Mechanizing Hypothesis Formation

Principles and Case Studies

Jan Rauch
Department of Information and Knowledge Engineering
Prague University of Economics and Business, Prague, Czechia

Milan Šimůnek
Department of Systems Analysis
Prague University of Economics and Business, Prague, Czechia

David Chudán
Department of Information and Knowledge Engineering
Prague University of Economics and Business, Prague, Czechia

Petr Máša
Department of Information and Knowledge Engineering
Prague University of Economics and Business, Prague, Czechia

CRC Press
Taylor & Francis Group
Boca Raton London New York

CRC Press is an imprint of the
Taylor & Francis Group, an **informa** business

A SCIENCE PUBLISHERS BOOK

First edition published 2022
by CRC Press
6000 Broken Sound Parkway NW, Suite 300, Boca Raton, FL 33487-2742

and by CRC Press
4 Park Square, Milton Park, Abingdon, Oxon, OX14 4RN

© 2022 Taylor & Francis Group, LLC

CRC Press is an imprint of Taylor & Francis Group, LLC

Library of Congress Cataloging-in-Publication Data (applied for)

ISBN: 978-0-367-54980-0 (hbk)
ISBN: 978-0-367-54982-4 (pbk)
ISBN: 978-1-003-09144-8 (ebk)

DOI: 10.1201/9781003091448

Typeset in Times New Roman
by Radiant Productions

To all users of the GUHA method.

Preface

The monograph *Mechanizing Hypothesis Formation (mathematical foundations for a general theory)* [41] starts with two questions:

Q1: *Can computers formulate and justify scientific hypotheses?*

Q2: *Can they comprehend empirical data and process them rationally, using the apparatus of modern mathematical logic and statistics to try to produce a rational image of the observed empirical world?*

Logic of discovery is developed in [41] as a set of answers to additional questions inspired by Q1 and Q2.

The question *"Are there methods for suggesting such a set of sentences, which is as interesting (important) as possible?"* is one of the questions inspired by Q1 and Q2. The GUHA method is introduced as an answer to this question. The method is realized by GUHA procedures working according to the schema introduced in Fig. 1.

An input of a GUHA procedure consists of an analyzed data matrix and parameters defining a large set of relevant patterns. The GUHA procedure generates all the defined patterns and verifies them in the given data matrix. The output of the procedure consists of all true relevant patterns.

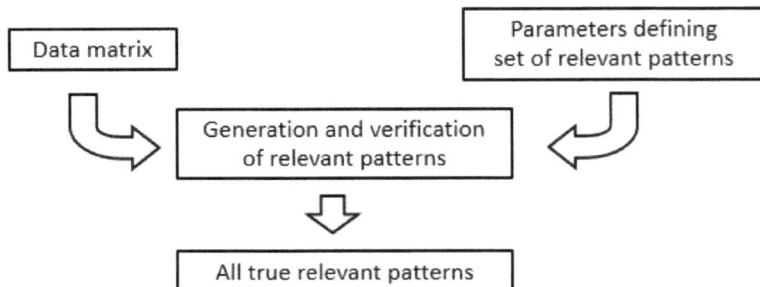

Figure 1: Schema of a GUHA procedure.

GUHA is an abbreviation of *General Unary Hypotheses Automaton.* *General* refers both to a wide field of applications and to the generality of results, *Unary* corresponds to the form of analyzed data. *Hypotheses* points to hypothesis formation and *Automaton* refers to the use of computers.

The GUHA procedures ASSOC and IMPL implemented in the 1970s deal with patterns, which can be understood as a generalization of association rules defined later in the 1990s. These procedures do not use the well-known apriori algorithm developed in the 1990s. Their implementation is based on a representation of analyzed data matrices by suitable strings of bits [101]. Early development and applications of the GUHA method are described in two special volumes of the International Journal of Man-Machine Studies—*Volume 10, Issue 1, January 1978* and *Volume 15, Issue 3, October 1981.*

The GUHA method is developed as a method of data mining since the advent of knowledge discovery in databases. An overview of activities and publications related to the GUHA method until 2010 is available in the paper *The GUHA method and its meaning for data mining, Journal of Computer and System Science*, 2010, Vol. 76, No. 1.

The representation of analyzed data by suitable strings of bits makes it possible to apply very fast basic computer instructions. They are used to get strings of bits representing derived Boolean attributes as well as to compute contingency tables concerning couples of Boolean and/or categorical attributes. Conditional contingency tables relevant to sub-matrices of a given data matrix defined by derived Boolean attributes are also used.

This led to the implementation of several additional GUHA procedures dealing with association rules, conditional association rules, conditional histograms, conditional patterns based on contingency tables of categorical attributes, couples of association rules, couples of conditional histograms, couples of conditional patterns based on contingency tables and action rules. All these procedures are implemented in the LISp-Miner software system developed at the *Faculty of Informatics and Statistics, Prague University of Economics and Business.*

We can summarize that the LISp-Miner system is intended to be used in descriptive data mining. The goal is to help to describe in detail a process producing a given data, not to predict the result of one particular step. The GUHA procedures implemented in the LISp-Miner can, among others, mine for a variety of exception rules and action rules as well as to solve tasks of subgroup discovery of various types.

A task to describe the behavior of tennis players under various circumstances in a way making possible to help a trainer lead his player to win is a typical task of descriptive data mining. A task of predictive data mining is to help to bettor to make his bet. Descriptive data mining is usually also the first step of predictive data mining. The patterns produced by particular GUHA procedures (rules, histograms, contingency tables, . . .) are understood as stones of a mosaic or items of a large puzzle. Our goal is to produce as much as possible complete mosaic or puzzle. Of course, this requires to use suitable formalized domain knowledge. Thus, research related to the LISp-Miner system concerns applications of various items of domain knowledge.

The LISp-Miner system has been applied in many practical tasks. Examples of applications are available in [5, 44, 85, 93, 121, 124, 125, 126, 127, 128, 131, 132, 137, 138, 139, 140, 155, 156, 161]. It has been and is still used in research and teaching at several universities. The LISp-Miner system proved to be a powerful tool to enhance the possibilities of the Business Intelligence tools [11].

Usefulness of the GUHA procedures implemented in the LISp-Miner system and achieved theoretical results led to the start of the development of a system of modules in the Python language which is very popular among data scientists. The goal is to make the GUHA procedures of the LISp-Miner system available to a large community of data scientists.

There are also various research activities related to the GUHA method and the LISp-Miner system. Let us mention the development and study of logical calculi formulas which correspond to patterns some of the GUHA procedures deal with [121, 139]. Both correct logical deduction rules and expert deduction rules are studied [124, 139]. Important research topics concern the generation of artificial data with defined hidden patterns. Such data are very useful when teaching data mining with the GUHA procedures and generally for testing and evaluation of data mining tools.

The goal of this book is to present

■ a brief overview of theoretical foundations of the GUHA method

■ principles of implementation of the GUHA procedures and the LISp-Miner system

■ examples of applications of the GUHA procedures of the LISp-Miner system to solve various tasks of descriptive data mining

■ examples of enhancing possibilities of the PowerBI software system by the GUHA procedures

■ main features of the CleverMiner system which is being developed as a system of modules of Python language implementing the GUHA procedures

■ an overview of important research results and research challenges.

The book is intended for both practitioners and researchers working in data mining, knowledge discovery in databases and data science. The book can also be used as a source of topics for seminar works and diploma theses for students of data mining. The book has four authors:

■ *Jan Rauch* authored the structure of the book, preface, Chapters 1–4, 6, 17, and 18, co-authored Chapters 7–13, and edited the whole book. He also authored theoretical principles of all the GUHA procedures 4ft-Miner, CF-Miner, KL-Miner, SD4ft-Miner, SDCF-Miner, SDKL-Miner and Ac4ft-Miner described in this book.

■ *Milan Šimůnek* authored Chapters 5 and 16 and co-authored Chapters 7–13. He is the author of the software architecture of the system LISp-Miner and the author of software of all the above mentioned GUHA procedures implemented in the LISp-Miner system.

■ *David Chudán* authored Chapter 14 concerning GUHA procedures and Business Intelligence.

■ *Petr Máša* authored Chapter 15 concerning the CleverMiner system that is a Python implementation of key GUHA procedures.

Acknowledgements

The work described here has been supported by funds VSE IP 400040 of institutional support for long-term conceptual development of science and research at the Faculty of Informatics and Statistics of the Prague University of Economics and Business.

Contents

SECTION I: THE GUHA PROCEDURES

SECTION II: APPLYING THE GUHA PROCEDURES

SECTION III: RELATED RESEARCH AND THEORY

Chapter 1

Introduction

This book deals with results achieved with mechanizing hypothesis formation introduced in monograph [41]. We use two types of results – observational calculi and the GUHA method. Observational calculi are special logical calculi formulas that correspond to statements on analyzed data. The acronym GUHA abbreviates **G**eneral **U**nary **H**ypotheses **A**utomaton. The GUHA method automatically generates statements on analysed data that are both true and interesting for data owners. Important features of mechanized hypothesis formation are summarized in Section 1.1.

We aim at applications of these results. Data mining is the main area of our applications. Relevant features of data mining are described in Section 1.2. However, we believe that Business Intelligence and Data science can also benefit from these results, see Section 1.3. Our applications deal with data matrices introduced in Section 1.4. We use items of domain knowledge described in Section 1.5. The goals of the book are introduced in Section 1.6 in more detail, together with a structure of the book and notes to using of the book.

1.1 Mechanizing Hypothesis Formation

Mechanizing hypothesis formation is an approach to exploratory data analysis introduced in the monograph [41]. The monograph starts with two questions:

Q1: *Can computers formulate and justify scientific hypotheses?*

Q2: *Can they comprehend empirical data and process them rationally, using the apparatus of modern mathematical logic and statistics to try to produce a rational image of the observed empirical world?*

A more detailed investigation of these questions resulted in the five questions of logic of discovery listed in Section 1.1.1. Answers to these five questions constitute a logic of discovery shortly introduced in Section 1.1.2. Observational calculi this book deals with are an important part of logic of discovery. The GUHA method introduced in Section 1.1.3 is a tool of logic of discovery. The research on mechanizing hypothesis formation has started in the 1960s. Notes to history and an overview of results relevant to this book are provided in Section 1.1.4.

1.1.1 Questions of logic of discovery

The rational image of the observed empirical world mentioned in question Q2 is understood as a set of all relevant statements concerning the observed empirical world and supported by the observed data. A schema of inductive inference

$$\frac{\textit{theoretical assumptions, observational statement}}{\textit{theoretical statement}}$$

is used to develop a theory answering questions Q1 and Q2. The schema means, that if we accept theoretical assumptions and verify a suitable statement about the observed empirical data, we accept a conclusion – a theoretical statement. The logic of discovery is introduced in [41] as a set of answers to questions (L0)–(L4):

(L0) In what languages does one formulate observational and theoretical statements? (What are the syntax and semantics of these languages? What is their relation to the classical first order predicate calculus?)

(L1) What are rational inductive inference rules bridging the gap between observational and theoretical sentences? (What does it mean that a theoretical statement is justified?)

(L2) Are there rational methods for deciding whether a theoretical statement is justified (based on given theoretical assumptions and observational statements)?

(L3) What are the conditions for a theoretical statement or a set of theoretical statements to be of interest (importance) with respect to the task of scientific cognition?

(L4) Are there methods for suggesting such a set of sentences, which is as interesting (important) as possible?

1.1.2 Logic of discovery and observational calculi

The goal of the monograph [41] is to develop a *mathematical logic of discovery* since only mathematically precise notions and results can be used as a basis for computer procedures. This leads to the application of notions and approaches used in mathematical logic and mathematical statistics. Both observational and theoretical calculi were developed and studied. Formulas of observational calculi correspond to statements on observed data. Sentences of theoretical calculi refer to systems of "possible worlds".

Observational and theoretical calculi are related by inductive inference rules corresponding to statistical hypothesis testing. Crucially, in many inductive inference rules, interesting theoretical statements are in one-to-one correspondence with suitable observational statements. This means that a task of finding all interesting theoretical statements can be formulated as a task of finding all interesting observational statements.

Answers to all questions (L0)–(L4) are developed in [41]. Answers to (L0)–(L2) constitute a *logic of induction*, answers to (L3)–(L4) constitute a *logic of suggestion*. Answers to (L0)–(L4) constitute a *logic of discovery*. However, only two of the answers to questions (L0)–(L4) are important to this book.

The first one is the answer to the question "*In what languages does one formulate observational and theoretical statements?*" included in (L0). **Observational calculi** are defined for this purpose. Formulas of observational calculi correspond to various assertions on analysed data. An association rule is a very simple example of a formula of observational calculus. Analytical GUHA procedures described in this book deal with patterns, which can be understood as formulas of suitable observational calculi. This brings useful possibilities both for implementation and for applications of these procedures, see further.

1.1.3 *GUHA method—tool of logic of discovery*

The answer to the question (L4) – "*Are there methods for suggesting such a set of sentences, which is as interesting (important) as possible?*" is the second answer important to this book. The **General Unary Hypotheses Automaton** method is introduced as the answer to this question. *General* refers both to a wide field of applications and to the generality of results, *Unary* corresponds to the form of analysed data. *Hypotheses* points to hypothesis formation and *Automaton* refers to the use of computers. The acronym GUHA is usually used for *General Unary Hypotheses Automaton*.

The method is realized by *GUHA procedures*. Each such procedure works according to the schema introduced in Fig. 1.1. An input of a GUHA procedure consists of an analysed data matrix and of parameters defining a large set of relevant patterns. The procedure generates all the defined patterns and verifies them in the given data matrix. The output of the procedure consists of all relevant patterns true in the data matrix.

The patterns must be formulas of observational calculi with clear syntax and semantics. This makes it possible to tune a definition of a set of relevant patterns according to the user's needs. In some cases, it is possible to use logical deduction

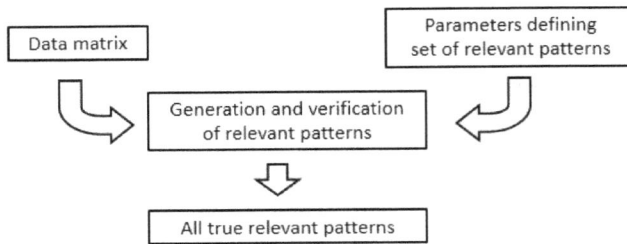

Figure 1.1: Schema of a GUHA procedure.

to optimize an algorithm of a GUHA procedure and/or output of the GUHA procedure. If suitable, only *prime patterns* are involved in the output. A pattern ω is prime if it is true and no more simple (i.e., shorter) true pattern ω' such that ω logically follows from ω' is included in the output.

1.1.4 Notes to history and overview of results

First considerations on the GUHA method come from the 1960s [36]. There is a series of papers [21, 22, 23, 24, 25, 26, 37, 38, 39] concerning the GUHA method and related logical questions. Statistical aspects of the GUHA method are studied in papers [51, 52, 53, 54]. Papers [34, 100] deal with missing information. Ideas and notions introduced in these papers are used to develop mathematical foundations for a general theory of mechanizing hypothesis formation presented in the monograph [41] (1978). Of course, various additional sources are used in [41]; let us mention [12, 15, 17, 90, 98].

Paper [40] (1978) introduces the GUHA method in 24 questions and answers. Various aspects of the implementation of the method are discussed in [91, 92, 101, 102]. An overview of software available in 1981 is presented in [60]. Early applications of the GUHA method are described in [61] (sociology, 1977), [141] (medicine, 1978), [157] (chemical engineering, 1981), and [71] (pedagogy, economics, and sports, 1981). Problems of interpretation of results of the GUHA procedures are discussed in [142]. A relation between the GUHA method and exploratory data analysis is discussed in [59].

Additional theoretical results related to logical and statistical features of the GUHA method are available in [27, 55, 56, 57, 58, 62, 63]. A new GUHA procedure ASSOC was designed in 1984 [29]. Results on the complexity of the GUHA procedures are introduced in [94, 95, 147]. There are also papers [28, 45, 35, 75] concerning using of artificial intelligence in applications of the GUHA method. They are inspired by the project *AM: An Artificial Intelligence Approach to Discovery in Mathematics as Heuristic Search* [83, 84]. Additional papers concerning the GUHA method are listed in the above introduced references, namely in [40, 41].

All the above-mentioned papers come from the period 1960s–1980s. Since the advent of data mining in the 1990s, the GUHA method is developed as a method of data mining [44]. A new implementation of the GUHA procedure ASSOC dealing with association rules was developed [47] and applied [48]. Various theoretical and practical aspects of development and applications of the GUHA method as well as related observational calculi were studied [30, 31, 32, 33, 42, 43, 46, 66, 67, 68, 72]. Possibilities of using fuzzy logic were also considered [65, 73, 74].

Research on mechanizing hypothesis formation as a data mining tool at *Prague University of Economics and Business* started in the second half of the 1990s [103]. The GUHA procedure 4ft-Miner dealing with enhanced association rules was implemented [125]. Development of the LISp-Miner system [158] started at the same time. Currently, the LISp-Miner system involves nine analytical procedures, which can be considered GUHA procedures. In addition, various tools for data transformations, interpretation of results of the GUHA procedures, and dealing with domain knowledge are available. Examples of applications of seven from the nine GUHA procedures of the LISp-Miner are used to achieve the goals of the book. There are dozens of relevant references. They are introduced in appropriate chapters of the book.

1.2   Data Mining

Data mining is a widely known discipline of informatics. It is very shortly introduced in Section 1.2.1. CRISP-DM outlined in Section 1.2.2 is the most used methodology of data mining. In this book, we tackle three important topics of data mining – rules discovery, exception and action rules mining, and subgroup discovery. Introductory remarks to these topics are presented in Sections 1.2.3, 1.2.4, and 1.2.5 respectively. Since the advent of data mining, the GUHA method is developed also as a method of data mining [44]. A role of the GUHA procedures in data mining is shortly discussed in Section 1.2.6.

1.2.1   Discipline of informatics

Data mining started at the beginning of the 1990s when owners of large databases realized assets hidden in their databases. Today, data mining is a matured discipline of informatics. It is largely applied and further developed. The development is driven by the need to analyze data produced by various new as well as traditional data sources.

There are plenty of monographs, conference proceedings, journal papers, textbooks, and courses concerning data mining. There are also various definitions of data mining. In the large textbook [1], data mining is understood as *the study of collecting, cleaning, processing, analyzing, and gaining useful in-*

sights from data. Our goal is neither to bring additional definitions nor to discuss features of known definitions. We will follow the point of view according to [1]. Note that the term *knowledge discovery in databases* is sometimes used instead of *data mining.* The term *data mining* is then understood as one of the steps in the process of knowledge discovery in databases. Let us point to `https://www.kdnuggets.com/` where the recent information relevant to data mining is available.

There are two branches of data mining – descriptive data mining and predictive data mining. Descriptive data mining uses given data to describe what has happened. Data mining with association rules, exception and action rules mining, and subgroup discovery introduced in Sections 1.2.3–1.2.5, as well as the GUHA method belong to descriptive data mining. Predictive data mining uses given data to predict future results. We are not interested in predictive data mining in this book.

1.2.2 CRISP-DM

Data mining is a complex process. The well-known CRISP-DM methodology [10] is usually used to describe the whole process. The CRISP-DM recognizes six mutually related phases of the data mining process. Mutual relations of the main phases are outlined in Fig. 1.2. The inner arrows correspond to main relations among particular phases. The outer circle emphasizes that the process has a cyclical character and that there are mutual relations and feedbacks among all the phases. A short description of the main features of particular phases follows.

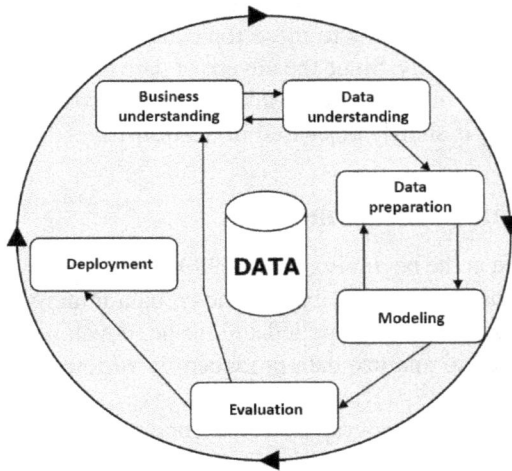

Figure 1.2: The CRISP–DM methodology.

1. **Business understanding**: This phase starts with a determination of business objectives, continues by assessing a situation, definition of data mining goals, and by preparation of the project plan. There is close feedback with the phase of data understanding.

2. **Data understanding**: Generally, this phase includes collecting, description and exploring initial data. It is important to verify data quality.

3. **Data preparation**: Selecting, cleaning, integrating, and formatting data are the key tasks of this phase.

4. **Modelling**: This phase starts with the selection modelling technique, continues by generation building and assessing of a model. In the case of the GUHA procedures application, this means the selection of a suitable GUHA procedure and calibrating its parameters to get a reasonable number of results.

5. **Evaluation**: A general goal of this phase is comparing results of the modelling phase with the evaluation criteria defined at the start of the project. This of course depends on the role of applied GUHA procedures in the whole project.

6. **Deployment**: Generally, this phase covers the planning of deployment, including monitoring and maintenance, producing a final report and reviewing the project. Again, the role of applied GUHA procedures in the whole project is important.

Note that this very short description is valid for both predictive and descriptive data mining.

1.2.3 Rules discovery

Association rules are studied to better understand the purchasing behavior of customers in supermarkets since the advent of data mining [2]. The association rule is an expression $X \rightarrow Y$, where X and Y are sets of items. The rule $X \rightarrow Y$ means that transactions containing a set X of items tend to contain a set Y of items. The expression {butter, cheese} $\rightarrow$ {bread} means that customers who buy butter and cheese often buy bread.

There are various measures of interestingness of association rules. The two most used measures of interestingness are confidence and support. *Confidence* is defined as a ratio

$$\frac{the\ number\ of\ transactions\ containing\ both\ X\ and\ Y}{the\ number\ of\ transactions\ containing\ X}.$$

Support is a ratio

$$\frac{the\ number\ of\ transactions\ containing\ both\ X\ and\ Y}{the\ number\ of\ all\ transactions}.$$

The association rule discovery task is a task of finding all association rules with confidence and support above given thresholds *minconf* and *minsup*. The association rules $X \to Y$, where X and Y are sets of items, were generalized. Association rules are further understood also as expressions $A \to S$ where both A and S are conjunctions of attribute-value pairs. The expression

$$Sex(male) \wedge Weight(very\ high) \wedge Smoking(yes) \to Infarction(yes)$$

is an example of such an association rule. The algorithm apriori [2] is usually used for association rules mining. There are also additional algorithms developed and used for this purpose [50, 164]. The arules package in the R-system [19] is an example of the apriori implementation. There is a very large number of publications on association rules. An overview of rules discovery is available in [18].

However, the concept of association rules was introduced and studied since the 1960s as an inherent part of the GUHA method research [36, 41]. Association rules are understood as relations $\varphi \approx \psi$ between general Boolean attributes φ and ψ derived from columns of an analysed data matrix. A general ASSOC procedure [29, 40] was developed to mine for association rules $\varphi \approx \psi$. Implementation of the ASSOC procedure uses an algorithm based on strings of bits [101].

The 4ft-Miner GUHA procedure [127] described in this book deals with association rules $\varphi \approx \psi$. Examples of association rules the 4ft-Miner procedure deals with are in Section 3.2. The 4ft-Miner procedure is described in detail in Chapter 7 together with several examples of its application. The general form $\varphi \approx \psi$ of association rules brings possibilities that are practically not available when using association rules $A \to S$ where both A and S are conjunctions of attribute-value pairs. The 4ft-Miner procedure deals also with conditional association rules. A comparison of the 4ft-Miner procedure and the arules package for R is available in [138] and also in Chapter 7.

1.2.4 Exception rules and action rules

Exception rules mining is a research challenge that has attracted many researchers. This has resulted in various approaches presented in many papers. The approaches to mine exception rules can be divided into two classes – directed and undirected [153].

Directed search of exception rules starts with getting user-specified beliefs in a form of suitable patterns. Discovered rules are then compared with given patterns and ranked according to their relation to given patterns. A user-expectation method is introduced in [86]. Two key measures of subjective interestingness are discussed - *unexpectedness* (patterns are interesting if they surprise the user) and *actionability* (patterns are interesting if they can help to achieve some current goals of the user). Various measures of actionability and degrees of beliefs are studied [146]. A domain ontology is used in order to define a measure of unexpectedness in [49]. An approach based on a definition of unexpectedness in terms of a logical contradiction of a rule and a belief is studied and implemented in [88]. It is crucial that if a knowledge discovery procedure mines for a given type of pattern, then patterns of the same type are used to express user beliefs. Thus, if rules are mined, then rules are also used to express user beliefs.

The undirected search does not use background knowledge. A large discussion on this approach is available in [151]. Undirected search for exception rules produces patterns pairs ⟨ strong rule; exception rule ⟩. A good example of such a result is presented also [151]: *using a seat belt is safe* is a strong rule, *using a seat belt is risky for a child* is a counterexample of the strong rule.

Various approaches to action rules are introduced and discussed in the monograph [13]. The notions of *stable* and *actionable attributes* are crucial in these approaches. An expression

$$Sex(\text{male}) \wedge BodyMassIndex(\text{high} \rightarrow \text{normal}) \Rightarrow BloodPressure(\text{high} \rightarrow \text{normal})$$

is a simple example of an action rule. It should be read as *"If a body mass index of male patients decreases from high to normal, then also blood pressure decreases from high to normal"*. Here *Sex* is a stable attribute. *BodyMassIndex* and *BloodPressure* are flexible attributes.

There are three GUHA procedures dealing with rules and they are described in this book. The first one is the 4ft-Miner procedure mentioned already at the end of the previous section. The second one is the SD4ft-Miner procedure dealing with couples of (possibly conditional) association rules of the form $\varphi \approx \psi$. An example of couples of association rules the procedure deals with is available in Section 3.5. The SD4ft-Miner procedure is described in detail in Chapter 10 together with several examples of its application.

The Ac4ft-Miner procedure dealing with action rules is the third GUHA procedure dealing with rules described in this book. An example of action rules the procedure deals with is available in Section 3.5. The Ac4ft-Miner procedure is described in detail in Chapter 13 together with examples of its application.

Let us emphasize that the 4ft-Miner, SDft-Miner, and Ac4ft-Miner procedures mine for various variants of association rules, exception rules and action rules. Their overview is in Chapter 6 and their detailed descriptions are in Chapters 7, 10, and 13.

1.2.5 Subgroup discovery

Subgroup discovery is an important method of descriptive data mining. It aims at finding interesting subsets of a given dataset. Interestingness criteria are based on a property of interest or a more complex quality function. Subgroups are locally interesting patterns, not global models. Found subgroups are described in a subgroup description language. An overview of subgroups discovery is available in [3]. Additional relevant information can be found in [9, 64].

The GUHA procedure CF-Miner mines for sub-matrices of a given data matrix for which a histogram of a given attribute has given shape. We call such histograms *conditional histograms*. We present a simple example of the application of the CF-Miner procedure to subgroup discovery.

We start with a decreasing histogram of accidents concerning one vehicle in the period 2005–2015 available in Fig. 1.3. The histogram concerns a data matrix

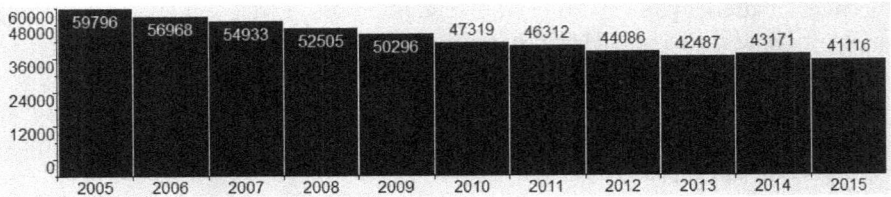

Figure 1.3: Decresing trend of all accidents in period 2005–2015.

Accidents derived from a dataset *UK Car Accidents* dataset and described in Section 2.3. The *Accidents* data matrix has 538 989 rows; each row corresponds to one accident.

Thus, a natural task arises to find all sub-matrices *Accidents*/χ such that a histogram of accidents belonging to this sub-matrix is increasing. Here χ is a Boolean attribute derived from columns of the *Accidents* data matrix corresponding to characteristics of accidents. *Accidents*/χ is a sub-matrix consisting of all rows satisfying χ.

This task can be solved by the GUHA procedure CF-Miner. Several simple parameters defining a set of relevant belong to an input of the CF-Miner. A run of the CF-Miner procedure resulted in 20 sub-matrices *Accidents*/χ such that histograms of accidents belonging to these sub-matrices are increasing. 61 059 sub-matrices were tested in 63 seconds. Details are available in Section 8.3.2.

A sub-matrix *Accidents*/(Type(Car)∧Speed_Limit(20)) is an example of a result of the CF-Miner. This sub-matrix has 6 351 rows and corresponds to a subgroup of accidents of cars on a road with a speed limit of 20 miles/hour. The corresponding histogram is available in Fig. 1.4.

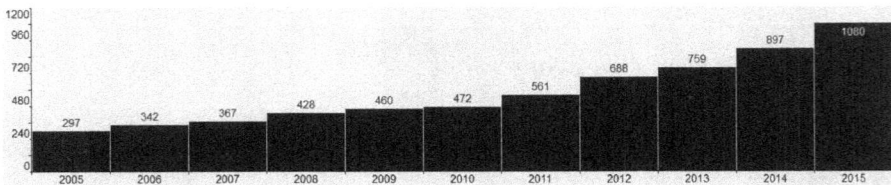

Figure 1.4: Histogram Year/(Type(Car)∧ Speed_Limit(20)).

The CF-Miner procedure is described in detail in Chapter 8 together with several examples of application. There are three additional GUHA procedures suitable for subgroup discovery. The SDCF-Miner procedure deals with couples of conditional histograms. An example of a couple of conditional histograms the procedure deals with is available in Section 3.6. The SDCF-Miner procedure is described in Chapter 11 together with several examples of its application.

The KL-Miner procedure mines for sub-matrices of a given data matrix for which a contingency table of two categorical attributes satisfies a given condition. An example of the results of the KL-Miner procedure is available in Section 3.4. The KL-Miner procedure is described in detail in Chapter 9. The SDKL-Miner procedure mines for couples of sub-matrices of a given data matrix for which contingency tables of two categorical attributes satisfy a given condition. An example of the results of the SDKL-Miner procedure is available in Section 3.7. The SDKL-Miner procedure is described in detail in Chapter 12.

1.2.6 Presented GUHA procedures and data mining

There are seven GUHA procedures described in this book. All of them are methods of descriptive data mining. A single result of one procedure is usually not the goal of the whole process of data mining as described in Section 1.2.2. Results produced by particular GUHA procedures should be understood as stones of a mosaic or items of a large puzzle. The application of the GUHA procedures aims to produce the complete puzzle as much as possible. Descriptive data mining is also the first step of predictive data mining.

All the described procedures have important common features. The common features are introduced in detail in Chapter 4. Boolean attributes are important parts of all patterns the described GUHA procedures deal with. The Boolean attribute Type(Car) $\wedge$ Speed_Limit(20) used in Fig. 1.4 is an example. Boolean attributes used in particular patterns, are derived from columns of analysed data matrices. This is described in Section 1.4.

Let us emphasize that interpretation of results of particular GUHA procedures requires the application of domain knowledge. There is research related to the formalization of particular items of domain knowledge and to application of formalized domain knowledge in post processing of results of the GUHA procedures. For more information see Chapter 17.

1.3 Business Intelligence and Data Science

Our goal is to show that Business Intelligence and Data science can benefit from mechanizing hypothesis formation. Basic ideas supporting this goal are presented in Sections 1.3.1 and 1.3.2.

1.3.1   Business Intelligence and GUHA procedures

There are many definitions of Business Intelligence (BI). In simple words, BI is a set of tools for the analysis of business information. In this book, we are interested in tools of self service BI dealing with smaller datasets than general BI. PowerBI (see https://powerbi.microsoft.com/en-us/desktop/) is a typical tool of self service BI. A more detailed introduction to BI and self service BI is available in Section 14.1.

PowerBI is usually used to manually find interesting and useful patterns in data. The histogram showing decreasing trend of all accidents in a period 2005–2015 similar to that introduced in Fig. 1.3 can be seen as a typical result of a PowerBI application. It is straightforward and natural to produce such a histogram as one of the first steps of an analysis of the *UK Car Accidents* dataset described in Section 2.3.

However, it is not reasonable to use PowerBI to solve a task to find all submatrices *Accidents/χ* such that a histogram of accidents belonging to this submatrix is increasing. A solution of this task by the CF-Miner GUHA procedure is shortly outlined in Section 1.2.5; for details see Section 8.3.2. The core of the problem is that it is not possible to manually test more than 60 000 relevant histograms.

We can conclude that the application of the GUHA procedures can significantly extend the results of PowerBI applications. Two scenarios of complementary usage of PowerBI and GUHA procedures are presented in Chapter 14. These scenarios can be of course applied also to additional self service BI tools.

1.3.2   Data science and mechanizing hypothesis formation

Data science is a very popular topic with several definitions. It can be understood as a unification of several disciplines (namely data mining, statistics, machine learning, and informatics) in order to understand processes producing data. Our goal is not to formulate an additional definition, let us only point to https://www.kdnuggets.com/topic/data-science where news concerning data science are available.

Python language is largely used by data scientists. This is the reason why the development of the CleverMiner project started. Its goal is to make GUHA procedures available to a community of the Python language. Scenarios of complementary usage of analytical tools of Python and GUHA procedures analogous to scenarios of complementary usage of BI tools and GUHA procedures will be easily realizable in the CleverMiner project. The CleverMiner project is introduced in Chapter 15.

Note that the Python users will be able to sort and filter patterns produced by the GUHA procedures. This way approaches to filter out uninteresting consequences of known items of domain knowledge will be easily applicable. These approaches are based on results concerning observational calculi. They are introduced in Chapters 17 and 18.

1.4 Data Matrix

The GUHA procedures described in this book deal with data matrices. An example of a data matrix is available in Section 1.4.1. A more formal definition of a data matrix is provided in Section 1.4.2. All the described GUHA procedures deal with Boolean attributes derived from columns of data matrices. Boolean attributes are introduced in Section 1.4.3. A very important notion of data submatrix is defined in Section 1.4.4.

1.4.1 Data matrix—an example

We use a well-known dataset *Adult* available from *UCI Machine Learning Repository* [14] to present an example of data matrix. The *Adult* dataset concerns 48 842 persons. Each person is described by 14 attributes introduced in detail in Section 2.2.

The *Adult* dataset can be written in a form of a data matrix outlined in Fig. 1.5. Each row corresponds to a person, and each column corresponds to an attribute. Only the first and last two rows as well as only five attributes are introduced. There are missing values, which are coded by the string "*NA*".

Adult	age	workclass	education	occupation	capital_gain	...
o_1	*39*	*State-gov*	*Bachelors*	*Adm-clerical*	*124*	...
o_2	*52*	*NA*	*HS-grad*	*NA*	*0*	...
$\vdots$	$\vdots$	$\vdots$	$\vdots$	$\vdots$	$\vdots$	$\ddots$
o_{48841}	*44*	*Private*	*Bachelors*	*Adm-clerical*	*11 678*	...
o_{48842}	*24*	*Private*	*11th*	*Craft-repair*	*0*	...

Figure 1.5: The *Adult* data matrix corresponding to the *Adult* dataset.

We see in Fig. 1.5, that the age of person o_1 is *39*, the workclass of this person is *State-gov* and that this person has education *Bachelors*. Note that we can get several data matrices from one initial dataset if we apply various data transformations.

1.4.2 Data matrix—definition

Rows of data matrices we deal with are in one-to-one correspondence with observed objects. Columns of data matrices correspond to attributes of observed objects. We will consider columns–attributes as functions assigning a value for each row. Each attribute has a set of its possible values.

Attributes age and capital_gain mentioned in the previous section assigns values – natural numbers. The set of possible values of attributes age is naturally limited. However, the possible values of attribute capital_gain can be seen as unlimited. Attribute workclass has eight possible values and attribute education has sixteen possible values introduced in Section 2.2.

We see in Fig. 1.5 that the value of attribute age for o_1 is *39*, the value of attribute workclass for row o_1 is *State-gov* etc. This can be written as age$(o_1,Adult) = 39$, workclass$(o_1,Adult) = State$-gov, etc. If there is no danger of misunderstanding, we can write only age$(o_1) = 39$, workclass$(o_1) = State$-gov, etc.

The GUHA procedures described in this book deal with data matrices with a final set of possible values for each attribute. We call possible values of such attributes as *categories*. A general form of a data matrix $\mathcal{M}$ we are interested in is shown in Fig. 1.6. We assume that the data matrix $\mathcal{M}$ has n rows $o_1,\ldots,o_n$ and K columns – categorical attributes $A_1,\ldots,A_K$. Particular categories of the attribute A_i can be coded by natural numbers $1,\ldots,t_i$ for $i = 1,\ldots,K$. We assume $t_i \geq 2$ for each i. This means that it holds $A_i(o_j,\mathcal{M}) = v_{j,i}$ and $v_{j,i} \in \{1,\ldots,t_i\}$ for $i = 1,\ldots,K$ and $j = 1,\ldots,n$. Again, we write only $A_i(o_j) = v_{j,i}$ if there is no danger of misunderstanding.

Of course, suitable strings of symbols can be used as names of attributes and their categories. The fact that only data matrices with a finite set of possible values for each column can be used in the GUHA procedures often leads to transformations of initial data matrices.

$\mathcal{M}$	A_1	A_2	A_3	$\ldots$	A_{K-1}	A_K
o_1	$v_{1,1}$	$v_{1,2}$	$v_{1,3}$	$\ldots$	$v_{1,K-1}$	$v_{1,K}$
o_2	$v_{2,1}$	$v_{2,2}$	$v_{2,3}$	$\ldots$	$v_{2,K-1}$	$v_{2,K}$
$\vdots$	$\vdots$	$\vdots$	$\vdots$	$\ddots$	$\vdots$	$\vdots$
o_{n-1}	$v_{n-1,1}$	$v_{n-1,2}$	$v_{n-1,3}$	$\ldots$	$v_{n-1,K-1}$	$v_{n-1,K}$
o_n	$v_{n,1}$	$v_{n,2}$	$v_{n,3}$	$\ldots$	$v_{n,K-1}$	$v_{n,K}$

Figure 1.6: A data matrix $\mathcal{M}$ with attributes $A_1,\ldots,A_K$ and rows $o_1,\ldots,o_n$.

1.4.3 Boolean attributes

The GUHA method uses Boolean attributes derived from attributes – columns of the analysed data matrices. *Basic Boolean attributes* are created first. A basic Boolean attribute is an expression $A(\alpha)$ where $\alpha \subsetneq \{a_1,\ldots a_t\}$ and $\{a_1,\ldots a_t\}$ is a set of all categories of the attribute A. The set α is a *coefficient* of the basic Boolean attribute $A(\alpha)$. A *basic Boolean attribute $A(\alpha)$ is true in a row o of $\mathcal{M}$*

if $A(o, \mathcal{M}) \in \alpha$; we write $A(\alpha)(o, \mathcal{M}) = 1$. If $A(o, \mathcal{M}) \notin \alpha$, then $A(\alpha)$ is *false in a row o*; we write $A(\alpha)(o, \mathcal{M}) = 0$. If there is no danger of misunderstanding, we write only $A(\alpha)(o) = 1$ or $A(\alpha)(o) = 0$. If $A(\alpha)(o) = 1$ then we also say that *o satisfies* $A(\alpha)$. If $A(\alpha)(o) = 0$ then we say that *o does not satisfy* $A(\alpha)$.

Each basic Boolean attribute is a *Boolean attribute*. If φ and ψ are Boolean attributes, then $\varphi \wedge \psi$, $\varphi \vee \psi$, and $\neg \varphi$ are *Boolean attributes*. Here $\wedge$, $\vee$, and $\neg$ are logical connectives of conjunction, disjunction and negation respectively. Semantics of these connectives is defined in the usual way. We write $\varphi(o, \mathcal{M}) = 1$ if φ is true in a row o of $\mathcal{M}$, analogously for $\varphi(o, \mathcal{M}) = 0$, $\varphi(o) = 1$ and $\varphi(o) = 0$. $\varphi \wedge \psi$, $\varphi \vee \psi$, and $\neg \varphi$ are called *derived Boolean attributes*.

In Fig. 1.7, there are examples of Boolean attributes derived from attributes A_1 and A_2. Attribute A_1 has categories $1, \ldots, 7$ and attribute A_2 has categories $1, 2, 3$. Basic Boolean attribute $A_1(6, 7)$ is true in a row o_1 because $A_1(o_1) = 6$ and $6 \in \{6, 7\}$. Basic Boolean attribute $A_1(6, 7)$ is false in row o_3 because and $A_1(o_3) = 1$ and $1 \notin \{6, 7\}$. Note, that pedantically we should write $A_1(\{6, 7\})$ and $A_2(\{1\})$ etc.; however, we will not do this.

The expression *True* is also considered as a Boolean attribute. It is identically true, it holds $True(o) = 1$ for all rows of all data matrices.

Let us emphasize that the theory developed for the GUHA method makes it possible to deal with missing information even for Boolean attributes derived from attributes columns with missing information. This is introduced in a precise manner in Section 4.5.

$\mathcal{M}$	A_1	A_2	$A_1(6,7)$	$A_2(1)$	$A_1(6,7) \wedge A_2(1)$	$A_1(1) \vee A_2(6,7)$	$\neg A_2(1)$
o_1	6	1	1	1	1	1	0
o_2	7	2	1	0	0	1	1
o_3	1	1	0	1	0	1	0
⋮	⋮	⋮	⋮	⋮	⋮	⋮	⋮
o_n	2	3	0	0	0	0	1

Figure 1.7: Basic and derived Boolean attributes.

1.4.4 Data sub-matrix

We are often interested not only in patterns concerning a whole data matrix. Recall the data matrix *Adult* introduced in Section 1.4.1. We can be interested in persons, whose income is large only. This means that we are interested in patterns hidden in all rows of the *Adult* data matrix such that a value of the attribute income is *Large*.

In other words, we are interested in all rows of the *Adult* data matrix for which the Boolean attribute income(*Large*) is true. We consider such rows as a sub-matrix *Adult* /income(*Large*) of the *Adult* data matrix.

This can be generalized. Let $\mathcal{M}$ be a data matrix and let χ be a Boolean data matrix created from columns of $\mathcal{M}$. Then $\mathcal{M}/\chi$ is a *sub-matrix* of $\mathcal{M}$ defined by χ. This is a data matrix consisting of all rows of $\mathcal{M}$ satisfying χ. Let us note that this definition can lead to an *empty data matrix*, i.e., a date matrix with no rows. This happens if χ is false for all rows of $\mathcal{M}$.

1.5 Data Matrix and Items of Domain Knowledge

One of the important outcomes of the business understanding phase is a set of relevant items of domain knowledge. We introduce four types of items of domain knowledge, which are used in Parts II and III of the book:

- groups of attributes, see Section 1.5.1.

- knowledge relevant to transformations of attributes, see Section 1.5.2.

- global properties of attributes, see Section 1.5.3.

- mutual dependence of attributes, see Section 1.5.4.

1.5.1 Groups of attributes

The attributes – columns of data matrices can be usually divided into naturally defined groups. Attributes of the *Adult* data matrix introduced in Section 2.2 can be divided into groups according to Table 1.1, see also [139].

Table 1.1: Groups of attributes of the *Adult* data matrix.

Group	Attributes
Personal	sex, age, education
Family	marital_status, relationship,
Society	native_country, workclass, race,
Employment	occupation, hours_per_week, income
Capital	capital_gain, capital_loss

Particular groups of attributes can be used to formulate reasonable analytical questions. A question *"Are there any strong relations between combinations of Boolean characteristics of the groups Personal and Employment and Boolean characteristics of the group Capital?"* is an example. Names of groups of attributes can be used to express this question concisely:

$$\mathcal{B}(\mathsf{Personal}) \wedge \mathcal{B}(\mathsf{Employment}) \ \Rightarrow^? \ \mathcal{B}(\mathsf{Capital}) \,.$$

Several analytical questions concerning the *Adult* data matrix formulated this way are presented in Section 7.3.

Groups of attributes tree	Attribute	Used	DBColumn	Categories	XCat	Sample categories
⊟ ⬚ Root group of attributes	marital_status	+	marital_status	7		Divorced, Married-AF-spous
⬚ Capital	relationship	+	relationship	6		Husband, Not-in-family, Othe
⬚ Employment						
⬚ Family						
⬚ Personal						
⬚ Society						

Figure 1.8: Groups of attributes of the *Adult* data matrix stored in the LISp-Miner.

Definitions of the groups of attributes can be stored in the LISp-Miner system and used in applications of particular GUHA procedures. The groups of attributes are presented in Table 1.1 are stored and presented by the LISp-Miner system in a way shown in Fig. 1.8.

1.5.2 Transformations of attributes

A set of categories of an attribute can be unsuitable for a given analytical question. We introduce two examples of items of domain knowledge concerning transformations of attributes.

The attribute age of the data matrix *Adult* has categories 17, ..., 90. These categories are too narrow to be used to solve most analytical questions. Important items of domain knowledge say which categories for the age attribute are suitable. We present two examples of suitable definitions of broader categories.

- We define a new attribute age_5 with 15 equidistant categories of length 5: $\langle 15; 20 \rangle, (20; 25 \rangle, (25; 30 \rangle, \ldots, (80; 85 \rangle, (85; 90 \rangle$. Since the last four intervals have too low frequencies, we join them into one interval $(70; 90 \rangle$. This way we get 12 intervals – categories $\langle 15; 20 \rangle, (20; 25 \rangle, (25; 30 \rangle, \ldots, (65; 70 \rangle, (70; 90 \rangle$ used to solve analytical questions.

- We define a new attribute age_exp with four categories – intervals defined by a domain expert: Young – $\langle 15; 25 \rangle$, Middle-aged – $\langle 26; 45 \rangle$, Senior – $\langle 46; 65 \rangle$, and Old – $\langle 66; 90 \rangle$.

Histograms of attributes age, age_5, and age_exp at the *Adult* data matrix are shown in Fig. 1.9.

1.5.3 Global properties of attributes

Histograms of attributes age, age_5, and age_exp at the *Adult* data matrix shown in Fig. 1.9 are examples of global properties of attributes. The histograms of attributes age_5, and age_exp can be characterized such as *"first increasing and then decreasing"*.

Thus, it is natural to ask if there are sub-matrices $Adult/\chi$ (i.e., subgroups of persons) such that a histogram of age_exp at $Adult/\chi$ is increasing or decreasing.

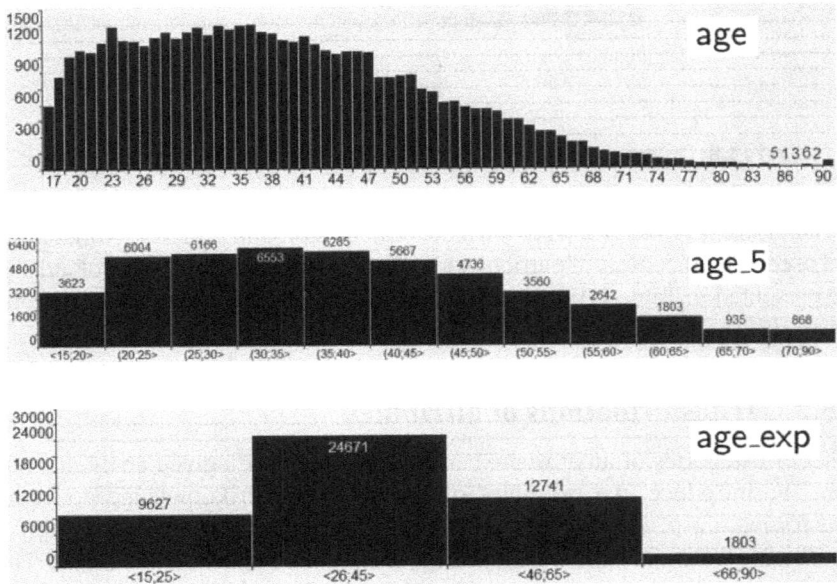

Figure 1.9: Histograms of attributes age, age_5, and age_exp.

An additional natural question is a question if there are sub-matrices $Adult/\chi$ such that a histogram of age_5 at $Adult/\chi$ is first decreasing and then increasing. These analytical questions are answered in Sections 8.2.1, 8.2.2, and 8.2.3. The procedure CF-Miner is used.

We can also ask if there are subgroups of persons, which can be understood as groups increasing or lowering columns of histograms. Such analytical questions are answered in Sections 7.4.1 and 7.4.2. The procedure 4ft-Miner and data matrix *Accidents* are used.

1.5.4 Mutual dependence of attributes

Additional items of domain knowledge resulting from the phase *Business understanding* are items concerning mutual relations of attributes. We present an example of an item of domain knowledge, including its symbolic expression:

■ *If education increases, then capital_gain increases as well.* Symbolically we write education ↑↑ capital_gain.

Such items of domain knowledge can be used to formulate analytical questions concerning exceptions and irregularities. They can be used also to filter out uninteresting patterns resulting from the analytical procedures. Examples of applications of such items of domain knowledge are available in Section 17.2, see also [122].

1.6 Goals, Structure, and Using the Book

1.6.1 Goals and structure

The main goal of this book is to present applications of two results related to mechanizing hypothesis formation – the GUHA method and observational calculi introduced in Section 1.1. To do this, we need to also present the necessary theoretical background. A first part of the theoretical background is already introduced in Sections 1.2–1.5. Let us emphasize that this first part is sufficient to understand all chapters except the last one. The second goal is to present related research. This is done in the last three chapters. The remaining part of the relevant theoretical background is available in the last chapter.

Four datasets are used to present applications of the GUHA procedures. They are introduced in Chapter 2. The rest of the book is divided into three sections:

I The GUHA procedures.

II Applying GUHA procedures.

III Related research and theory.

The first goal of **Section I—The GUHA procedures** is to give an overview of all seven GUHA procedures the book deals with. Examples of all seven types of patterns particular procedures mine for are presented in Chapter 3. All seven GUHA procedures have common features concerning namely contingency tables, sets of relevant Boolean attributes to be generated, and dealing with missing information. The second goal of part I is to present the common features in one place. This is done in Chapter 4. The third goal is to introduce the main features of the LISp-Miner system which implements all the GUHA procedures described in this book. This is done in Chapter 5.

Section II—Applying GUHA procedures is a core of the book. About forty short case studies and examples of applications of the GUHA procedures introduced in Section I are presented. Section II starts with Chapter 6 where an overview of all the presented applications of the GUHA procedures is available. Examples of applications of each of seven GUHA procedures are available in Chapters 7–13. Let us note that a detailed comparison of the GUHA approach and apriori approach to association rules is among others available in Chapter 7. Chapter 14 is devoted to a relation of the GUHA procedures and Business Intelligence (BI). Two scenarios of possible complementary usage of self-service BI and the ability to generate and verify a large set of relevant patterns by GUHA procedures are introduced. First experiences with the implementation of the GUHA procedures in Python language are described in Chapter 15.

Section III—Related research and theory gives an overview of important research topics, achieved results, and research challenges related to the presented

GUHA procedures. The needs, advantages and possibilities of the generation of artificial data are discussed in Chapter 16. Three examples of applications of formalized items of domain knowledge in data mining with association rules and histograms are presented in Chapter 17. Recall that observational calculi are defined to partly answer the question "*In what languages does one formulate observational and theoretical statements?*", see Section 1.1.2. Main features of observational calculi are presented in Chapter 18. Results on logical deduction rules and expert deduction rules are presented together with related research challenges.

1.6.2   Using the book

There are various ways how to use the book. We do not assume that the book will be studied step by step from Chapter 1 to Chapter 18 even if it is also possible. Let us note that continuous reading of all details concerning analysed datasets introduced in Chapter 2 will probably be extremely boring. Several hints on how to use the book are available below.

- It is possible to get a basic understanding of each of the described GUHA procedures in Chapter 3 and then to use an available link to a chosen procedure to get detailed information.

- A practitioner solving a particular problem of data mining can start with an overview of examples in Chapter 6 where links are available to detailed descriptions of applications of corresponding GUHA procedures.

- A user of a business intelligence software system can draw inspiration in Chapter 14 on how to enhance the possibilities of his system.

- A Python programmer can get information about the implementation of the GUHA procedures in Chapter 15.

The following topics can be useful when using the book.

- At least one Boolean attribute derived from columns of analysed data matrices is an inherent part of each pattern the described GUHA procedures deal with. An important part of the input of each GUHA procedure is thus a definition of at least one set of relevant Boolean attributes. A detailed definition of all possibilities of a definition of a set of relevant Boolean attributes is available in Section 4.4.

- Patterns the described GUHA procedures deal with are evaluated using several types of contingency tables. Overviews of patterns and types of contingency tables are available in Sections 4.1 and 4.2. Principles of evaluation of particular patterns are described in Section 4.3.

■ All the GUHA procedures use a secured approach to missing information when dealing with Boolean attributes. This approach is combined with additional possibilities depending on a particular procedure. Introductory information on dealing with missing information is available in Section 4.5. Additional information is in Chapters 7–13 concerning particular procedures.

■ Three Chapters 16–18, in Section III can be used as a source of research results and challenges related to the generation of artificial data, applications of formalized items of domain knowledge, logical deduction rules and expert deduction rules related to the presented GUHA procedures.

Chapter 2

Datasets

We use four datasets to demonstrate the possibilities of the GUHA procedures. Which datasets and why are they used are explained in Section 2.1. Details on particular datasets are presented in Sections 2.2–2.5.

2.1 Which Datasets and Where are they Used

The first dataset we use is the *Adult* dataset mentioned already in Section 1.4.1. It is used in Section 7.2 to compare the well known algorithm apriori and the 4ft-Miner GUHA procedure that concerns datamining with association rules. This dataset is used also in Section 7.3 to present additional features of the 4ft-Miner GUHA procedure. The *Adult* dataset is further used in Section 8.2 to present several examples of applications of the CF-Miner GUHA procedure in data mining with increasing or decreasing conditional histograms. It is used also in Sections 17.1.1 and 17.3 in considerations on applications of domain knowledge in data mining with association rules and conditional histograms. The *Adult* dataset is used also in Chapter 3 to present principles of the GUHA procedures. The *Adult* dataset is described in Section 2.2 in more detail.

The second dataset we use is the *UK Car Accidents* dataset. It comes from a very popular source KAGGLE of datasets available at https://www.kaggle.com/. Transformations of the *UK Car Accidents* dataset resulted in a data matrix *Accidents* with 538 989 rows. The *Accidents* data matrix is used in Sections 7.4 and 8.3 to present additional examples of data mining with association rules and conditional histograms. Examples of applications of the GUHA procedures SD4ft-Miner and SDCF-Miner presented in Sections 10.2 and 11.2 are also based on the *Accidents* data matrix. The SD4ft-

Miner GUHA procedure mines for interesting couples of conditional association rules. The SDCF-Miner GUHA procedure mines for interesting couples of conditional histograms rules. The *Accidents* data matrix is introduced in Section 2.3.

The medical dataset *STULONG* is the third dataset we use. It is used in Sections 9.2 and 12.2 to present possibilities of the KL-Miner and SDKL-Miner GUHA procedures. The KL-Miner GUHA procedure mines for interesting conditional relations of two categorical attributes. The SDKL-Miner procedure mines for couples of data matrices interesting what concerns relations of two categorical attributes. The *STULONG* dataset is also used in Section 13.2 to demonstrate possibilities of the Ac4ft-Miner procedure to mine with action rules. The *STULONG* dataset is described in Section 2.4.

The fourth dataset we use is an artificial dataset *Hotel* concerning a fictive hotel. This dataset was created with a help of the ReverseMiner procedure introduced in Chapter 16. The *Hotel* dataset is used to present possibilities of the KL-Miner and Ac4ft-Miner procedures in Sections 9.3 and 13.3 respectively. The *Hotel* dataset is described in Section 2.5.

2.2   *Adult* **Dataset**

The *Adult* dataset is available from *UCI Machine Learning Repository* (https://archive.ics.uci.edu/ml/datasets/Adult) [14]. The basic information concerning this dataset is presented in Section 2.2.1 together with details on particular attributes involved in the source dataset. Several additional derived attributes presented in Section 2.2.2 are used to create the *Adult* data matrix. Items of domain knowledge relevant to the *Adult* dataset are introduced in Section 2.2.3.

2.2.1   *Adult* **Dataset—basic info**

The *Adult* dataset concerns 48 842 persons. Each person is described by fifteen attributes. We do not use the attributes fnlwgt and education-num. The attribute fnlwgt is a weight calculated by the creators of the dataset from control data provided by the Population Division of the U.S. census bureau. The attribute education-num is a numeric representation of the attribute education.

We decided not to use the attribute fnlwgt because this attribute is also removed in the model application of the apriori algorithm described in [20]. The apriori application described in [20] is used in [138] to compare the apriori algorithm and the 4ft-Miner GUHA procedure. Removing the fnlwgt attribute ensures that we will deal with the same dataset as used in [20, 138]. There are nine categorical attributes introduced in Table 2.1.

Table 2.1: Categorical attributes.

	Attribute	Categories/frequencies
1	education 16 categories	*Preschool*/83, *1st-4th*/247, *5th-6th*/509, *7th-8th*/955 *9th*/756, *10th*/1 389, *11th*/1 812, *12th*/657, *HS-grad*/15 784, *Some-college*/10 878, *Assoc-acdm*/1 601, *Assoc-voc*/2 061, *Bachelors*/8 025, *Masters*/2 657, *Prof-school*/834, *Doctorate*/594
2	income missings: 16 281	*Small*/24 720, *Large*/7 841
3	sex	*Female*/16 192, *Male*/32 650
4	marital_status 7 categories	*Divorced*/6 633, *Married-AF-spouse*/37, *Married-civ-spouse*/22 379, *Married-spouse-absent*/628, *Never-married*/16 117, *Separated*/1 530, *Widowed*/1 518
5	relationship 6 categories	*Husband*/19 716, *Not-in-family*/12 583, *Other-relative*/1 506, *Own-child*/7 581, *Unmarried*/5 125, *Wife*/2 331
6	workclass 8 categories missings: 2 799	*Federal-gov*/1 432, *Local-gov*/3 136, *Without-pay*/21, *Never-worked*/10, *Private*/33 906, *Self-emp-inc*/1 695, *Self-emp-not-inc*/3 862, *State-gov*/1 981
7	occupation 14 categories missings: 2 809	*Adm-clerical*/5 611, *Craft-repair*/6 112, *Exec-managerial*/6 086, *Machine-op-inspct*/3 022, *Other-service*/4 923, *Prof-specialty*/6 172, *Sales*/5 504, *+ 7 occupations with frequency* < 3 000, *their total frequency* = 8 603
8	race 5 categories	*Amer-Indian-Eskimo*/470, *Asian-Pac-Islander*/1 519, *Black*/4 685, *Other*/406, *White*/41 762
9	native_country 41 categories missings: 857	*United-States*/43 832, *Mexico*/951, *Philippines*/295, *Germany*/206, *Puerto-Rico*/184, *Canada*/182, *El-Salvador*/155, *India*/151, *+ 33 countries* with frequency < 150, their total frequency = 1 172

The expression "*Preschool*/83" in the first row of Table 2.1 means that there are 83 persons with preschool education. This is the same for additional categories and attributes. There are missing values for the attributes income, workclass, occupation, and native_country introduced in Table 2.1.

There are four attributes with possible values – positive integers listed in Table 2.2. They are used to create derived attributes according to Section 2.2.2.

2.2.2 *Adult* **Dataset—derived attributes**

New categories are defined for the attributes hours_per_week, capital_gain, and capital_loss in Table 2.3. A defining interval and a frequency of a category are introduced. The definitions of categories are the same as in [20] and [138].

The attribute age is used to define two derived attributes age_5, and age_exp see Section 1.5.2 and also Fig. 1.9.

Table 2.2: Attributes with possible values – natural integers.

	Attribute	Values
1	age	17, ..., 99, see the histogram in Fig. 1.9
2	hours_per_week	1, ..., 99
3	capital_gain	0, ..., 99 999
4	capital_loss	0, ..., 4 356

Table 2.3: New categories for hours_per_week, capital_gain, and capital_loss.

Attribute	Categories/frequencies
hours_per_week	*Part-time* – $\langle 1;25\rangle$/5 913, *Full-time* – $\langle 26;40\rangle$/28 577 *Over-time* – $\langle 41;60\rangle$/12 676, *Workaholic* – $\langle 61;168\rangle$/1 676
capital_gain	*None*/44 807, *Low* – $\langle 1;7\,268\rangle$/2 345, *High* – $\langle 7\,269;99\,999\rangle$/1 690
capital_loss	*None*/46 560, *Low* – $\langle 1;1\,887\rangle$/1 166, *High* – $\langle 1\,888;4\,500\rangle$/1 116

The *Adult* data matrix has attributes introduced in Table 2.1, Table 2.3, and attributes age_5 and age_exp.

2.2.3 *Adult* **Dataset—items of domain knowledge**

There are four types of items of domain knowledge for the *Adult* dataset:

■ groups of attributes

■ knowledge relevant to transformations of attributes

■ mutual dependence of attributes

■ global properties of attributes.

All these items of domain knowledge have already been introduced in Section 1.5.

2.3 *UK Car Accidents* **Dataset**

The *UK Car Accidents* dataset we use is available at https://www.kaggle.com/silicon99/dft-accident-data. It comes from the UK Department for Transport and concerns car accidents in UK in the period 2005–2015. We use this dataset because it is well understood. However, our goal is not to present an exhaustive analysis of this dataset. This requires more specific knowledge than we have and that can be assumed by readers.

UK Car Accidents dataset consists of three files – Accidents0515.csv, Casualties0515.csv, and Vehicles0515.csv. Each row of the file Accidents0515.csv corresponds to one accident in the period from January 1st, 2005 to December

31st, 2015. There are 1 780 653 accidents. 538 989 accidents concern one vehicle, 1 058 932 accidents concern two vehicles, 142 219 accidents concern three vehicles, etc. The highest number of vehicles involved in one accident is 67.

We limit ourselves to accidents concerning one vehicle only to decrease the complexity of analysis. There are 538 989 accidents concerning only one vehicle. Thus, we use the *Accidents* data matrix with 538 989 rows prepared in a way described in Section 2.3.1. Six groups of columns – attributes of this data matrix are defined. They are described in Sections 2.3.2–2.3.7.

2.3.1 Accidents *data matrix and groups of attributes*

We imported files Accidents0515.csv and Vehicles0515.csv into tables Accidents0515 and Vehicles0515 of an MS-ACCESS database. Then we prepared a table Accidents_1 from table Accidents0515 such that all rows corresponding to accidents related to more than one vehicle were filtered out. Column Number_of_Vehicles was used. After that, we joined this table with table Vehicles0515 using the identifier Accident_Index. The resulting table Accidents_1_Vehicle contains all available information relevant to accidents of one vehicle. It was used to create an *Accidents* data matrix .

Either actual values or codes – natural numbers are used in files Accidents0515.csv and Vehicles0515.csv. Labels for codes are given in several *.csv files. Missing values are coded as -1 in the original data.

The columns of the *Accidents* data matrix were used to create attributes for further analysis. Each column corresponds to one attribute. However, we use only some of the attributes which can be created in this way. The attributes – columns of the Accidents_1 table were divided into naturally defined groups. The definitions of groups of attributes are stored also in the LISp-Miner system. A way of presenting groups by the LISp-Miner system is shown in Fig. 2.1.

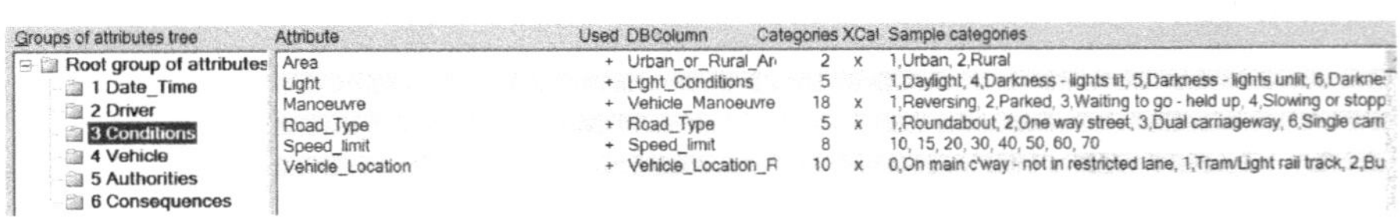

Figure 2.1: *Accidents* groups of attributes.

2.3.2 Group Date_Time

There are four attributes in the group Date_Time: Year, Month, DayOfWeek, and Hour. Histograms showing distribution of accidents among categories of these attributes are available in Fig. 2.2–Fig. 2.5.

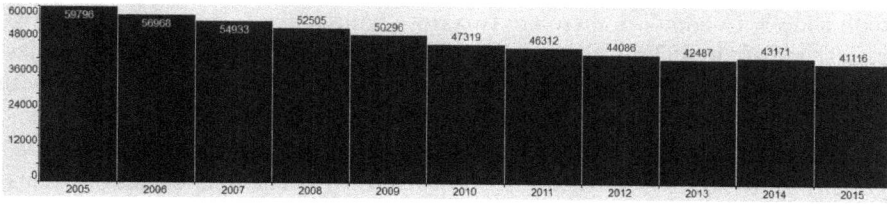

Figure 2.2: Attribute Year – the numbers of accidents in years 2005–2015.

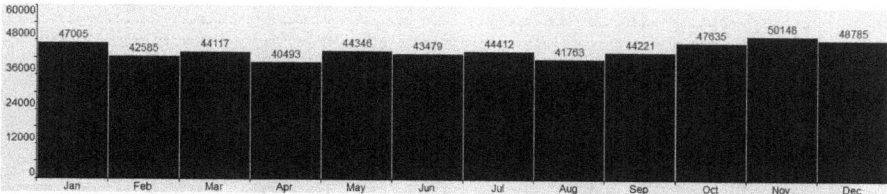

Figure 2.3: Attribute Month – the numbers of accidents in particular months.

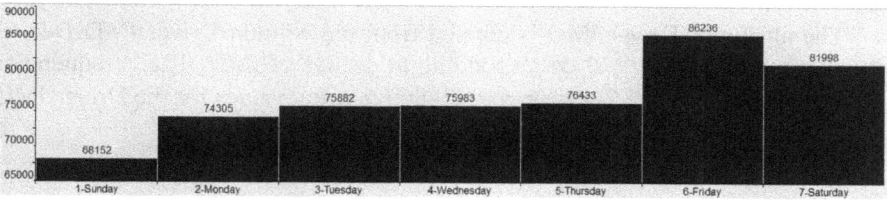

Figure 2.4: Attribute DayOfWeek – the numbers of accidents in days of week.

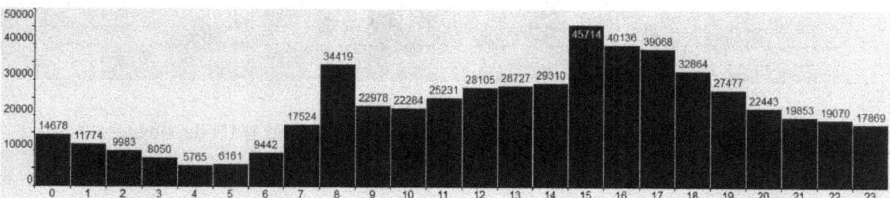

Figure 2.5: Attribute Hour – the numbers of accidents in particular hours of the day.

2.3.3 Group Driver

There are four attributes in the group Driver: Sex, Age, Driver_IMD, and Journey.

The attribute Sex is created from the column Sex_of_Driver. There is a special code 3 with the label *Not known* for this column. There are 36 451 rows with code 3. In addition, there are two rows with code -1. We joined these two codes, and all 36 453 rows (i.e., 6.7%) are further considered as rows with missing values for the attribute Sex. The Sex attribute has two categories: *Male* with frequency 365 266, i.e., 67.8% and *Female* with frequency 137 270, i.e., 25.5%.

The attribute Age is created from the column Driver_Age_Band. It has categories *0–5, 6–10, 11–15, 16–20, …, 66–75, Over 75*. The frequencies of the first three categories are too low – 16, 110, 1011 respectively. We deleted these three categories, and we work further with eight categories *16–20, …, 66–75, Over 75*. Their frequencies are available in Fig. 2.6.

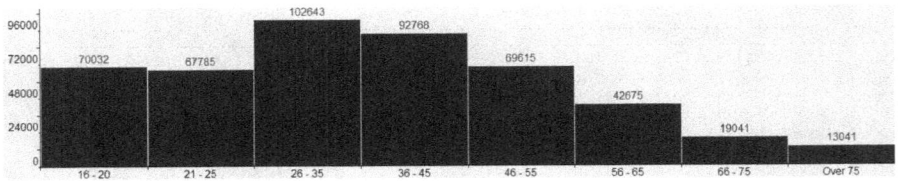

Figure 2.6: The numbers of accidents for particular categories of the Age attribute.

The attribute Driver_IMD is created from the column Driver_IMD_Decile. It has categories *1, …, 10* corresponding to deciles of IMD. Their frequencies are introduced in Fig. 2.7. There are 199 869 missing values for the Driver_IMD attribute.

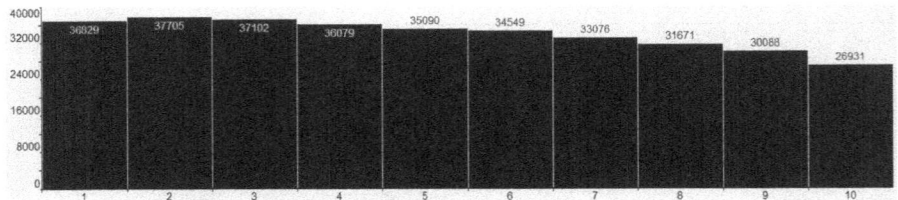

Figure 2.7: The numbers of accidents in particular IMD deciles.

The attribute Journey is created from the column Journey_Purpose. There are 3 697 rows with the code 5 – *Other*, 147 970 rows with the code 6 – *Not known*, 218 981 rows with the code 15 – *Other/Not known (2005-10)*, and 8 101 rows with the code -1 – *Data missing or out of range* for the column Journey_Purpose. We joined all these rows and we consider them as rows with a value (a category) *Other/Not known*. The attribute Journey has five categories: *Other/Not known* with frequency 378 749, i.e., 70.3%, *Journey as part of work* with frequency 104 270, i.e., 19.3%, *Commuting to/from work* with frequency 47 450, i.e., 8.8%,

Taking pupil to/from school with frequency 7 183, i.e., 1.3%, and *Pupil riding to/from school* with frequency 1 337, i.e., 0.2%.

2.3.4   Group Conditions

There are six attributes in the group Conditions: Area, Road_Type, Speed_Limit, Light, Vehicle_Location, and Manoeuvre.

The attribute Area is created from the column Urban_or_Rural_Area. The attribute has two categories: *Urban* with frequency 340 200, i.e., 63.1% and *Rural* with frequency 198 743, i.e., 36.9%. There are 46 missing values.

The attribute Road_Type is created from the column Road_Type. The attribute has five categories: *Single carriageway* with frequency 427 848, i.e., 79.4%, *Dual carriageway* with frequency 65 947, i.e., 12.2%, *Roundabout* with frequency 18 275, i.e., 3.4%, *One way street* with frequency 17 665, i.e., 3.3%, and *Slip road* with frequency 4 897, i.e., 0.9%. There are 4 357 missing values coded as 9 – *Unknown*, this corresponds to 0.8%.

The attribute Speed_Limit is created from the column Speed_Limit. It has eight categories: *30* with frequency 350 099, i.e., 65.0%, *60* with frequency 100 795, i.e., 18.7%, *70* with frequency 33 864, i.e., 6.3%, *40* with frequency 31 335, i.e., 5.8%, *50* with frequency 13 234, i.e., 2.5%, *20* with frequency 9 636, i.e., 1.8%, *10* with frequency 16, i.e., 0.0%, and *15* with frequency 10, i.e., 0.0%.

The attribute Light is created from the column Light_Conditions. It has eight categories: *Daylight* with frequency 358 282, i.e., 66.5%, *Darkness—lights lit* with frequency 118 168, i.e., 21.9%, *Darkness—no lighting* with frequency 52 923, i.e., 9.8%, *Darkness—lighting unknown* with frequency 6 682, i.e., 1.2%, and *Darkness—lights unlit* with frequency 2 934, i.e., 0.5%.

The attribute Vehicle_Location is created from the column Vehicle_Location-_Restricted_Lane. It has ten categories. We state details on the three most frequent categories: *On main c'way—not in the restricted lane* with frequency 518 451, i.e., 96.2%, *Footway (pavement)* with frequency 11 445, i.e., 2.1%, and *Bus lane* with frequency 4 169, i.e., 0.8%. They represent 99.1% of all accidents. There are 43 missing values.

The attribute Manoeuvre is created from the column Vehicle_Manoeuvre. It has 18 categories. We state details on the five most frequent categories: *Going ahead other* with frequency 295 473, i.e., 54.8%, *Going ahead right-hand bend* with frequency 54 050, i.e., 10.0%, *Going ahead left-hand bend* with frequency 43 449, i.e., 8.1%, *Slowing or stopping* with frequency 25 722, i.e., 4.8%, and *Turning right* with frequency 25 477, i.e., 4.7%. They represent 82.4% of all accidents. There are 323 missing values.

2.3.5   Group Vehicle

There are two attributes in the group Vehicle: Type and Vehicle_Age.

The attribute Type is created from the column Vehicle_Type. It has 21 categories. We state details on the five most frequent categories: *Car* with frequency 386 346, i.e., 71.7%, Bus or coach (17 or more pass seats), i.e., *Bus_coach_17+* with frequency 40 935, i.e., 7.6%, *Van/Goods 3.5 tonnes mgw or under* with frequency 20 869, i.e., 3.9%, *Motorcycle over 500cc* (sometimes denoted as Motorcycle 500cc+) with frequency 20 639, i.e., 3.8%, and *Motorcycle 125cc and under* with frequency 15 383, i.e., 2.9%. They represent 89.8% of all accidents.

The attribute Vehicle_Age is created from the column Age_of_Vehicle. The column Age_of_Vehicle has 82 values. We joined values from 16 to 20 to one category and all values over 20 to one additional category. This resulted in 17 categories of the attribute Vehicle_Age, see Fig. 2.8.

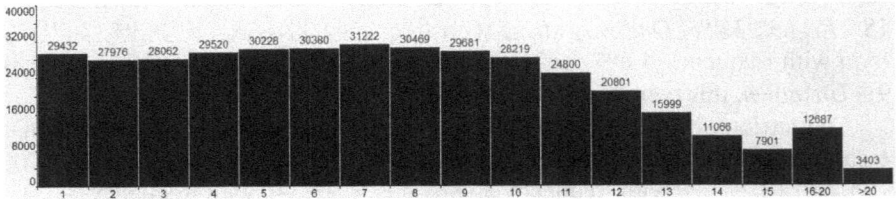

Figure 2.8: Frequencies of categories of attribute Vehicle_Age.

2.3.6 Group Authorities

There are four attributes in the group Authorities: District, District2+, Highway, and Police.

The attribute District is created from the column Local_Authority_District. It has 416 categories. We state details on the five most frequent categories: *Birmingham* with frequency 9 461, i.e., 1.8%, *Leeds* with frequency 7 365, i.e., 1.4%, *Westminster* with frequency 6 636, i.e., 1.2%, *Glasgow City* with frequency 6 425, i.e., 1.2%, and *Edinburgh, City* of with frequency 5 495, i.e., 1.0%. They represent 6.6% of all accidents.

The attribute District2+ is created from the column Local_Authority_District such that only districts with at least 2 000 accidents are considered. The five most frequent categories are the same as for the attribute District.

The attribute Highway is created from the column Local_Authority_Highway. It has 207 categories. We state details on the five most frequent categories: *E10000016, Kent* with frequency 14 777, i.e., 2.7%, *E10000030, Surrey* with frequency 12 909, i.e., 2.4%, *E10000017, Lancashire* with frequency 11 760, i.e., 2.2%, *E10000012, Essex* with frequency 10 976, i.e., 2.0%, and *E10000014, Hampshire* with frequency 10 538, i.e., 2.0%. They represent 11.3% of all accidents.

The attribute Police is created from the column Police_Force. It has 51 categories. We state details on the five most frequent categories: *Metropolitan Police* with frequency 79 135, i.e., 14.7%, *West Yorkshire* with frequency 20 608, i.e., 3.8%, *Strathclyde* with frequency 20 261, i.e., 3.8%, *West Midlands* with frequency 20 241, i.e., 3.8%, and *Thames Valley* with frequency 18 322, i.e., 3.4%. They represent 29.4% of all accidents.

2.3.7   Group Consequences

We use only the Severity attribute from the group Consequences. The attribute Severity is created from the column Accident_Severity. It has 3 categories: *Fatal* with frequency 10 349, i.e., 1.9%, *Serious* with frequency 105 754, i.e., 19.6%, and *Slight* with frequency 422 886, i.e., 78.5%.

2.4   *STULONG* Dataset

The medical dataset *STULONG*[1] was one of the datasets offered at Discovery Challenge workshops of the ECML/PKDD conferences 2002, 2003, and 2004 [8]. The *STULONG* dataset is a result of a longitudinal study with several goals related to atherosclerosis risk factors prevalence in a population of middle aged men. It consists of several mutually related files. We use here only a part of the *Entry* data matrix describing results of entry examination of observed men.

Our goal is not to present a qualified application of statistical analysis or even to get new medical knowledge. Our only goal is to demonstrate some of the possibilities of three GUHA procedures implemented in the LISp-Miner system.

The *Entry* data matrix is introduced in Section 2.4.1 together with groups of attributes. Attributes of particular groups are described in Sections 2.4.2–2.4.8.

2.4.1   Entry data matrix and groups of attributes

1 417 men have been examined during the entry examination. 244 attributes have been surveyed with each patient. Values of 64 attributes are either codes or results of size measurements of different variables or results of transformations of the rest of the attributes. Values of all these 64 attributes are stored in the data matrix *Entry*. We use here the *Entry* data matrix with 27 attributes divided into 7 groups introduced in Fig. 2.9.

---

[1]The study (STULONG) was realized at the 2nd Department of Medicine, 1st Faculty of Medicine of Charles University and Charles University Hospital, U nemocnice 2, Prague 2 (head. Prof. M. Aschermann, MD, SDr, FESC), under the supervision of Prof. F. Boudík, MD, ScD, with collaboration of M. Tomeǎková, MD, PhD and Ass. Prof. J. Bultas, MD, PhD. The data were transferred to the electronic form by the European Centre of Medical Informatics, Statistics and Epidemiology of Charles University and Academy of Sciences (head. Prof. RNDr. J. Zvárová, DrSc).

Groups of attributes tree	Attribute	Used	DBColumn	Categories	XCal	Sample categories
⊟ 📁 Root group of attributes	FamilyAnamnesisRrisk	+	RARISK	2	x	yes, no
📁 1 Personal	HypertensionRisk	+	HTRISK	2	x	no, yes
📁 2 Anamnesis	CholesterolRisk	+	CHOLRISK	2	x	yes, no
📁 **3 Risks**	ObesityRisk	+	OBEZRISK	2	x	yes, no
📁 4 Measurement	SmokingRisk	+	KOURRISK	2	x	yes, no
📁 5 Alcohol consumptic						
📁 6 Blood pressure						
📁 7 Biochemical examir						

Figure 2.9: *STULONG* groups of attributes.

2.4.2 Group Personal

There are four attributes in the group Personal: Age, Education, MaritalStatus, and Responsibility.

The attribute Age has two categories: ≤ 45 with frequency 561, i.e., 39.6% and > 45 with frequency 856, i.e., 60.4%.

The attribute Education has four categories: *basic* with frequency 151, i.e., 10.7%, *apprentice* with frequency 405, i.e., 28.6%, *secondary* with frequency 444, i.e., 31.3%, and *university* with frequency 397, i.e., 28.0%. There are 20 missing values corresponding to 1.4%.

The attribute MaritalStatus has four categories: *married* with frequency 1 207, i.e., 85.2%, *divorced* with frequency 104, i.e., 7.3%, *single* with frequency 95, i.e., 6.7%, and *widower* with frequency 10, i.e., 0.7%. There is one missing value.

Attribute Responsibility concerns responsibility in a job. It has four categories: manager with frequency 286, i.e., 20.2%, partly independent with frequency 435, i.e., 30.7%, others with frequency 636, i.e., 44.9%, and pensioner with frequency 25, i.e., 1.8%. There are 35 missing values corresponding to 2.5%.

2.4.3 Group Anamnesis

There are five Boolean attributes in the group Anamnesis: Diabetes, Hyperlipidemia, Hypertension, Ictus, and Myocardial infarction, see Table 2.4.

Table 2.4: Boolean attributes of group Anamnesis.

Attribute	yes		no		missing	
	%	Freq.	%	Freq.	%	Freq.
Diabetes	2.1	30	97.2	1 378	0.6	9
Hyperlipidemia	3.8	54	57.5	815	38.7	548
Hypertension	15.5	220	84.1	1 192	0.4	5
Ictus	0.1	2	99.4	1 408	0.5	7
Myocardial infarction	2.4	34	97.2	1 378	0.4	5

2.4.4   Group Risks

There are five Boolean attributes in the group Risks: FamilyAnamnesisRrisk, HypertensionRisk, CholesterolRisk, ObesityRisk, and SmokingRisk, see Table 2.5.

Table 2.5: Boolean attributes of group Risk.

Attribute	yes		no		missing	
	%	Freq.	%	Freq.	%	Freq.
FamilyAnamnesisRisk	81.7	1 158	17.6	250	0.6	9
HypertensionRisk	24.7	350	72.6	1 029	2.7	38
CholesterolRisk	26.9	381	72.1	1 021	1.1	15
ObesityRisk	21.5	305	77.9	1 104	0.6	8
SmokingRisk	43.8	620	56.2	796	0.1	1

2.4.5   Group Measurement

There are three attributes in the group Measurement: BMI, Subscapularis, and Triceps.

The attribute BMI has six categories defined using value $val = \frac{weight\ in\ kg}{(height\ in\ m)^2}$: *underweight* ($val < 18.5$) with frequency 4, i.e., 0.3%, *normal* ($val \in \langle 18.5; 25 \rangle$) with frequency 534, i.e., 37.7%, *overweight* ($val \in \langle 25; 30 \rangle$) with frequency 723, i.e., 51.0%, *obese class I* ($val \in \langle 30; 35 \rangle$) with frequency 127, i.e., 9.0%, *obese class II* ($val \in \langle 35; 40 \rangle$) with frequency 19, i.e., 1.3%, and *obese class III* ($val \geq 40$) with frequency 2, i.e., 0.1%. There are 8 missing values correspond to 0.6%.

The attribute Subscapularis has three categories defined using the height of a skinfold above musculus subscapularis in mm denoted as val_S: *low* ($val_S < 15$) with frequency 392, i.e., 27.7%, *average* ($val_S \in \langle 15; 21 \rangle$) with frequency 460, i.e., 32.5%, and *high* ($val_S \geq 21$) with frequency 428, i.e., 30.2%. There are 137 missing values corresponding to 9.7%.

The attribute Triceps has three categories defined using the height of a skinfold above musculus triceps in mm denoted as val_T. *low* ($val_T < 8$) with frequency 400, i.e., 28.2%, *average* ($val_T \in \langle 8; 11 \rangle$) with frequency 443, i.e., 31.3%, and *high* ($val_T \geq 11$) with frequency 438, i.e., 30.9%. There are 136 missing values corresponding to 9.6%.

2.4.6   Group Alcohol consumption

There are four attributes in the group Alcohol consumption: Alcohol, Beer, Wine, and Liquors.

The attribute Alcohol has three categories: *no* with frequency 131, i.e., 9.2%, *occasionally* with frequency 748, i.e., 52.8%, and *regularly* with frequency 428, i.e., 30.2%. There are 76 missing values corresponding to 5.4%.

The attribute Beer has three categories: *no* with frequency 465, i.e., 32.8%, *up to 1 litre/day* with frequency 777, i.e., 54.8%, and *more than 1 litre/day* with frequency 157, i.e., 11.1%. There are 18 missing values corresponding to 1.3%.

The attribute Wine has three categories: *no* with frequency 675, i.e., 47.6%, *up to half a litre/day* with frequency 36, i.e., 2.5%, and *more than half a litre/day* with frequency 36, i.e., 2.5%. There are 17 missing values corresponding to 1.2%.

Attribute Liquors has three categories: *no* with frequency 759, i.e., 53.6%, *up to 100 cc/day* with frequency 574, i.e., 40.5%, and *more than 100 cc/day* with frequency 76, i.e., 5.4%. There are 8 missing values corresponding to 0.6%.

2.4.7 Group Blood pressure

There are two attributes in the group Blood pressure: Diastolic and Systolic.

The Diastolic attribute resulted from two consecutive measurements of diastolic blood pressure. Results of these measurements are denoted as $Diastolic_1$ and $Diastolic_2$. There are five categories of the Diastolic attribute determined using value $val_i = \frac{Diastolic_1 + Diastolic_2}{2}$: *very low* ($val_i < 75$) with frequency 254, i.e., 17.9%, *lower* ($val_i \in \langle 75; 80 \rangle$) with frequency 159, i.e., 11.2%, *average* ($val_i \in \langle 80; 85 \rangle$) with frequency 359, i.e., 25.3%, *higher* ($val_i \in \langle 85; 90 \rangle$) with frequency 190, i.e., 13.4%, and *very high* ($val_i \geq 90$) with frequency 449, i.e., 31.7%. There are 6 missing values corresponding to 0.4%.

The Systolic attribute resulted from two consecutive measurements of systolic blood pressure. Results of these measurements are denoted as $Systolic_1$ and $Systolic_2$. There are five categories of the Systolic attribute determined using value $val_Y = \frac{Systolic_1 + Systolic_2}{2}$: *very low* ($val_Y < 115$) with frequency 209, i.e., 14.7%, *lower* ($val_Y \in \langle 115; 125 \rangle$) with frequency 321, i.e., 22.7%, *average* ($val_Y \in \langle 125; 132 \rangle$) with frequency 273, i.e., 19.3%, *higher* ($val_i \in \langle 85; 90 \rangle$) with frequency 309, i.e., 21.8%, and *very high* ($val_Y \geq 145$) with frequency 299, i.e., 21.1%. There are 6 missing values corresponding to 0.4%.

Let us note, that we are aware of additional possibilities of definition of categories of attributes Diastolic and Systolic. However, our goal is not a deep analysis of the *STULONG* data. Our goal is only to demonstrate the possibilities of the GUHA procedures of the LISp-Miner system.

2.4.8 Group Biochemical examination

There are three attributes in the group Biochemical examination: Urine, Cholesterol, and Triglycerides.

The Urine attribute has three categories: *normal* with frequency 1 351, i.e., 95.3%, *sugar positive* with frequency 31, i.e., 2.2%, and *albumen positive* with frequency 17, i.e., 1.2%. There are 18 missing values corresponding to 1.3%.

The Cholesterol attribute has five categories defined using the level of cholesterol in mg% denoted as val_{Ch}: *very low* ($val_{Ch} < 196$) with frequency 276, i.e., 19.5%, *lower* ($val_{Ch} \in \langle 196; 220 \rangle$) with frequency 264, i.e., 18.6%, *average* ($val_{Ch} \in \langle 220; 243 \rangle$) with frequency 299, i.e., 21.1%, *higher* ($val_{Ch} \in \langle 243; 270 \rangle$) with frequency 265, i.e., 18.7%, and *very high* ($val_{Ch} \geq 270$) with frequency 298, i.e., 21.0%. There are 15 missing values corresponding to 1.1%.

The Triglycerides attribute has five categories defined using the level of triglycerides in mg% denoted as val_T: *very low* ($val_T < 100$) with frequency 218, i.e., 15.4%, *lower* ($val_T \in \langle 100; 128 \rangle$) with frequency 222, i.e., 15.7%, *average* ($val_T \in \langle 128; 162 \rangle$) with frequency 299, i.e., 21.1%, *higher* ($val_T \in \langle 162; 209 \rangle$) with frequency 231, i.e., 16.3%, and *very high* ($val_T \geq 209$) with frequency 229, i.e., 16.2%. There are 288 missing values corresponding to 20.3%.

2.5 Fictive *Hotel* Dataset

The artificial dataset *Hotel* concerning a fictive hotel was created with the help of the ReverseMiner procedure introduced in Chapter 16. The set consists of three data matrices – *Hotel*, *Meteo*, and *Forex*. The *Hotel* data matrix has 2 000 rows. Each row corresponds to one stay of a guest in the hotel. The stays come from the period January 1st, 2012–December 31st, 2013.

Each of the *Meteo* and *Forex* data matrices has 731 rows, one row for each day. The *Meteo* data matrix contains weather information. The *Forex* data matrix contains information concerning the exchange rate CZK/EUR.

These three data matrices were joined into one data matrix *HotelPlusExternal* which is used in several applications of the GUHA procedures. The *HotelPlusExternal* data matrix is introduced in Section 2.5.1 together with groups of attributes. Attributes of particular groups are described in Sections 2.5.2–2.5.8. Note that names of some of the attributes are used without prefixes used in Chapter 16. This means that we use Month instead of VMonth, etc. There are no missing values in the *HotelPlusExternal* data matrix.

2.5.1 HotelPlusExternal *data matrix and groups of attributes*

Dozens of attributes can be derived from the *HotelPlusExternal* data matrix. We consider 26 attributes divided into seven groups. The definitions of groups of attributes are stored also in the LISp-Miner system, see Fig. 2.10.

The group Domicile is a subgroup of the group Guest. The groups Check-in and Price are subgroups of the group of Stay.

Groups of attributes tree	Attribute	Used DBColumn	Categories XCat	Sample categories
⊟ ☐ Root group of attributes	Age	+ GAge	4	below 21, 21-27, 28-59, 60+
⊟ ☐ Guest	Sex	+ GSex	2	female, male
☐ Domicile				
☐ Meteo				
☐ Questionnaire				
⊟ ☐ Stay				
☐ Check-in				
☐ Price				

Figure 2.10: *HotelPlusExternal* groups of attributes.

2.5.2 Group Guest

There are two attributes in the group Guest: Age and Sex. In addition, the group Domicile described in Section 2.5.3 is a subgroup of the group Guest.

The attribute Age has four categories: *below 21* with frequency 120, i.e., 6.0%, *21–27* with frequency 230, i.e., 11.5%, *28–59* with frequency 1020, i.e., 51.0%, and *60+* with frequency 630, i.e., 31.5%.

The attribute Sex has two categories: *female* with frequency 1010, i.e., 50.5%, and *male* with frequency 990, i.e., 49.5%.

2.5.3 Group Domicile

The Domicile group is a sub-group of the group Guest. There are three attributes in the group Domicile: State, City, and Foreigner.

The attribute State has five categories: *Austria* with frequency 515, i.e., 25.8%, *Czechia* with frequency 913, i.e., 45.6%, *Germany* with frequency 356, i.e., 17.8%, *Poland* with frequency 71, i.e., 3.5%, and *Slovakia* with frequency 145, i.e., 7.3%.

The attribute City has 28 categories. We state details on the five most frequent categories: *Linz* with frequency 254, i.e., 12.7%, *Budweis* with frequency 152, i.e., 7.6%, *Wien* with frequency 145, i.e., 7.3%, *Prague* with frequency 141, i.e., 7.0%, and *Muenchen* with frequency 141, i.e., 7.0%. They represent 41.6% of all stays.

The attribute Foreigner is related to Czechia. It has two categories: *yes* with frequency 1 087, i.e., 54.4%, and *no* with frequency 913, i.e., 45.6%.

2.5.4 Group Meteo

There are two attributes in the group Meteo: Sky and Temperature.

The attribute Sky has three categories: *cloudy* with frequency 660, i.e., 33.0%, *rain/snow* with frequency 500, i.e., 25.0%, and *sunny* with frequency 840, i.e., 42.0%.

The attribute Temperature has five categories: *extremely cold* with frequency 46, i.e., 2.3%, *cold* with frequency 757, i.e., 37.9%, *neutral* with frequency 528, i.e., 26.4%, *warm* with frequency 658, i.e., 32.9%, and *extremely hot* with frequency 11, i.e., 0.6%.

2.5.5 Group Questionnaire

There are five attributes in the group Questionnaire: QSatisfaction, QRoom, QFood, QService, and QCulture.

The attribute QSatisfaction concerns the overall satisfaction of guests with their stays in the hotel. It has three categories: *low* with frequency 570, i.e., 28.5%, *average* with frequency 950, i.e., 47.5%, and *high* with frequency 480, i.e., 24.0%.

The attribute QRoom concerns the satisfaction of guests with their room during their stays in the hotel. It has five categories: *very low* with a frequency of 388, i.e., 19.4%, *lower* with frequency 152, i.e., 7.6%, *average* with frequency 377, i.e., 18.9%, *higher* with frequency 414, i.e., 79.0%, and *very high* with frequency 412, i.e., 21.1%.

The attribute QFood concerns the satisfaction of guests with their food during their stays in the hotel. It has five categories: *very low* with frequency 377, i.e., 18.9%, *lower* with frequency 397, i.e., 19.9%, *average* with frequency 404, i.e., 20.2%, *higher* with frequency 416, i.e., 20.8%, and *very high* with frequency 406, i.e., 20.3%.

The attribute QService concerns the satisfaction of guests with the hotel service during their stays. It has five categories: *very low* with frequency 372, i.e., 18.6%, *lower* with frequency 392, i.e., 19.6%, *average* with frequency 418, i.e., 20.9%, *higher* with frequency 414, i.e., 20.7%, and *very high* with frequency 404, i.e., 20.2%.

The attribute QCulture concerns the satisfaction of guests with the cultural life in the hotel during their stays. It has five categories: *very low* with a frequency of 377, i.e., 18.9%, *lower* with frequency 422, i.e., 21.1%, *average* with frequency 389, i.e., 19.4%, *higher* with frequency 393, i.e., 19.6%, and *very high* with frequency 419, i.e., 20.9%.

2.5.6 Group Stay

There are five attributes in the group Stay: Nights, SaturdayNight, Persons, PersonNights, and TypeOfVisits.

The attribute Nights concerns the length of stays in nights. It has ten categories, see Fig. 2.11.

The attribute SaturdayNight has two categories saying if stays contain Saturday night or not: *yes* with frequency 1 390, i.e., 69.5%, and *no* with frequency 610, i.e., 30.5%.

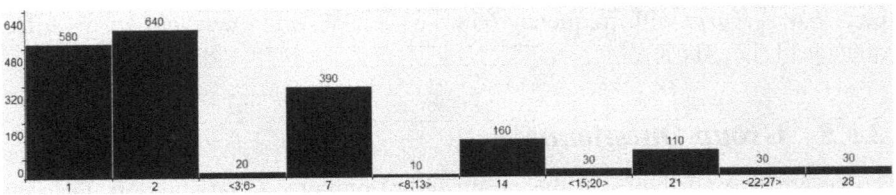

Figure 2.11: Frequencies of categories of attribute Nights.

The attribute Persons indicates the number of persons taking part in stays. It has four categories: *1* with frequency 540, i.e., 27.0%, *2* with frequency 770, i.e., 38.5%, *3* with frequency 280, i.e., 14.0%, and *4* with frequency 410, i.e., 20.5%.

The attribute PersonNights concerns a product val_{PN} of a category of the attribute Persons and a length of stay in nights. It has five categories: *very low* ($val_{PN} = 1$) with frequency 173, i.e., 8.7%, *lower* ($val_{PN} \in \langle 2; 4 \rangle$) with frequency 469, i.e., 23.4%, *average* ($val_{PN} \in \langle 4; 8 \rangle$) with frequency 533, i.e., 26.6%, *higher* ($val_{PN} \in \langle 8; 21 \rangle$) with frequency 362, i.e., 18.1%, and *very high* ($val_{PN} \geq 21$) with frequency 463, i.e., 23.1%.

The attribute TypeOfVisits attribute has two categories: *business* with frequency 410, i.e., 20.5%, and *private* with frequency 1 590, i.e., 79.5%.

2.5.7 Group Check-in

There are five attributes in the group Check-in: Year, Month, Season, Day-OfWeek, and WeekEnd.

The attribute Year has two categories: *2012* with frequency 944, i.e., 47.2%, and *2013* with frequency 1 056, i.e., 52.8%.

The attribute Month indicates a month of check-in. It has twelve categories, see Fig. 2.12.

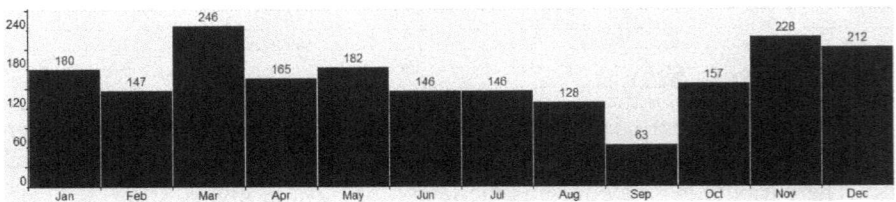

Figure 2.12: Categories of attribute Month.

The attribute Season has two categories: *yes* with frequency 959, i.e., 48.0%, and *no* with frequency 1 041, i.e., 52.0%. Yes means that a check-in is in *Jan, Feb, Jun, Jul, Aug*, or *Dec*.

The attribute DayOfWeek indicates a day of week of check-in. It has seven categories, see Fig. 2.13.

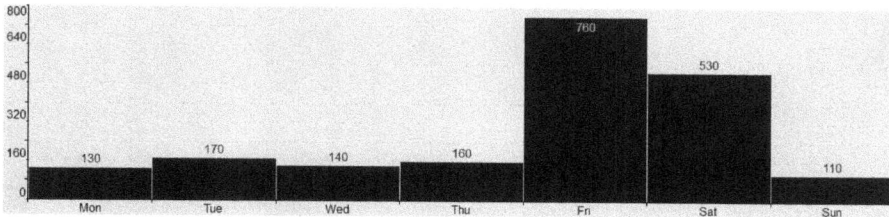

Figure 2.13: Categories of attribute DayOfWeek.

The WeekEnd attribute has two categories: *yes* with frequency 1 290, i.e., 64.5%, and *no* with frequency 710, i.e., 35.5%. *Yes* means that a check-in is in *Fri* or *Sat*.

2.5.8 Group Price

There are three attributes in the group Price: PriceRoom, PriceFood, and Price-Total.

The attribute PriceRoom indicates a price of a room in CZK. It has five categories: *very low* with frequency 371, i.e., 18.6%, *lower* with frequency 271, i.e., 13.6%, *average* with frequency 539, i.e., 26.9%, *higher* with frequency 412, i.e., 20.6%, and *very high* with frequency 407, i.e., 20.4%.

The attribute PriceFood indicates the price of food in CZK. It has five categories: *very low* with frequency 1 328, i.e., 66.4%, *lower* with frequency 10, i.e., 0.5%, *average* with frequency 34, i.e., 1.7%, *higher* with frequency 217, i.e., 10.9%, and *very high* with frequency 411, i.e., 20.6%.

The attribute PriceTotal indicates a total price of a stay in CZK. It has five categories: *very low* with frequency 398, i.e., 19.9%, *lower* with frequency 19.8 395, i.e., 0.5%, *average* with frequency 387, i.e., 19.4%, *higher* with frequency 413, i.e., 10.9%, and *very high* with frequency 407, i.e., 20.4%.

THE GUHA PROCEDURES

I

First, examples of patterns the all seven described GUHA procedures deal with are introduced in Chapter 3. All described procedures have common features described in Chapter 4. The LISp-Miner system is introduced in Chapter 5.

Chapter 3

Principle and Simple Examples

The main features of all seven GUHA procedures this book deals with are introduced. All the procedures use the same simple principle described in Section 3.1. The first three procedures mine for simple patterns – rules, histograms, and pairs of categorical attributes. The procedure 4ft-Miner dealing with association rules is presented in Section 3.2. Histograms and the procedure CF-Miner are introduced in Section 3.3. Pairs of categorical attributes are shortly described in Section 3.4 together with the KL-Miner procedure. All the simple patterns can be evaluated on sub-matrices of an analysed data matrix. A set of relevant sub-matrices can be defined in various ways.

Additional three procedures deal with couples of simple patterns. Couples of association rules and the procedure SD4ft-Miner are introduced in Section 3.5. Couples of histograms are presented in Section 3.6 together with the SDCF-Miner procedure. Couples of pairs of categorical attributes and the SDKL-Miner procedure are briefly described in Section 3.7. The Ac4ft-Miner procedure mines for action rules, see Section 3.8.

Note that the introduced patterns present only a form of patterns, which can be mined by the presented GUHA procedures. Real practical interestingness depends on many circumstances related to real application. Very often, the found patterns are widely known or they can be seen as clear consequences of various patterns of domain knowledge. More information relevant to this topic is available in Chapter 17.

Examples of applications of the above mentioned procedures GUHA procedures are available in [5, 44, 85, 93, 121, 124, 125, 126, 127, 128, 131, 132, 137, 138, 139, 140, 155, 156].

3.1 GUHA Procedures Principle

All the GUHA procedures work according to the schema shown in Fig. 3.1 introduced already in Fig. 1.1.

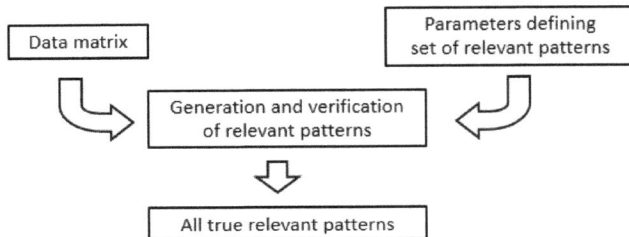

Figure 3.1: Schema of a GUHA procedure.

Particular procedures deal with different types of patterns informally introduced in Sections 3.2–3.8. Association rules, couples of association rules and action rules consist of Boolean attributes introduced in Section 1.4.3. Additional patterns consist of Boolean attributes, and categorical attributes introduced in Section 1.4.2.

The input of each GUHA procedure includes an analysed data matrix and at least one definition of a set of relevant Boolean attributes or of a set of relevant attributes. All used definitions have common features described in Chapter 4. All the patterns the GUHA procedures deal with are evaluated by conditions concerning contingency tables or couples of contingency table. Three types of contingency tables are used. They are informally introduced in Sections 3.2–3.8. Formal definitions of contingency tables are available in Chapter 4.

Each GUHA procedure has its specific set of conditions concerning contingency tables. These conditions are called quantifiers, and they are described in chapters concerning particular procedures. Examples of quantifiers are presented in Sections 3.2–3.8.

Note that the Boolean attributes particular procedures deal with have rich syntax. This follows from applications of basic Boolean attributes in a form $A(\alpha)$ where α is a subset of a set of all categories of the attribute A, see Section 1.4.3. Various types of subsets are automatically generated by particular procedures as described in Chapter 4. This makes the GUHA procedures different from the well known apriori algorithm [2] where only conjunctions of couples $A(a)$ are used where a is a category of an attribute A.

3.2 Association Rules and 4ft-Miner

The pattern

$$\text{education}(\textit{Masters, Prof-school}) \sim_{2.47} \text{income}(\textit{Large}) \qquad (3.1)$$

is a simple example of association rule concerning the data matrix *Adult* introduced in Section 2.2. It says that a relative frequency of persons with large income among persons satisfying education(*Masters, Profs-chool*) is 2.47 times higher than a relative frequency of persons with large income in the whole data matrix. A contingency table 4*ft*(education(*Masters, Profs-chool*), income(*Large*), *Adult*) introduced in Fig. 3.2 can be used to verify this assertion.

Adult	income(*Large*)	¬income(*Large*)
education(*Masters, Prof-school*)	1 382	2 109
¬ education(*Masters, Prof-school*)	6 459	38 892

Figure 3.2: 4*ft*(education(Masters, Profs-chool), income(Large), Adult).

Tables like 4*ft*(education(*Masters, Prof-school*), income(*Large*), *Adult*) are called *4ft-tables*. The 4ft-table in Fig. 3.2 says that there are 1 382 rows of the *Adult* data matrix satisfying both education(*Masters, Prof-school*) and income(*Large*), 2 109 rows satisfying education(*Masters, Prof-school*) and not satisfying income(*Large*), etc.

There are 1 382 + 2 109 persons satisfying education(*Masters, Prof-school*), thus a relative frequency of persons with large income among persons satisfying education(*Masters, Prof-school*) is $\frac{1382}{1382+2109} = 0.3959$. There are 1 382 + 6 459 persons satisfying income(*Large*), thus a relative frequency of persons with large income in the whole data matrix is $\frac{1382+6459}{1382+2109+6459+38882} = 0.1605$. It holds $\frac{0.3959}{0.1605} = 2.47$ which confirms the assertion of the rule (3.1).

A pattern (3.2) is a general form of association rules.

$$\varphi \approx \psi \qquad (3.2)$$

Here φ and ψ are Boolean attributes, φ is an *antecedent*, ψ is a *succedent* (*consequent*) of the rule (3.2). The symbol $\approx$ is a *4ft-quantifier*. The symbol $\sim_{2.47}$ in the rule (3.1) is a *4ft-quantifier of lift*.

The pattern

$$\text{occupation}(\textit{Adm-clerical}) \rightarrow_{0.96} \text{hours}(\textit{Part-Time,Full-time})/\text{age}(\textit{Old}) \qquad (3.3)$$

is a simple example of a conditional association rule concerning also the data matrix *Adult*. It says that if we consider a data sub-matrix *Adult*/age(*Old*), then 96 percent of persons satisfying occupation(Adm-clerical) satisfy also hours(Part-Time,Full-time).

This can be verified using a contingency table $4ft$(occupation(*Adm-clerical*), hours(*Part-Time,Full-time*), *Adult*/age(*Old*)) shown in Fig. 3.3. We write "*Adm-clerical*" only instead of "occupation(*Adm-clerical*)" in Fig. 3.3; similarly for "*Part-Time,Full-time*".

Adult/age(*Old*)	Part-Time,Full-time	¬(Part-Time,Full-time)
Adm-clerical	151	6
¬ Adm-clerical	231	1 646

Figure 3.3: $4ft$((*Adm-clerical*), (*Part-Time,Full-time*), *Adult*/age(*Old*)).

There are $151 + 6$ persons satisfying occupation(Adm-clerical) and 151, i.e., 96 percent of them satisfy also hours(Part-Time,Full-time). This confirms the assertion of the rule (3.3). Let us note that the symbol $\rightarrow_{0.96}$ is a *4ft-quantifier of confidence*.

A pattern (3.4) is a general form of conditional association rules.

$$\varphi \approx \psi/\chi \qquad\qquad (3.4)$$

Here χ is a Boolean attribute, it is a condition of the conditional rule (3.4).

More details on association rules and the 4ft-Miner procedure dealing with association rules and conditional association rules are available in Chapter 7.

3.3 Histograms and CF-Miner

There are two examples of histograms in Fig. 3.4. The first one is a conditional histogram age_exp/[hours(*Part-time*),*Abs*] presented in the upper part of Fig. 3.4. It shows frequencies of particular categories of the attribute age_exp for a sub-matrix *Adult*/hours(*Part-time*). The second one is a histogram age_exp/[hours(*Part-time*),*Cat*] showing a percentage of the rows of the data sub-matrix *Adult*/hours(*Part-time*) from a frequency of each category the age_exp attribute for the whole data matrix *Adult*.

A CF-table CF(age_exp,*Adult*/hours(*Part-time*)) introduced in Fig. 3.5 defines both histograms. We see in the the first column of the second row of CF(age_exp,*Adult*/hours(*Part-time*)), that there are 2 819 young persons satisfying hours(*Part-time*) and 9 627 young persons in the whole data matrix *Adult*. This means that a percentage of rows of the data sub-matrix *Adult*/hours(*Part-time*) from a frequency of the Young category is $100\frac{2819}{9627} = 29$. This is the same for additional categories of the age_exp attribute.

Histogram age_exp/[hours(*Part-time*), *Abs*]

Histogram age_exp/[hours(*Part-time*), *Cat*]

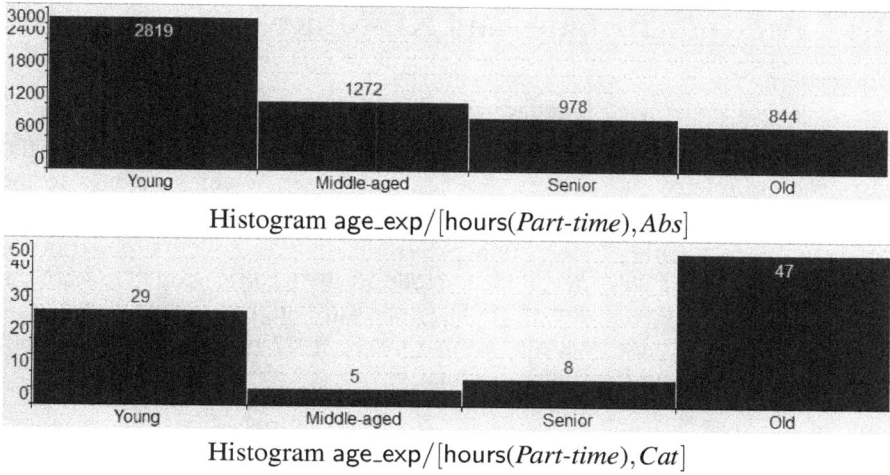

Figure 3.4: Examples of histograms.

Adult	Young	Middle-aged	Senior	Old
hours(*Part-time*)	2 819	1 272	978	844
True	9 627	24 671	12 741	1 803

Figure 3.5: CF-table $CF(\text{age\_exp}, Adult/\text{hours}(Part\text{-}time))$.

Generally speaking, we deal with CF-patterns of a form

$$\approx A/[\chi, Type]. \tag{3.5}$$

Here $\approx$ is a CF-quantifier, it defines a condition concerning CF-tables. χ is a Boolean attribute. *Type* is a type of histogram. *Abs* and *Cat* are examples of possible values of *Type*.

The expression $[StDn \geq 3]$ is an example of CF-quantifier and the expression

$$[StDn \geq 3]\ \text{age\_exp/}\,[\text{hours}(Part\text{-}time), Abs] \tag{3.6}$$

(3.6) is an example of CF-pattern. The CF-quantifier $[StDn \geq 3]$ means that a histogram has at least three steps down. A step down in a histogram is a couple of neighbouring columns such that the second column is lower than the first one. Note that the CF-pattern (3.6) is true in the data matrix *Adult* since it holds $2819 > 1272 > 978 > 844$.

A more precise and detailed description of the CF-Miner procedure is available in Chapter 8 together with examples of applications.

3.4   Pairs of Attributes and KL-Miner

The pattern

$$\text{Systolic} \approx^K_{0.82} \text{Diastolic}/\chi \tag{3.7}$$

is an example of KL-pattern expressing that categorical attributes Systolic and Diastolic are related. It concerns the data matrix *Entry* which belongs to the *STULONG* dataset introduced in Section 2.4. The KL-pattern (3.7) is justified by a contingency table of the attributes Systolic and Diastolic presented in the left part of Fig. 3.6. Categories of the Systolic attribute correspond to rows of the contingency table. Categories of the Diastolic attribute correspond to columns of the contingency table. We can see that there are 12 patients satisfying both Systolic(*very low*) and Diastolic(*very low*), etc.

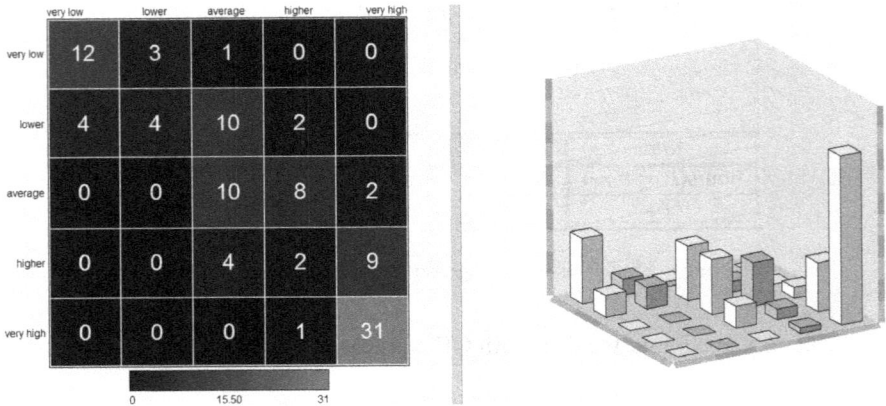

	very low	lower	average	higher	very high
very low	12	3	1	0	0
lower	4	4	10	2	0
average	0	0	10	8	2
higher	0	0	4	2	9
very high	0	0	0	1	31

0   15.50   31

Figure 3.6: Contingency table justifying the KL-pattern Systolic $\approx^K_{0.82}$ Diastolic$/\chi$.

The contingency table in the left part of Fig. 3.6 concerns a data sub-matrix *Entry*$/\chi$ where

$\chi = \text{Hyperlipidemia}(no) \land \text{CholesterolRisk}(no) \land \text{Alcohol}(regularly) \land$

$\land \text{Beer}(up\ to\ 1\ litre\ /\ day,\ more\ than\ 1\ litre\ /\ day)$.

Evaluation of this contingency table results in the value 0.82 of the Kendall's coefficient. The symbol $\approx^K_{0.82}$ is called a *KL-quantifier*.

Generally speaking, we deal with KL-patterns of a form

$$R \approx C/\chi\ . \tag{3.8}$$

Here R and C are categorical attributes, χ is a Boolean attribute. KL-quantifier $\approx$ defines a condition concerning contingency tables of two categorical attributes. If the KL-pattern (3.8) is evaluated in a data matrix $\mathcal{M}$, then the condition given by $\approx$ is applied to a contingency table of R and C in a data matrix $\mathcal{M}/\chi$.

A more precise and detailed description of the KL-Miner procedure is available in Chapter 9 together with examples of applications.

3.5 Couples of Association Rules and SD4ft-Miner

A pattern

$$\text{District}(Ryedale) \bowtie \text{District}(Leicester) :$$
$$\text{DayOfWeek}(Thursday) \rightarrow_{*4.7} \text{Severity}(Fatal, Serious) \tag{3.9}$$

is an example of SD4ft-pattern corresponding to a couple of association rules. It concerns the *Accidents* data matrix introduced in Section 2.3. It means that a confidence of a conditional association rule

$$\text{DayOfWeek}(Thursday) \rightarrow_{0.359} \text{Severity}(Fatal, Serious)/\text{District}(Ryedale) \tag{3.10}$$

is 4.7 times higher than a confidence of a conditional association rule

$$\text{DayOfWeek}(Thursday) \rightarrow_{0.076} \text{Severity}(Fatal, Serious)/\text{District}(Leicester) \tag{3.11}$$

This can be verified using 4ft-tables presented in Fig. 3.7 where we write only "(*Fatal, Serious*)" instead of "Severity(*Fatal, Serious*)".

Accidents / District(*Ryedale*)	(*Fatal, Serious*)	¬ (*Fatal, Serious*)
DayOfWeek(*Thursday*)	101	180
¬ DayOfWeek(*Thursday*)	514	1 119

$4ft(\text{DayOfWeek}(Thursday), (Fatal, Serious), Accidents/\text{District}(Ryedale))$

Accidents/District(*Leicester*)	*Fatal, Serious*	¬ (*Fatal*),*Serious*)
DayOfWeek(*Thursday*)	117	1 419
¬ DayOfWeek(*Thursday*)	802	8 421

$4ft(\text{DayOfWeek}(Thursday), (Fatal, Serious), Accidents/\text{District}(Leicester))$

Figure 3.7: 4ft-tables for SD4ft-pattern (3.9).

A confidence of the rule (3.10) is equal to $\frac{101}{101+180} = 0.359$ and a confidence of the rule (3.11) is equal to $\frac{117}{117+1419} = 0.076$. We can conclude that the confidence of the rule 3.10 is 4.7 times higher than a confidence of the rule 3.11. In other words, a chance that an accident on Thursday is fatal or serious in district Ryedale is 4.7 times higher than in district Leicester. The symbol $\rightarrow_{*4.7}$ is an *SDft-quantifier*.

Generally speaking, we deal with SD4ft-patterns in one of the forms:

$$\alpha \bowtie \beta : \varphi \approx \psi/\chi, \quad \alpha \bowtie \alpha \wedge \beta : \varphi \approx \psi/\chi, \quad \alpha \bowtie \neg\alpha : \varphi \approx \psi/\chi . \tag{3.12}$$

Here $\alpha, \beta, \varphi, \psi$ and χ are Boolean attributes, $\approx$ is an SD4ft-quantifier. Attribute χ can be omitted. A more precise and detailed description of the SD4ft-Miner procedure is available in Chapter 10 together with examples of applications.

3.6 Couples of Histograms and SDCF-Miner

An example of an interesting couple of histograms is shown in Fig. 3.8. Both histograms concern the data matrix *Accidents* introduced in Section 2.3.

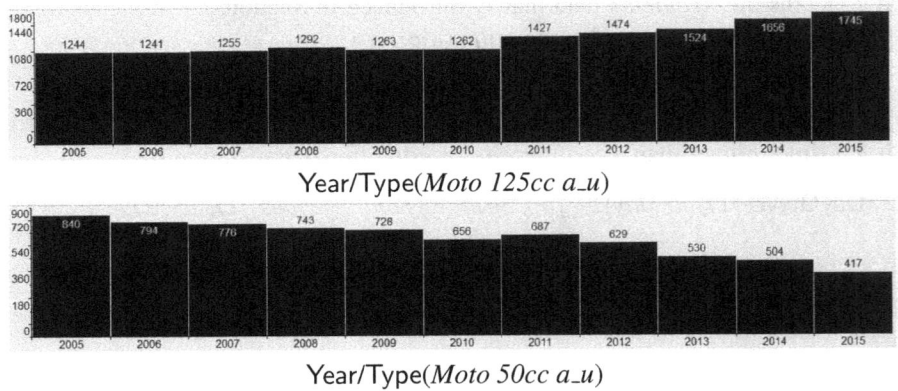

Figure 3.8: An example of couple of interesting histograms.

The first one is a histogram Year/Type(*Moto 125cc a_u*) available in the upper part of Fig. 3.8. It shows frequencies of particular categories of the attribute Year for a sub-matrix *Accidents*/Type(*Moto 125cc a_u*). We see that there is an increasing trend of accidents of motorcycles 125cc and under.

The second one is a histogram Year/Type(*Moto 50cc a_u*) available in the bottom part of Fig. 3.8. It shows the frequencies of particular categories of the attribute Year for a sub-matrix *Accidents*/Type(*Moto 50cc a_u*). We see that there is a decreasing trend of accidents of motorcycles 50cc and under.

This contrast of trends makes the couple of histograms interesting. The data mining process does not result in particular interesting histograms. The process results in couples of histograms interestingness of which follows from comparison of both histograms. In the above example, a result can be written as an SDCF-pattern

$$\text{Type}(\textit{Moto 125cc a\_u}) \bowtie \text{Type}(\textit{Moto 50cc a\_u}) : [StUp = 7; StDn = 9]\text{Year}.$$
$$(3.13)$$

The expression $[StUp, 7; StDn, 9]$ is a SDCF-quantifier. It means that the first histogram Year/Type(*Motorcycle 125cc a_u*) has 7 *steps up* and that the second histogram Year/Type(*Motorcycle 50cc a_u*) has 9 *steps down*. A step up in a histogram is a couple of neighbouring columns such that the second column is higher than the first one. A step down is defined analogously.

Generally speaking, we deal with SDCF-patterns in one of the forms:

$$\alpha \bowtie \beta :\approx A/\chi, \qquad \alpha \bowtie \alpha \wedge \beta :\approx A/\chi, \qquad \alpha \bowtie \neg \alpha :\approx A/\chi . \qquad (3.14)$$

Here α, β, and χ are Boolean attributes, A is an attribute. A symbol $\approx$ is an SDCF-quantifier. A more precise and detailed description of the SDCF-Miner procedure is available in Chapter 11 together with examples of applications.

3.7 Couples of Pairs of Attributes and SDKL-Miner

An example of an interesting couple of pairs of categorical attributes is presented in Fig. 3.9 by contingency tables KL(Wine,Beer,*Entry*/Spirits(*No*)) and KL(Wine,Beer,*Entry*/Spirits(≤ 0.1) visualised in the bottom of Fig. 3.9.

Entry / Spirits(*No*)		Beer				*Entry /* Spirits(≤ 0.1)		Beer		
		No	1-2	> 2				*No*	1-2	> 2
	No	6	4	2			*No*	2	1	0
Wine	≤ 0.5	5	4	0		Wine	≤ 0.5	0	19	0
	> 0.5	1	0	0			> 0.5	0	0	0

KL(Wine,Beer,*Entry*/Spirits(*No*)) KL(Wine,Beer,*Entry*/Spirits(≤ 0.1)

Figure 3.9: Contingency tables for a couple of pairs of categorical attributes.

Recall that the attributes Spirits, Wine, and Beer concern consumption of alcohol, see Section 2.4. Contingency table KL(Wine,Beer,*Entry*/Spirits(*No*)) in the left upper part of Fig. 3.9 says that there are six patients that do not drink spirits, no wine, and no beer, there are five patients that drink neither spirits nor beer, but they drink not more than 0.5 litre of wine per day. This is analogous for additional combinations of categories of the attributes Wine and Beer in both contingency tables.

Interestingness of this couple of pairs of categorical attributes is in the different distribution of frequencies in both contingency tables. This means that the relation of drinking wine and beer among patients who do not drink spirits differs from this relation among patients who drink not more than 0.1 litre of spirits per day.

This interestingness can be expressed by SDKL-pattern

$$\text{Spirits}(No) \bowtie \text{Spirits}(\leq 0.1) : \text{Wine} \approx^{K}_{-0.18;0.80} \text{Beer}. \qquad (3.15)$$

This means that the values of the Kendall's coefficient for the contingency tables $KL(\text{Wine}, \text{Beer}, Entry/\text{Spirits}(No))$ and $KL(\text{Wine}, \text{Beer}, Entry/\text{Spirits}(\leq 0.1)$ are -0.18 and 0.80 respectively.

Generally speaking, we deal with SDKL-patterns in one of the forms:

$$\alpha \bowtie \beta : R \approx C/\chi, \quad \alpha \bowtie \alpha \wedge \beta : R \approx C/\chi, \quad \alpha \bowtie \neg\alpha : R \approx C/\chi \ . \quad (3.16)$$

Here α, β, and χ are Boolean attributes, R and C are attributes. A symbol $\approx$ is an SDKL-quantifier. A more precise and detailed description of the SDKL-Miner procedure is available in Chapter 12 together with examples of applications.

3.8 Action Rules and Ac4ft-Miner

A pattern

$$\text{Height}(160; 175) \wedge \text{BMI}(\textit{obese class I} \rightarrowtail \textit{normal}) \rightarrow_{/2.7} \text{Syst}\langle 140; 180\rangle \quad (3.17)$$

is an example of action rule. It concerns the *Entry* data matrix belonging to the *STULONG* dataset introduced in Section 2.4. It means that a change of a value of BMI from *obese class I* to *normal* causes that a chance to have systolic blood pressure in the interval $\langle 140; 180\rangle$ (i.e., high blood pressure) decreases 2.7 times for patients satisfying Height(160; 175).

The action rule (3.17) can be understood as a couple of association rules

$$\text{Height}(160; 175) \wedge \text{BMI}(\textit{obese class I}) \rightarrow_{0.53} \text{Syst}\langle 140; 180\rangle \quad (3.18)$$

and

$$\text{Height}(160; 175) \wedge \text{BMI}(\textit{normal}) \rightarrow_{0.2} \text{Syst}\langle 140; 180\rangle \ . \quad (3.19)$$

The first rule (3.18) is called a *before rule* (or an *initial rule*). Its confidence is given by the 4ft-table

$$4ft(\text{Height}(160; 175) \wedge \text{BMI}(\textit{obese class I}), \text{Syst}(160; 175), Entry)$$

in the upper part of Fig. 3.10. The value of confidence is $\frac{49}{49+43} = 0.53$. The second rule (3.19) is called an *after rule* (or a *final rule*). Its confidence is given by the 4ft-table

$$4ft(\text{Height}(160; 175) \wedge \text{BMI}(\textit{normal}), \text{Syst}(160; 175), Entry)$$

in the bottom part of Fig. 3.10. The value of confidence is $\frac{76}{76+305} = 0.20$. We can conclude that a change of a value of BMI from *obese class I* to *normal* decreases a probability to have systolic blood pressure in the interval $\langle 140; 180 \rangle$ $\frac{49}{49+43} / \frac{76}{76+305} = 2.67$ times for patients satisfying Height$\langle 160; 175 \rangle$.

	Entry	Syst$\langle 160; 175 \rangle$	¬Syst$\langle 160; 175 \rangle$
Height$\langle 160; 175 \rangle \wedge$ BMI(*obese class I*)		49	43
¬(Height$\langle 160; 175 \rangle \wedge$ BMI(*obese class I*))		378	947

$4ft$(Height$\langle 160; 175 \rangle \wedge$ BMI(*obese class I*), Syst$\langle 160; 175 \rangle$, *Entry*)

	Entry	Syst$\langle 160; 175 \rangle$	¬Syst$\langle 160; 175 \rangle$
Height$\langle 160; 175 \rangle \wedge$ BMI(*normal*)		76	305
¬(Height$\langle 160; 175 \rangle \wedge$ BMI(*normal*))		351	685

$4ft$(Height$\langle 160; 175 \rangle \wedge$ BMI(*normal*), Syst$\langle 160; 175 \rangle$, *Entry*)

Figure 3.10: 4ft-tables for action rule (3.17).

The Boolean attribute Height$\langle 160; 175 \rangle$ in the action rule (3.17) is called a *stable part of antededent*, the expression BMI(*obese class I* $\rightarrowtail$ *normal*) is a *change of antecedent*, the symbol $\rightarrow_{/2.7}$ is an *Ac4ft-quantifier*, and the Boolean attribute Syst$\langle 140; 180 \rangle$ is a *stable part of succedent*.

Generally speaking, we deal with action rules of a form

$$\varphi \wedge \Phi \approx \psi \wedge \Psi / \chi \,. \tag{3.20}$$

Each action rule can be seen as a couple of association rules

$$\varphi \wedge \varphi_B \approx \psi \wedge \psi_B / \chi \quad \text{and} \quad \varphi \wedge \varphi_A \approx \psi \wedge \psi_A / \chi \tag{3.21}$$

where φ, φ_B, φ_A, ψ, ψ_B, and χ are Boolean attributes. φ_B is an *antecedent before*, φ_A is an *antecedent after*, ψ_B is a *succedent before*, ψ_A is a *succedent after* and χ is a *condition*. φ_B and φ_A can be derived from Φ, which is called a *change of antecedent* ψ_B and ψ_A can be derived from Ψ, which is called a *change of succedent*. The symbol $\approx$ is an *Ac4ft quantifier*, it corresponds to a condition concerning a couple of 4ft-tables.

If the action rule (3.20) is evaluated in a data matrix $\mathcal{M}$, then the condition corresponding to the Ac4ft quantifier $\approx$ is applied to a couple of 4ft-tables $4ft(\varphi \wedge \varphi_B, \psi \wedge \psi_B, \mathcal{M} / \chi)$ and $4ft(\varphi \wedge \varphi_A, \psi \wedge \psi_A, \mathcal{M} / \chi)$. If the condition is satisfied, then the action rule (3.20) is true in $\mathcal{M}$ and it is listed in the output of the procedure.

A more precise and detailed description of the Ac4ft-Miner procedure is available in Chapter 13 together with examples of applications.

Chapter 4

Common Features

The seven GUHA procedures described in this book deal with various types of patterns. Their examples are presented in the previous chapter. Even if the procedures deal with various types of patterns, they have important common features. Formal similarities can be seen in Section 4.1, where a short overview of procedures and patterns is available. The patterns are evaluated using three types of contingency tables introduced in Section 4.2. Principles of evaluation of particular types of patterns are introduced in Section 4.3.

Boolean attributes are very important elements of all patterns the seven GUHA procedures deal with. The usefulness of the GUHA procedures depends on the possibilities to define a set of relevant patterns in a sufficiently precise way. Relevant Boolean attributes the GUHA procedures deal with are not only conjunctions of attribute-value pairs often used in association rules produced by the apriori algorithm. The GUHA procedures deal with Boolean attributes of substantially more general syntax introduced in Section 4.4. Possibilities of a definition of a set of relevant Boolean attributes are also described in Section 4.4.

All the described GUHA procedures can handle data with missing information. A special approach to missing information – *secured completion* developed in [41] is applied in all procedures. Principles of dealing with missing information are described in Section 4.5. Note that additional approaches to missing information are available for the 4ft-Miner procedure, see Section 7.1.4.

4.1 Overview of Procedures and Patterns

An overview of GUHA procedures described in this book and the patterns particular procedures dealt with is available in Table 4.1. Additional information on particular procedures are provided in chapters introduced in the last column of Table 4.1.

Table 4.1: Overview of GUHA procedures.

Procedure	Patterns	Name of Pattern	Chapter
4ft-Miner	$\varphi \approx \psi$	enhanced association rule, association rule	7
	$\varphi \approx \psi/\chi$	conditional association rule	
CF-Miner	$\approx A/\chi$	histogram, CF-pattern, conditional histogram	8
KL-Miner	$R \approx C/\chi$	KL-pattern	9
SD4ft-Miner	$\alpha \bowtie \beta : \varphi \approx \psi$	SD4ft-pattern	10
	$\alpha \bowtie \alpha \wedge \beta : \varphi \approx \psi$		
	$\alpha \bowtie \neg \alpha : \varphi \approx \psi$		
	$\alpha \bowtie \beta : \varphi \approx \psi/\chi$	conditional SD4ft-pattern	
	$\alpha \bowtie \alpha \wedge \beta : \varphi \approx \psi/\chi$		
	$\alpha \bowtie \neg \alpha : \varphi \approx \psi/\chi$		
SDCF-Miner	$\alpha \bowtie \beta : \approx A/\chi$	SDCF-pattern	11
	$\alpha \bowtie \alpha \wedge \beta : \approx A/\chi$		
	$\alpha \bowtie \neg \alpha : \approx A/\chi$		
SDKL-Miner	$\alpha \bowtie \beta : R \approx C/\chi$	SDKL-pattern	12
	$\alpha \bowtie \alpha \wedge \beta : R \approx C/\chi$		
	$\alpha \bowtie \neg \alpha : R \approx C/\chi$		
Acft-Miner	$\varphi \wedge \Phi \approx \psi \wedge \Psi$	action rule	13
	$\varphi \wedge \Phi \approx \psi \wedge \Psi/\chi$	conditional action rule	

Here $\alpha, \beta, \varphi, \chi, \psi$ are Boolean attributes introduced in Section 1.4.3; A, R, C are categorical attributes introduced in Section 1.4.2 and Φ, Ψ are changes of Boolean attributes introduced later in Section 13.1. The symbol $\approx$ is a quantifier specifying relations among Boolean attributes, categorical attributes and changes of Boolean attributes occurring in patterns.

Note that the patterns the particular procedures deal with have various names introduced in Table 4.1. Additional names are introduced together with detailed descriptions of particular procedures. All the procedures work according to the schema introduced in Fig. 3.1.

4.2 Contingency Tables

Three types of contingency tables are used to evaluate patterns that the described GUHA procedures deal with – 4ft-tables, CF-tables and KL-tables.

The first one is a 4ft-table $4ft(\kappa,\lambda,\mathcal{M})$ of Boolean attributes κ and λ in a data matrix $\mathcal{M}$. It is a quadruple $4ft(\kappa,\lambda,\mathcal{M}) = \langle a,b,c,d \rangle$, where a is the number of rows of $\mathcal{M}$ satisfying both κ and λ, b is the number of rows satisfying κ and not satisfying λ, etc., see Fig. 4.1.

$\mathcal{M}$	λ	$\neg\lambda$
κ	a	b
$\neg\kappa$	c	d

Figure 4.1: 4ft-table $4ft(\kappa,\lambda,\mathcal{M}) = \langle a,b,c,d \rangle$.

4ft-tables are used by the procedures 4ft-Miner, SD4ft-Miner, and Ac4ft-Miner.

The second type of contingency tables used by the described GUHA procedures is a CF-table $CF(A,\mathcal{M}/\chi)$ of a categorical attribute A and a Boolean attribute χ in a data matrix $\mathcal{M}$. We assume that the attribute A has categories $a_1,\ldots,a_t$.

The CF-table $CF(A,\mathcal{M}/\chi)$ is a vector of $2t$ non-negative integers $\langle n_1,\ldots,n_t,m_1,\ldots,m_t \rangle$. Here, n_i denotes the number of rows o of a data matrix $\mathcal{M}/\chi$ for which $A(o) = a_i$ and m_i denotes the number of rows o of a data matrix $\mathcal{M}$ for which $A(o) = a_i$; we assume $i = 1,\ldots,t$. The CF-table $CF(A,\mathcal{M}/\chi)$ is often written in a form of table presented in Fig. 4.2. the sums $n = n_1 + \cdots + n_t$ and $m = m_1 + \cdots + n_t$ are also introduced.

$\mathcal{M}$	a_1	...	a_t	Σ
χ	n_1	...	n_t	n
$True$	m_1	...	m_t	m

Figure 4.2: CF-table $CF(A,\mathcal{M}/\chi) = \langle n_1,\ldots,n_t,m_1,\ldots,m_t \rangle$.

CF-tables are used by the procedures CF-Miner and SDCF-Miner.

The third type of contingency tables used by the described GUHA procedures is a KL-table $KL(R,C,\mathcal{M}/\chi)$ of categorical attributes R,C and a Boolean attribute χ in a data matrix $\mathcal{M}$. We assume that the attribute R has categories $r_1,\ldots,r_K$ and that the attribute C has categories $c_1,\ldots,c_L$.

The KL-table $KL(R,C,\mathcal{M}/\chi)$ is a data matrix $\{n_{i,j}\}_{i=1,\ldots,K}^{j=1,\ldots,L}$ of non-negative integers. Here, $n_{i,j}$ denotes the number of rows o of a data matrix $\mathcal{M}/\chi$ satisfying $R(o) = r_i$ and $C(o) = c_j$ where $i = 1,\ldots,K$ and $j = 1,\ldots,L$; see Fig. 4.3. KL-tables are used by the procedures KL-Miner and SDKL-Miner.

M/χ	c_1	$\cdots$	c_L
r_1	$n_{1,1}$	$\cdots$	$n_{1,L}$
$\vdots$	$\vdots$	$\ddots$	$\vdots$
r_K	$n_{K,1}$	$\cdots$	$n_{K,L}$

Figure 4.3: KL-table $KL(R,C,\mathcal{M}/\chi) = \{n_{i,j}\}_{i=1,\dots,K}^{j=1,\dots,L}$.

4.3 Principles of Patterns Evaluation

A symbol $\approx$ occurring in patterns that a particular GUHA procedure deals with is a *generalised quantifier*. A condition concerning one or two contingency tables is assigned to each generalised quantifier. The condition is defined such that it expresses an interesting relation among Boolean attributes and attributes involved in patterns.

A pattern Ω with a generalised quantifier $\approx$ is true in a data matrix $\mathcal{M}$ if a condition assigned to $\approx$ is satisfied in a contingency table or in a couple of contingency tables corresponding to the pattern Ω and the data matrix $\mathcal{M}$. We then write $Val(\Omega,\mathcal{M}) = 1$. If a condition assigned to $\approx$ is not satisfied in a contingency table or in a couple of contingency tables corresponding to the pattern and the data matrix $\mathcal{M}$, then the pattern Ω is false in a data matrix $\mathcal{M}$. We then write $Val(\Omega,\mathcal{M}) = 0$.

Patterns the procedure 4ft-Miner, CF-Miner, and KL-Miner deal with are evaluated using a contingency table introduced in Table 4.2. Patterns, the procedures SD4ft-Miner, SDCF-Miner, and SDKL-Miner deal with, are evaluated using couples of contingency table introduced in Tables 4.3, 4.4, and 4.5 respectively. Couples of suitable 4ft-tables are used to verify action rules $\varphi \wedge \Phi \approx \psi \wedge \Psi$ and conditional action rules $\varphi \wedge \Phi \approx \psi \wedge \Psi/\chi$, see Section 13.1.

Table 4.2: Contingency tables used by the 4ft-Miner, CF-Miner, and Kl-Miner.

Procedure	Pattern	Contingency table
4ft-Miner	association rules $\varphi \approx \psi$	$4ft(\varphi,\psi,\mathcal{M})$
	conditional association rule $\varphi \approx \psi/\chi$	$4ft(\varphi,\psi,\mathcal{M}/\chi)$
CF-Miner	conditional histogram $\approx A/\chi$	$CF(A,\mathcal{M}/\chi)$
KL-Miner	KL-pattern $R \approx C/\chi$	$KL(R,C,\mathcal{M}/\chi)$

4.4 Set of Relevant Boolean Attributes

A definition of a set of relevant patterns includes one or more definitions of sets of relevant Boolean attributes $\alpha, \beta, \varphi, \chi, \psi$ introduced in Section 4.1. All definitions of sets of relevant Boolean attributes have the same, slight complex structure.

Table 4.3: Couples of contingency tables used by the SD4ft-Miner procedure.

Pattern	Used contingency tables
$\alpha \bowtie \beta : \varphi \approx \psi$	$4ft(\varphi, \psi, \mathcal{M}/\alpha), \quad 4ft(\varphi, \psi, \mathcal{M}/\beta)$
$\alpha \bowtie \alpha \wedge \beta : \varphi \approx \psi$	$4ft(\varphi, \psi, \mathcal{M}/\alpha), \quad 4ft(\varphi, \psi, \mathcal{M}/\alpha \wedge \beta)$
$\alpha \bowtie \neg\alpha : \varphi \approx \psi$	$4ft(\varphi, \psi, \mathcal{M}/\alpha), \quad 4ft(\varphi, \psi, \mathcal{M}/\neg\alpha)$
$\alpha \bowtie \beta : \varphi \approx \psi/\chi$	$4ft(\varphi, \psi, \mathcal{M}/\chi \wedge \alpha), \quad 4ft(\varphi, \psi, \mathcal{M}/\chi \wedge \beta)$
$\alpha \bowtie \alpha \wedge \beta : \varphi \approx \psi/\chi$	$4ft(\varphi, \psi, \mathcal{M}/\chi \wedge \alpha), \quad 4ft(\varphi, \psi, \mathcal{M}/\chi \wedge \alpha \wedge \beta)$
$\alpha \bowtie \neg\alpha : \varphi \approx \psi/\chi$	$4ft(\varphi, \psi, \mathcal{M}/\chi \wedge \alpha), \quad 4ft(\varphi, \psi, \mathcal{M}/\chi \wedge \neg\alpha)$

Table 4.4: Couples of contingency tables used by the SDCF-Miner procedure.

Pattern	Used contingency tables
$\alpha \bowtie \beta : \approx A/\chi$	$CF(A, \mathcal{M}/\chi \wedge \alpha), \quad CF(A, \mathcal{M}/\chi \wedge \beta)$
$\alpha \bowtie \alpha \wedge \beta : \approx A/\chi$	$CF(A, \mathcal{M}/\chi \wedge \alpha), \quad CF(A, \mathcal{M}/\chi \wedge \alpha \wedge \beta)$
$\alpha \bowtie \neg\alpha : \approx A/\chi$	$CF(A, \mathcal{M}/\chi \wedge \alpha), \quad CF(A, \mathcal{M}/\chi \wedge \neg\alpha)$

Table 4.5: Couples of contingency tables used by the SDKL-Miner procedure.

Pattern	Used contingency tables
$\alpha \bowtie \beta : R \approx C/\chi$	$KL(R, C, \mathcal{M}/\chi \wedge \alpha), \quad KL(R, C, \mathcal{M}/\chi \wedge \beta)$
$\alpha \bowtie \alpha \wedge \beta : R \approx C/\chi$	$KL(R, C, \mathcal{M}/\chi \wedge \alpha), \quad KL(R, C, \mathcal{M}/\chi \wedge \alpha \wedge \beta)$
$\alpha \bowtie \neg\alpha : R \approx C/\chi$	$KL(R, C, \mathcal{M}/\chi \wedge \alpha), \quad KL(R, C, \mathcal{M}/\chi \wedge \neg\alpha)$

However, this makes it possible to tune definitions of sets of relevant patterns in a very fine way.

Relevant Boolean attributes are called *cedents*. This comes from association rules $\varphi \approx \psi$. Here φ is called *antecedent* and ψ is called *succedent*. Thus, both φ and ψ are *cedents*. Naturally, also α, β, and χ are cedents. Each cedent is a conjunction of *partial cedents* which correspond to the Boolean characteristics of groups of attributes. Each partial cedent is a conjunction or a disjunction of *literals*.

Each literal is a basic Boolean attribute $A(\alpha)$ or its negation $\neg A(\alpha)$. Recall that α is a *coefficient* of $A(\alpha)$. There are seven types of coefficients introduced in Section 4.4.1 together with details on literals. Possibilities of a definition of a set of relevant literals are introduced in Section 4.4.2.

An example of partial cedents is available in Section 4.4.3. A definition of a set of relevant partial cedents is presented in Section 4.4.4. A definition of a set of relevant cedents together with an example are available in Section 4.4.5.

Examples of input screens presented in Sections 4.4.2–4.4.5 are based on the LISp-Miner system. We are going to implement these procedures also in the Python language, see Chapter 15. This is the reason why we use a special simple language to express all parameters defining sets of relevant literals, partial cedents and cedents.

4.4.1   Literals and types of coefficients

Literals are created from basic Boolean attributes $A(\alpha)$ introduced in Section 1.4.3. Recall that A is an attribute – a column of an analysed data matrix $\mathscr{M}$. We assume that $\alpha \subsetneq \{a_1,\ldots,a_t\}$ where $\{a_1,\ldots,a_t\}$ is a set of all categories (possible values) of A. A set α is a *coefficient* of a basic Boolean attribute $A(\alpha)$. If $\alpha = \{a_{i_1},\ldots,a_{i_k}\}$, then k is a *length of the coefficient* α.

A *literal* is a basic Boolean attribute $A(\alpha)$ or its negation $\neg A(\alpha)$. A literal $A(\alpha)$ is also called as a *positive literal*; $\neg A(\alpha)$ is a *negative literal*.

There are seven *types of coefficients* – *Subsets, One category, Sequences, Left cuts, Right cuts, Cuts,* and *Cyclical*. A type of coefficient together with a minimal and a maximal length belong to input parameters defining a set of literals to be automatically generated from a given attribute. We use attributes race and education introduced in Section 2.2.1 and attribute Month introduced in Section 2.3.2 to present examples of literals.

Each coefficient α of a Boolean attribute $A(\alpha)$ is of type *Subsets*. The attribute race has five categories – *White, Asian-Pac-Islander, Amer-Indian-Eskimo, Other,* and *Black.* Thus race(*White*) and $\neg$race(*Other*) are examples of literals with coefficients of type *Subsets* of length 1. A basic Boolean attribute race(*Asian-Pac-Islander, Amer-Indian-Eskimo*) is an example of a positive literal with a coefficient of type *Subsets* of length 2. Let us note that the attribute race is nominal, there is no reasonable ordering of particular categories. A set of relevant coefficients of type *Subsets* of a given attribute is specified by a minimal and maximal length of coefficients.

A set of relevant coefficients of type *One category* of a given attribute is specified by a concrete particular category. The category *White* specifies one positive literal race(*White*) and one negative literal $\neg$race(*White*).

The remaining types of coefficients are suitable for ordinal attributes. An attribute is *ordinal* if there is a meaningful sequence or ranking between particular categories. The attribute education is ordinal; his 16 categories have a following natural ranking: *Preschool, 1st-4th, 5th-6th, 7th-8th, 9th, 10th, 11th, 12th, HS-grad, Some-college, Assoc-acdm, Assoc-voc, Bachelors, Masters, Prof-school, Doctorate.*

Let us further assume that an ordinal attribute A has t categories ranked as follows: $a_1, a_2, \ldots, a_{t-1}, a_t$. Then a coefficient α of a basic Boolean attribute $A(\alpha)$ is of type *Sequences* if it holds $A(\alpha) = A(a_u, a_{u+1}, \ldots, a_{v-1}, a_v)$ where $1 \leq u \leq v \leq t$. Note that $v - u + 1$ is a length of the coefficient of $A(a_u, a_{u+1}, \ldots, a_{v-1}, a_v)$. Boolean attributes education(*1st-4th, 5th-6th, 7th-8th*) and education(*HS-grad, Some-college, Assoc-acdm, Assoc-voc*) are examples of Boolean attributes with coefficients of type *Sequences*. A coefficient of the Boolean attribute education(*Preschool, Doctorate*) is not of type *Sequences*.

A coefficient α of a basic Boolean attribute $A(\alpha)$ is of type *Left cuts* if it holds $A(\alpha) = A(a_1, a_2, \ldots, a_{v-1}, a_v)$ where $1 \leq v < t$. Note that v is a length of the coefficient of $A(a_1, a_2, \ldots, a_{v-1}, a_v)$. Boolean attributes education(*Preschool, 1st-4th*) and education(*Preschool, 1st-4th, 5th-6th*) are examples of Boolean attributes with coefficients of type *Left cuts*. A coefficient of the Boolean attribute education(*1st-4th, 5th-6th*) is not of type *Left cuts*. We can say that a coefficient of type *Left cuts* of a Boolean attribute $A(\alpha)$ contains the first v categories of A.

A coefficient α of a basic Boolean attribute $A(\alpha)$ is of type *Right cuts* if it holds $A(\alpha) = A(a_{t-v+1}, a_{t-v+2}, \ldots, a_{t-1}, a_t)$ where $v \geq 1$. Note that v is a length of the coefficient of $A(a_{t-v+1}, a_{t-v+2}, \ldots, a_{t-1}, a_t)$. Boolean attributes education(*Masters, Prof-school, Doctorate*) and education(*Assoc-acdm, Assoc-voc, Bachelors, Masters, Prof-school, Doctorate*) are examples of Boolean attributes with coefficients of type *Right cuts*. A coefficient of the Boolean attribute education(*Bachelors, Masters, Prof-school*) is not of type *Right cuts*. We can say that a coefficient of type *Right cuts* of a Boolean attribute $A(\alpha)$ contains the last v categories of A.

A coefficient α of a Boolean attribute $A(\alpha)$ is of type *Cuts* if it is of type *Left cuts* or *Right cuts*.

The last type *Cyclical* is applicable for cyclical attributes. An ordinal attribute A with t categories ranked as $a_1, a_2, \ldots, a_{t-1}, a_t$ is a *cyclical attribute* if we can say that the category a_1 follows the category a_t. Attribute Month with categories *Jan, Feb, Mar, ..., Oct, Nov, Dec* is a cyclical attribute since we can say that *Jan* follows *Dec*. A coefficient α of a basic Boolean attribute $A(\alpha)$ is of type *Cyclical* if it is of type *Sequence* or if it holds $A(\alpha) = A(a_u, \ldots, a_t, a_1, \ldots, a_v)$ where $2 \leq u \leq t$ and $1 \leq v < u$. Note that $t - u + 1 + v$ is a length of the coefficient of $A(a_u, \ldots, a_t, a_1, \ldots, a_v)$. Coefficients of basic Boolean attributes Month(*Oct, Nov,Dec*), Month(*Nov, Dec, Jan*), Month(*Dec, Jan, Feb*) are of type *Cyclical*. Coefficient of a basic Boolean attribute Month(*Nov, Dec, Feb*) is not of type *Cyclical* since the category of *Jan* is missing.

4.4.2 Set of relevant literals

An example of a window for a definition of a set of relevant literals to be derived from attribute education is shown in Fig. 4.4.

Input screens of the LISp-Miner system (see Chapter 5) and the input of the GUHA procedures implemented in the Python language (see Chapter 15) are fundamentally different. Thus, an $\mathscr{RL}$-expression $\mathscr{RL}(A)$ in a form

$$\langle A; Coef, Min, Max, Gace, B/R, Class \rangle \tag{4.1}$$

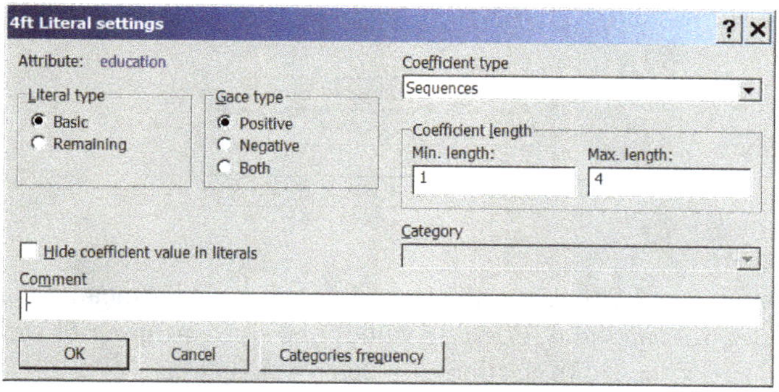

Figure 4.4: Example of a definition of a set of relevant literals.

is used to define a set of literals to be automatically generated from attribute A. The parameters $Coef, Min, Max$ define a set of relevant coefficients. The parameter $Gace$ determines a set of literals to be generated from relevant coefficients. The parameters B/R and $Class$ interact with the same parameters from additional $\mathscr{RL}$-expression used in a definition of a set of relevant partial cedents. Possible values of the $Coef$ parameter are introduced in Table 4.6 together with their meaning specified in the previous section.

Table 4.6: Values of the *Coef* parameters.

Value	Generated coefficients α of $A(\alpha)$
Subsets	coefficients of type *Subsets*
Sequences	coefficients of type *Sequences*
Cyclical	coefficients of type *Cyclical*
Left cuts	coefficients of type *Left cuts*
Right cuts	coefficients of type *Right cuts*
Cuts	coefficients of type *Cuts*
One category	$\{a\}$ where a is a category given as a value of parameter *Min*. A value of the *Max* parameter is undefined. Thus, one basic Boolean attribute $A(a)$ is defined in this case.

A length $\mathscr{L}(\alpha)$ of a coefficient α of a basic Boolean attribute $A(\alpha)$ is the number of categories belonging to α, see the previous section. The parameters *Min* and *Max* specify the minimal length and maximal length of generated coefficients α. It must hold:

$$1 \leq Min \leq \mathscr{L}(\alpha) \leq Max < \text{ the number of categories of } A . \quad (4.2)$$

The parameter *Gace* has three possible values specifying literals to be generated from the basic Boolean attributes $A(\alpha)$ defined by the parameters *Coef, Min,* and *Max*, see Table 4.7.

Table 4.7: Values of the *Gace* parameters.

Value	Generated literals
Pos	$A(\alpha)$
Neg	$\neg A(\alpha)$
Both	both $A(\alpha)$ and $\neg A(\alpha)$

The parameter B/R has possible values B and R. Their meaning is explained in Sections 4.4.3 and 4.4.4. Possible values and their meaning of the parameter *Class* are also explained in Sections 4.4.3 and 4.4.4.

We can summarize that a set of relevant literals to be generated from an attribute A is a subset of a set of all literals defined by parameters *Coef*, *Min*, *Max*, and *Gace* of an $\mathscr{RL}$-expression $\mathscr{RL}(A) = \langle A; Coef, Min, Max, Gace, B/R, Class \rangle$. It is crucial that each $\mathscr{RL}$-expression is a part of an $\mathscr{RP}$-expression defining a set of relevant partial cedents – conjunctions or disjunctions of relevant literals. Which relevant literal can be used in relevant partial cedents depends on configurations of values of corresponding parameters B/R and *Class*.

If $\mathscr{RL}(A) = \langle A; Coef, Min, Max, Gace, B/R, Class \rangle$ is an $\mathscr{RL}$-expression, then $\mathscr{SRL}(A)$ denotes a set of all literals defined by parameters *Coef*, *Min*, *Max*, and *Gace* of $\mathscr{RL}(A)$. We can summarize:

- If *Gace = Pos*, then $\mathscr{SRL}(A)$ is a set of all $A(\alpha)$

- If *Gace = Neg*, then $\mathscr{SRL}(A)$ is a set of all $\neg A(\alpha)$

- If *Gace = Both*, then $\mathscr{SRL}(A)$ is a set of all $A(\alpha)$ and $\neg A(\alpha)$,

where α satisfies conditions given by parameters *Coef*, *Min* and *Max*. This means that α is given by a value of *Coef* according to Table 4.6, and that $Min \leq \mathscr{L}(\alpha) \leq Max$. If *Coef = One category*, then $\mathscr{SRL}(A) = \{A(a)\}$ where a is a given category of A.

4.4.3 *Example of partial cedents*

An example of a definition of a set Personal of partial cedents in the LISp-Miner system is available in Fig. 4.5. There are four rows corresponding to four $\mathscr{RL}$-expressions

$\langle \text{sex}; Subsets, 1, 1, pos, B, - \rangle$,

$\langle \text{age\_exp}; Subsets, 1, 1, pos, R, Age \rangle$,

$\langle \text{age\_5}; Sequences, 1, 3, pos, R, Age \rangle$, and

$\langle \text{education}; Sequences, 1, 4, pos, R, - \rangle$.

These $\mathscr{RL}$ expressions define four sets $\mathscr{SRL}(\text{sex})$, $\mathscr{SRL}(\text{age\_exp})$, $\mathscr{SRL}(\text{age\_5})$, and $\mathscr{SRL}(\text{education})$ of relevant literals.

4ft Antecedent Partial cedent Settings								
Basic parameters								
Name: Personal								
Min. length: 0	Max. length: 3		Literals boolean operation type: Conjunction					
Comment: -								
Options								
Allow only a consecutive sequence of literals in cedent (only neighbouring literals):				No				
Linked coefficents (all literals must have the same coefficient as in the first one):				No				
Literals Settings								
Underlying attribute		Categories	X-cat	Coefficient type	Length	+/-	B/R	Class of equiv.
sex		2	No	Subsets	1 - 1	pos	Basic	-
age_exp		4	No	Subsets	1 - 1	pos	Remaining	Age
age_ed5		15	No	Sequences	1 - 3	pos	Remaining	Age
education		16	No	Sequences	1 - 4	pos	Remaining	-

Figure 4.5: Definition of set Personal of partial cedents.

The definition of a set Personal of partial cedents shown in Fig. 4.5 defines a set of Boolean characteristics of a group of attributes ⟨sex, age_exp,age_5, education⟩ concerning personal characteristics. It is assumed that the set Personal will be used to define a set Clients of cedents – Boolean characteristics of clients using sets Personal and Employment, see Section 4.4.5. The set Personal of partial cedents is defined according to the following requirements:

RQ1: A partial cedent from the Personal set needs not be used to define each cedent from the set Clients.

RQ2: Each partial cedent from the Personal set is either a single Boolean attribute or a conjunction of two or three Boolean attributes belonging to the sets $\mathscr{SRL}$(sex), $\mathscr{SRL}$(age_exp), $\mathscr{SRL}$(age_5), and $\mathscr{SRL}$(education). Of course, maximally one of attributes from each of these sets can be used.

RQ3: Each partial cedent from the Personal set must contain a literal created from attribute sex.

RQ4: There is no partial cedent in the Personal set containing literals both from the set $\mathscr{SRL}$(age_exp) and from the set $\mathscr{SRL}$(age_5).

These requirements are met in the following way.

RQ1: A minimal length of partial cedents is set to 0, see expression "Min. length: 0" in the second row of the frame Basic parameters.

RQ2: A maximal length of partial cedents is set to 3, see expression "Max. length: 3". Type of operations is set to conjunction see expression "Literals boolean operation type: Conjunction". The LISp-Miner automatically uses maximally one literal from each of sets of relevant literals in question.

RQ3: The LISp-Miner generates only partial cedents with at least one attribute for which a value of parameter B/R is equal to B (B abbreviates Basic and R abbreviates Remaining). Note that attribute sex is the only attribute with value B for parameter B/R.

RQ4: The LISp-Miner generates only partial cedents with maximally one attribute from each class of equivalence. Note that values of parameter *Class* for attributes age_exp and age_5 are *Age*. This means that both attributes belong to the class of equivalence *Age*.

4.4.4 *Set of relevant partial cedents*

Each partial cedent is a conjunction or a disjunction of literals. A length of a partial cedent is the number of literals occurring in the partial cedent. A set of relevant partial cedents is given by an $\mathscr{RP}$-expression

$$\langle Name_P, Type_P, Min_P, Max_P; \mathscr{RL}(A_1),\ldots,\mathscr{RL}(A_u)\rangle . \tag{4.3}$$

Here, $Name_P$ is a name of the set of relevant partial cedents, Min_P is a minimal length, Max_P is a maximal length of partial cedents, $Type_P$ is a type of partial cedents and $\mathscr{RL}(A_1),\ldots,\mathscr{RL}(A_u)$ are $\mathscr{RL}$ expressions where $A_1,\ldots,A_u$ are mutually different attributes and $u \geq 1$. We assume $\mathscr{RL}(A_i) = \langle A_i; Coef_i, Min_i, Max_i, Gace_i, B/R_i, Class_i\rangle$ where $i = 1,\ldots,u$. An $\mathscr{RP}$-expression is usually written in a tabular form, see Table 4.8. Each $\mathscr{RP}$-expression satisfies conditions PC1 and PC2.

Table 4.8: $\mathscr{RP}$-expression in a tabular form.

Name: *Name*		Type: *Type*		Length: $Min_P - Max_P$	
Attribute	Coefficients	Length	Gace	B/R	Class
A_1	$Coef_1$	$Min_1 - Max_1$	$Gace_1$	B/R_1	$Class_1$
⋮	⋮	⋮	⋮	⋮	⋮
A_u	$Coef_u$	$Min_u - Max_u$	$Gace_u$	B/R_u	$Class_u$

PC1: *Name* is a string, Min_P, Max_P, u are integer numbers satisfying $u > 0$, $0 \leq Min_P \leq Max_P \leq u$. It holds $Type =$ "Conjunction" or $Type =$ "Disjunction".

PC2: It either holds $Class_i = $ " $-$ " for $i = 1,\ldots,u$ or there are mutually disjoint subsets $Cl_1,\ldots,Cl_v$ of the set $\{A_1,\ldots,A_u\}$ such that it holds $Class_i = $ " $-$ " or $Class_i = "Cl_j"$ for $i = 1,\ldots,u$ where $j \in \{1,\ldots,v\}$. The subsets $Cl_1,\ldots,Cl_v$ are called *classes of equivalence*.

Note that, according to the condition PC2, classes of equivalence can be omitted, or a class of equivalence can be assigned to some or to all attributes occurring in an $\mathscr{R}\mathscr{P}$-expression. Of course, it is reasonable that the number v of classes of equivalence is smaller than the number of attributes.

Recall the definition of the set Personal of partial cedents according to Fig. 4.5. It is shown again in the upper part of Fig. 4.6. The set Personal is defined by an $\mathscr{R}\mathscr{P}$-expression Personal. A tabular form of this $\mathscr{R}\mathscr{P}$-expression is in the bottom part of Fig. 4.6.

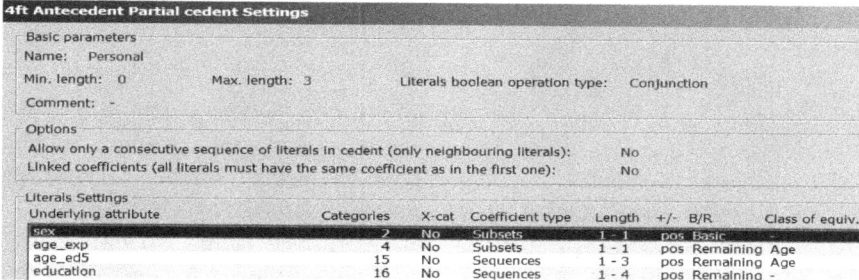

4ft Antecedent Partial cedent Settings						
Basic parameters						
Name: Personal						
Min. length: 0	Max. length: 3		Literals boolean operation type: Conjunction			
Comment: -						
Options						
Allow only a consecutive sequence of literals in cedent (only neighbouring literals):				No		
Linked coefficients (all literals must have the same coefficient as in the first one):				No		

Literals Settings Underlying attribute	Categories	X-cat	Coefficient type	Length	+/-	B/R	Class of equiv.
sex	2	No	Subsets	1 - 1	pos	Basic	
age_exp	4	No	Subsets	1 - 1	pos	Remaining	Age
age_ed5	15	No	Sequences	1 - 3	pos	Remaining	Age
education	16	No	Sequences	1 - 4	pos	Remaining	-

Name: Personal		Type: *Conjunction*		Length: $0-3$	
Attribute	Coefficients	Length	Gace	B/R	Class
sex	*Subsets*	$1-1$	*Pos*	B	—
age_exp	*Subsets*	$1-1$	*Pos*	R	*Age*
age_5	*Sequences*	$1-3$	*Pos*	R	*Age*
education	*Sequences*	$1-4$	*Pos*	R	—

Figure 4.6: $\mathscr{R}\mathscr{P}$-expression Personal in the tabular form.

A set $Name_P$ of all partial cedents defined by a $\mathscr{R}\mathscr{P}$-expression $\langle Name_P, Min_P, Max_P, Type_P; \mathscr{R}\mathscr{L}(A_1), \ldots, \mathscr{R}\mathscr{L}(A_u)\rangle$ is defined using the sets $\mathscr{S}\mathscr{R}\mathscr{L}(A_i)$ of relevant literals given by $\mathscr{R}\mathscr{L}$-expression $\mathscr{R}\mathscr{L}(A_1), \ldots, \mathscr{R}\mathscr{L}(A_u)$. First, let us assume $Type_P = Conjunction$. Then a set $Name_P$ consists of all conjunctions of literals $\lambda_{j_1} \wedge \cdots \wedge \lambda_{j_t}$ such that

RP1) $\max\{1, Min_P\} \leq t \leq Max_P$, $1 \leq j_1 < j_2 < \cdots < j_{t-1} < j_t \leq u$ and $\lambda_{j_i} \in \mathscr{S}\mathscr{R}\mathscr{L}(A_{j_i})$ for $i = 1, \ldots, t$

RP2) there is $1 \leq z \leq t$ such that $B/R_{j_z} = B$

RP3) there are no $k \neq l$ such that $1 \leq k \leq t$, $1 \leq l \leq t$ and $Class_{j_k} = Class_{j_l}$.

Note that RP2 means that there is at least Basic attribute and RP3 means that no two attributes belong to the same class of equivalence.

If $Type_P = Disjunction$, then a set $Name_P$ consists of all disjunctions of literals $\lambda_{j_1} \vee \cdots \vee \lambda_{j_t}$ satisfying RP1, RP2 and RP3.

An $\mathscr{R}\mathscr{P}$-expression $\langle Name_P, Type_P, 0, Max_P; \mathscr{R}\mathscr{L}(A_1), \ldots, \mathscr{R}\mathscr{L}(A_u)\rangle$ with $Min_P = 0$ means that the partial cedents from the set $Name_P$ need not be used in cedents defined by a $\mathscr{RC}$-expression in which this $\mathscr{RP}$-expression occurs, see next section.

4.4.5 Set of relevant cedents

Each cedent is a conjunction of partial cedents. A length of a cedent is the number of literals occurring in the cedent. A set of relevant cedents is given by an $\mathscr{RC}$-expression

$$\langle Name_C, Min_C, Max_C; NameP_1, \ldots, NameP_r \rangle . \tag{4.4}$$

Here $Name_C$ is a name of the set of relevant cedents, Min_C is a minimal length, Max_C is a maximal length of relevant cedents, it holds $0 \leq Min_C \leq Max_C$. $NameP_1, \ldots, NameP_r$ are names of sets of relevant partial cedents. We assume that these sets are defined by $\mathscr{RP}$-expressions $\langle NameP_i, Min_{P_i}, Max_{P_i}, Type_i; ListRL_i \rangle$ for $i = 1, \ldots, r$. Here $ListRL_i$ is a list of $\mathscr{RL}$-expressions used in $\mathscr{RP}$-expression defining the set $NameP_i$ of relevant partial cedents. An $\mathscr{RC}$-expression can be written in a short tabular form, see Table 4.9.

Table 4.9: $\mathscr{RC}$-expression in a short tabular form.

Name: $Name_C$	Length: $Min_C - Max_C$
Set of partial cedents	Length
$NameP_1$	$Min_{P_1} - Max_{P_1}$
$\vdots$	$\vdots$
$NameP_r$	$Min_{P_r} - Max_{P_r}$

A set $Name_C$ of all relevant cedents defined by an $\mathscr{RC}$-expression $\langle Name_C, Min_C, Max_C; NameP_1, \ldots, NameP_r \rangle$ is determined by sets $NameP_1, \ldots,$ $NameP_r$ of relevant partial cedents according to conditions RC1 and RC2.

RC1) If $Min_C = 0$, then $True \in Name_C$.

RC2) Each conjunction $\kappa_{j_1} \wedge \cdots \wedge \kappa_{j_t}$ satisfying the conditions

- $\max\{1, Min_C\} \leq t \leq Max_C$
- $1 \leq j_1 < j_2 < \cdots < j_{t-1} < j_t \leq r$
- $\kappa_{j_i} \in NameP_{j_i}$ for $i = 1, \ldots, t$

belongs to $Name_C$.

Let us have an $\mathscr{RP}$-expression Employment available in Table 4.10. This $\mathscr{RP}$-expression defines a set Employment of partial cedents. The attributes occupation and hours_per_week are introduced in Sections 2.2.1 and 2.2.2.

Table 4.10: $\mathscr{RP}$-expression Employment.

Name: Employment		Type: *Conjunction*		Length: $0-3$	
Attribute	Coefficients	Length	Gace	B/R	Class
occupation	*Subsets*	$1-3$	*Both*	B	–
hours_per_week	*Subsets*	$1-1$	*Pos*	B	–

The set Employment of partial cedents can be used together with a set Personal of partial cedents introduced in Fig. 4.6 to define a set Clients of cedents – Boolean characteristics of potential clients. The Clients can be defined by $\mathscr{RC}$-expression Clients:

$$\langle \text{Clients}, 1, 5; \text{Personal}, \text{Employment} \rangle . \tag{4.5}$$

A short tabular form of this $\mathscr{RC}$-expression is available in Table 4.11.

Table 4.11: $\mathscr{RC}$-expression Clients in short tabular form.

Name: Clients	Length: $1-5$
Personal	$0-3$
Employment	$0-2$

Each $\mathscr{RC}$-expression can be written also in a long tabular form. A long tabular form of $\mathscr{RC}$-expression Clients is available in Table 4.12.

Table 4.12: $\mathscr{RC}$-expression Clients in a long tabular form.

Name: Clients		Length: $1-5$			
Name: Personal		Type: *Conjunction*		Length: $0-3$	
sex	*Subsets*	$1-1$	*Pos*	B	–
age_exp	*Subsets*	$1-1$	*Pos*	R	*Age*
age_5	*Sequences*	$1-3$	*Pos*	R	*Age*
education	*Sequences*	$1-4$	*Pos*	R	–
Name: Employment		Type: *Conjunction*		Length: $0-3$	
occupation	*Subsets*	$1-3$	*Both*	B	–
hours_per_week	*Subsets*	$1-1$	*Pos*	B	–

4.5 Missing Information

Missing information is a serious problem of data mining. The GUHA procedures described in this book can handle data matrices with missing information introduced in Section 4.5.1. Attributes can get a special value denoted as X and interpreted as *unknown*. A specific approach called *secured X-extension* is developed in [41]. This approach is applied to get values both of basic and derived Boolean attributes, see 4.5.2. Note that this approach is equivalent to Kleene logics [78] when dealing with derived Boolean attributes. This leads to Boolean attributes with possible values *true, false* and X meaning *unknown*. Such attributes are called *three-valued attributes*.

This way, we have to deal with contingency tables in data matrices with missing values. Then, deleting all frequencies concerning missing information is applied to verify all patterns the described GUHA procedures deal with.

Let us note that the *secured X-extension* is applied also to association rules the 4ft-Miner procedure deals with. Also optimistic approach to missing information as well as ignoring missing information can be used when dealing with the 4ft-Miner procedure. See Section 7.1 for details.

4.5.1 Data matrices with missing information

We deal with missing information. This means that each attribute can get a value X in addition to its regular values. The X value is interpreted as *the value of the attribute is not known*. Thus we have to deal with a data matrix $\mathscr{M}^X$ with missing information. An example of such a data matrix is in the left part of Fig. 4.7.

$\mathscr{M}^X$	A_1	A_2	A_3	...	A_K		$\mathscr{M}_1$	A_1	A_2	A_3	...	A_K
o_1	2	X	6	...	2		o_1	2	3	6	...	2
o_2	X	4	X	...	1		o_2	4	4	2	...	1
o_3	X	X	2	...	3		o_3	2	1	2	...	3
⋮	⋮	⋮	⋮	⋱	⋮		⋮	⋮	⋮	⋮	⋱	⋮
o_n	1	1	1	...	X		o_n	1	1	1	...	2

Data matrix $\mathscr{M}^X$ Data matrix $\mathscr{M}_1$ – a completion of $\mathscr{M}^X$

Figure 4.7: Data matrix $\mathscr{M}^X$ with missing information and its completion $\mathscr{M}_1$.

A *completion of a data matrix with missing values* is a data matrix in which all X values are replaced by possible regular values. A data matrix $\mathscr{M}_1$ in the right part of Fig. 4.7 is an example of a completion of $\mathscr{M}^X$.

4.5.2 Secured completion of missings and Boolean attributes

Recall that $A(o,\mathscr{M})$ is a value of an attribute A in a row o of data matrix $\mathscr{M}$. We use this notation also for data matrices with missing information. Thus, we have $A_1(o_1,\mathscr{M}^X) = 2$ and $A_2(o_1,\mathscr{M}^X) = X$. Values of Boolean attributes are clearly defined for data matrices without missing information, see Section 1.4.3. $A(1)(o,\mathscr{M})$ denotes a value of a basic Boolean attribute $A(1)$ in a row o of a data matrix $\mathcal{M}$ without missing information. Analogously, $\varphi(o,\mathcal{M})$ denotes a value of a Boolean attribute φ in a row o of $\mathcal{M}$.

We need to define values of Boolean attributes in data matrices with missing information. We start with following observations concerning data matrices $\mathcal{M}^X$ and $\mathcal{M}$ from Fig. 4.7. It holds

- $A_1(o_1,\mathcal{M}^X) = 2$, thus $A_1(2)(o_1,\mathcal{M}) = 1$ in each completion $\mathcal{M}$ of $\mathcal{M}^X$

- from the same reason $A_1(1)(o_1,\mathcal{M}) = 0$ in each completion $\mathcal{M}$ of $\mathcal{M}^X$

- $A_2(o_1,\mathcal{M}_1) = 3$, thus $A_2(5)(o_1,\mathcal{M}_1) = 0$. There sure is a completion $\mathcal{M}_2$ of $\mathcal{M}^X$ such that $A_2(o_1,\mathcal{M}_2) = 5$ and thus $A_2(5)(o_1,\mathcal{M}_2) = 1$.

It also holds

- $A_1(2) \vee A_2(5)(o_1,\mathcal{M}) = 1$ in each completion $\mathcal{M}$ of $\mathcal{M}^X$ since $A_1(2)(o_1,\mathcal{M}) = 1$ in each completion $\mathcal{M}$ of $\mathcal{M}^X$

- $A_1(1) \wedge A_2(5)(o_1,\mathcal{M}) = 0$ in each completion $\mathcal{M}$ of $\mathcal{M}^X$ since $A_1(1)(o_1,\mathcal{M}) = 0$ in each completion $\mathcal{M}$ of $\mathcal{M}^X$

- $A_1(2) \wedge A_2(5)(o_1,\mathcal{M}_1) = 0$ since $A_2(5)(o_1,\mathcal{M}_1) = 0$ and also $A_1(2) \wedge A_2(5)(o_1,\mathcal{M}_2) = 1$ where $\mathcal{M}_2$ is a completion of $\mathcal{M}^X$ such that $A_2(o_1,\mathcal{M}_2) = 5$.

From these and additional simple examples, we can conclude that there are three possibilities for each Boolean attribute ω and a row o of a data matrix $\mathcal{M}^X$ with missing information:

1. $\omega(o,\mathcal{M}) = 1$ for each completion $\mathcal{M}$ of a data matrix $\mathcal{M}^X$

2. $\omega(o,\mathcal{M}) = 0$ for each completion $\mathcal{M}$ of a data matrix $\mathcal{M}^X$

3. There are completions $\mathcal{M}$ and $\mathcal{M}'$ of $\mathcal{M}^X$ such $\omega(o,\mathcal{M}) = 1$ and $\omega(o,\mathcal{M}') = 0$.

This observation leads to the following definition of values of Boolean attributes for data matrices with missing information. The definition is called *secured X-extension of handling with missing values* for Boolean attributes. Shortly we can say *secured handling with missing values*. If $\mathcal{M}^X$ is a data matrix with missing information, φ is a Boolean attribute and o is a row of $\mathcal{M}^X$, then:

- $\varphi(o,\mathcal{M}^X) = 1$ if it holds $\varphi(o,\mathcal{M}) = 1$ for each completion $\mathcal{M}$ of a $\mathcal{M}^X$. We say that φ is true for a row o of $\mathcal{M}^X$

- $\varphi(o,\mathcal{M}^X) = 0$ if it holds $\varphi(o,\mathcal{M}) = 0$ for each completion $\mathcal{M}$ of a $\mathcal{M}^X$. We say that φ is false for a row o of $\mathcal{M}^X$

■ $\varphi(o, \mathscr{M}^X) = X$ if there are completions $\mathscr{M}_1$ and $\mathscr{M}_2$ of $\mathscr{M}^X$ such that $\varphi(o, \mathscr{M}_1) = 1$ and $\varphi(o, \mathscr{M}_2) = 0$. We say that φ has X-value for a row o of $\mathscr{M}^X$ or that a value of φ for a row o of $\mathscr{M}^X$ is unknown.

We refer to such Boolean attributes as *Boolean attributes with missing values* or *three-valued attributes*.

It is easy to prove that the above given definition of values of three-valued attributes in data matrices with missing information can also be formulated in an inductive way. First, we again use a secured handling with missing values for basic Boolean attributes. Then we use a usual inductive definition of values of $\varphi \wedge \psi$, $\varphi \vee \psi$ and $\neg \varphi$. However, we use the truth tables according to Fig. 4.8 instead of usual truth tables for Boolean attributes.

$\varphi \wedge \psi$		ψ			$\varphi \vee \psi$		ψ				φ	$\neg \varphi$
		1	X	0			1	X	0		1	0
	1	1	X	0		1	1	1	1		X	X
φ	X	X	X	0	φ	X	1	X	X		0	1
	0	0	0	0		0	1	X	0			

Figure 4.8: Truth tables for $\wedge$, $\vee$, and $\neg$ for three-valued attributes.

Examples of three valued attributes are in Fig. 4.9.

$\mathscr{M}^X$	attributes				three-valued attributes			
row	A_1	A_2	...	A_K	$A_1(2)$	$A_2(5)$	$A_1(1) \wedge A_2(5)$	$A_1(1) \vee A_2(5)$
o_1	2	X	...	2	1	X	X	1
o_2	X	4	...	1	X	0	0	X
o_3	X	X	...	3	X	X	X	X
⋮	⋮	⋮	⋱	⋮	⋮	⋮	⋮	⋮
o_n	1	1	...	X	0	0	0	0

Figure 4.9: Three-valued attributes in data matrix $\mathscr{M}^X$ with missing information.

Chapter 5

LISp-Miner System

This chapter describes the *LISp-Miner* system, which is developed under an academic project of the same name for nearly 25 years. The main concept of the system is described in the following text as a reflection of initial and still valid requirements. Both the system architecture and its implementation layers were derived from the main concept. The most important achievement is the long-term adaptability of the system, as new functionalities were continuously added and seamlessly integrated into a concept designed long ago (e.g., data analysis on a grid/cloud or an automated analysis by scripting).

5.1 Overview of LISp-Miner

The LISp-Miner system is an academic system for data analysis developed in 1995 at the Prague University of Economics and Business [158, 125, 127, 159, 160, 136, 138, 139]. LISp-Miner system as one of several implementations of the GUHA method since the 1960s (see Section 1.1.3), is the most complex and the most feature rich implementation of this method so far. It is still important to remember that the LISp-Miner is an experimental software with UI tailored both to educational purposes and to research projects. Priorities of the system development are influenced by its academic nature and available funds.

5.1.1 Teaching and research tool

One of the main goals of the LISp-Miner development was its deployment as a tool for teaching data mining related courses. An in-house tool allows for a demonstration of inner details of used algorithms, better understanding of techniques used and it offers a possibility to add new features both as a result of research grants or as a result of students' diploma and dissertation thesis.

LISp-Miner is (or was) used at these universities: Prague University of Economics and Business, Faculty of Mathematics and Physics, Charles University in Prague, Czech Technical University in Prague, Technical University of Tampere, Université Lumière Lyon 2, the State University of New York at Binghamton, and Florida State University.

There were tens of seminar-, bachelor-, diploma- and dissertation-theses written, which relate to the LISp-Miner system. Several research projects are based on or use the system: project *AR2NL* — automatic translation of associational rules into natural language [150, 149], project *SEWEBAR* — presentation of found patterns in a readable form as web applications [111, 132, 79, 82], EU-wide research grants — e.g., [16] and in research grants in the medical domain. Additional applications of the LISp-Miner are presented, e.g., in [155, 93]. The LISp-Miner system is also related to research directions described in Chapters 16–18.

5.1.2 Home page

The LISp-Miner home page is available at `https://lispminer.vse.cz` and it is primarily used for a free download of the system and its modules. But it also contains an overview of the LISp-Miner system and the project, a list of important publications and links to other web pages concerning data mining and GUHA procedure and usage of the system, prepared in the wiki-form mainly.

An interesting record of the system development is a complete log of changes going back to version 3.00.00 from August of 17th 2001 and available at `https://lispminer.vse.cz/download/relnotes.html`.

5.2 Requirements and Prerequisites

The most important aspect of the LISp-Miner system development was (and still is) an academic environment in which it is developed. Suitability to test new research ideas and simultaneously to be ready for use in the day-to-day education of students in the field of data mining had to be preserved all the time. LISp-Miner is free to use and its installation requirements are kept very low to comply with a level of hardware and software equipment of university computer labs.

Regardless of its academic nature, the system was designed for the maximal speed of solving complex analytical tasks (mainly of generation and verification of GUHA hypotheses). So the user interaction is kept as smooth as possible, even for students to iterate several task solutions during one teaching lesson.

Last, but not least, we aimed for a robust system architecture with easy maintenance and adaptability, without any major redesign necessary for the longest possible time. After nearly 25 years of continuous development and many new data-mining modules and features successfully integrated, we could conclude that this aim was achieved.

5.3 Main Concept

The LISp-Miner is not the first implementation of the GUHA method (see Preface), and it is even not its last (see Chapter 15). The history of GUHA implementation goes back to 1960 (see Section 1.1.3). The design of the LISp-Miner system was based on experiences with previous GUHA implementations but also brought a new modular architecture and an integration with the just rightly released CRISP-DM methodology (see Section 1.2). Formerly, only the data mining phases of data preparation, modelling and data interpretations were supported in the system. Later, we also enhanced the coverage of the data mining phases of data understanding and of analytical questions formulations.

5.3.1 Context diagram

Context diagram of the LISp-Miner system is in Fig. 5.1.

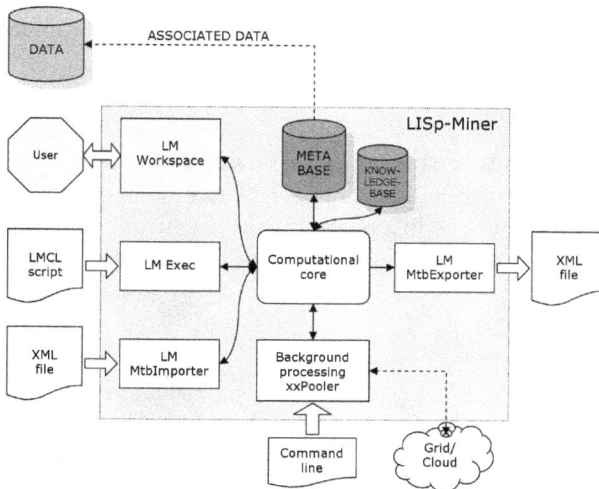

Figure 5.1: Context diagram of the LISp-Miner system.

Input of the system consists of data to be analysed in form of relational database tables (see Section 5.3.2). Output consists of interesting patterns found in the analysed data, described, e.g., in terms of association rules, conditional histograms and contingency-tables, action rules, trees or clusters. There is a central storage place for both meta-data and the found patterns called Metabase (see Section 5.3.3).

There are several modules in the LISp-Miner system, mainly the LM Workspace module, LM Exec module (see Section 5.3.7), all of them sharing the computational core implementing GUHA procedures and meta-data manipulation methods.

The LM Workspace is a basic module for common users and for an interactive communication with the system. It offers all the necessary support for all the phases of data mining process from data exploration and preprocessing to modeling and interpretation/evaluation of results.

There are two minor modules – LMMtbImporter and LMMtbExporter, for communications with other systems (usually in the PMML format, but other XML/Text formats are possible by user-defined templates), for detail see, e.g., [111, 132, 81, 82]. The last important group of modules supports remote solutions of data mining tasks on the cloud, or respectively on distributed solutions nodes of a computer grid. These modules could be launched from the operating system command prompt, so therefore could be easily integrated into a higher-level system.

5.3.2 *Analysed data*

Analysed data must be in the table format and it is usually imported into a relational database (beforehand by other means, or by using a tailored function of the LISp-Miner system). There is no restriction for the database system used, apart from ODBC connectivity available, which is a common standard from 1990s till now. So any of the main-stream database systems are suitable – PostgreSQL, Oracle, MS SQL Server, MySQL, . . . , also MS Access is good enough for small data, and even a plain text with a proper formatting works. There are no changes written into analysed data during analysis by the LISp-Miner system.

5.3.3 *Metabase*

All the user data are stored in a special relational database called Metabase. User data consists of a description of analysed data (aka "meta-data database", and thus Metabase), every form of preprocessing of data (e.g., transformations, discretizations, grouping), data-mining tasks descriptions, and finally, all the found patterns for each of the tasks and their annotations.

Metabase is thus a central storage place of meta-data for all the modules in the LISp-Miner system. Every module reads all the necessary information from the metabase and stores its outputs there. The open nature of the metabase allows even other systems to directly access information stored there. Every metabase is associated with just single analysed data. Single analysed data could be but associated with many metabases.

By using a relational database for both meta-data and found results, was a novel approach in 1990ies, when most of then used data-mining systems stored tasks in a proprietary format and a plain text was mainly used for outputs. An open-access relational database allowed for independent modules to both read and write task descriptions or to access found results and display them in a suitable way.

5.3.4   Knowledgebase

The original structure of the LISp-Miner system was enriched later by a new database-like entity called knowledgebase, as a consequence of coverage of the Domain understanding phase of CRISP-DM methodology (see Section 1.2). Knowledgebase stores primarily domain knowledge – generally accepted facts – from a given domain (see Chapter 17). Further, it includes tips for data pre-processing (e.g., granularity of discretization, domain specific threshold values, mutual influence of attributes). Knowledgebase helps with analysis of data from a given domain, so it is a source of domain knowledge for many metabases. Therefore, it was separated from metabase-entity to solve this one-to-many cardinality.

5.3.5   Context diagram of GUHA-procedure

All the data-mining procedures implemented into the LISp-Miner system so far are GUHA-procedures in the sense of [41]. There is a context diagram of a GUHA-procedure adapted for LISp-Miner system architecture in Fig. 5.2.

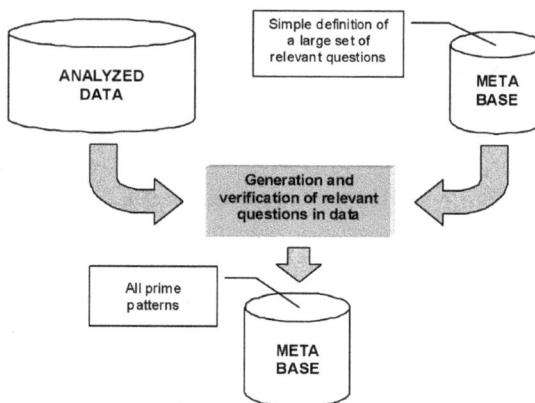

Figure 5.2: Context diagram of a GUHA-procedure adapted for LISp-Miner system.

5.3.6   LM Workspace module

As mentioned above, the LM Workspace is the main module for interactive work with the LISp-Miner system. Most of the users use this module only and do not need to take care of other modules.

The main screen of LM Workspace module is in Fig. 5.3.

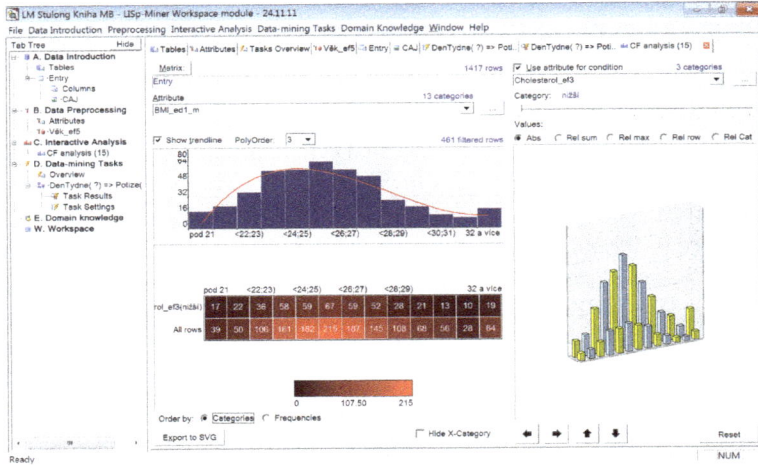

Figure 5.3: Main window of the LM Workspace module.

Interface accommodates a commonly used tab-format with a hierarchical tree of opened tabs in the left part of the screen. The tree is organized correspondingly to data-mining phases as defined in the CRISP-DM methodology (see Section 1.2). Therefore it helps users with an orientation and keeps them on track of data mining analysis process.

Majority of actions in the LM Workspace are logged locally in the metabase, see picture Fig. 5.4.

Activities are tagged by their importance and could be exported as a plain text file. Analysis of logs improves teaching process and in evaluation of students' works. A large amount of logs concerning the analysis of predefined data with the same goals helps improve the LISp-Miner system interface and functionality.

An automated way of validation of created data-mining tasks and found patterns is another feature that highly simplifies the evaluation of students' works.

As an introductory lesson, each student follows prepared instructions for the analysis of artificially prepared data (see Chapter 16). There are validation templates prepared for each step of this analysis. Before proceeding to the next step, students are required to use the validation function to self-check the current content of theirs metabase against the corresponding template. Validation automatically spots every difference (bar for lower/upper case mismatch in naming and

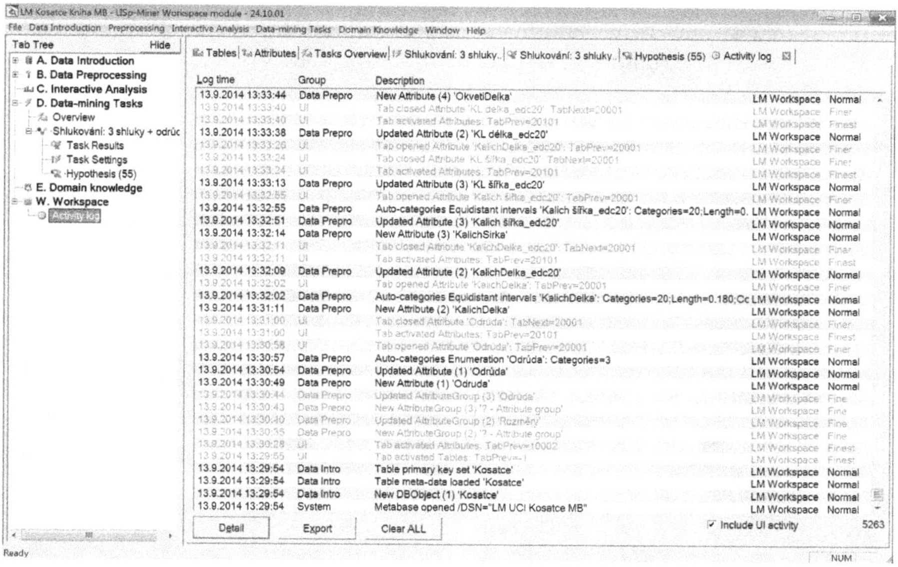

Figure 5.4: Detailed activity log, including even user switching among tabs.

other not important discrepancies) and summarizes them in a report. The most serious are labeled as errors and minor ones as warnings. Although, it is possible to continue with the analysis, it is always easier to correct errors sooner than later. And students know that the validation of metabase will be part of evaluation done by a teacher.

Validation templates could be prepared for any analysed data and every phase of the data-mining analysis process. They are successfully used also for debugging of developed modules for data-mining automation (see the next section). Results from each automated step could be automatically checked against a template prepared in advance to verify flawless functionality of the automation code.

5.3.7 Data-mining automation module

The LM Exec module allows for a semi-automated interaction with the LISp-Miner system using scripting language LMCL (LISp-Miner Control Language). LMCL was introduced in [159] and some results of automation are in [160, 136].

Data-mining process automation is a long term research goal (see [163]). It seems possible to describe algorithmically at least some of the steps of data analysis, and therefore to automate them. An example of means for such automation is LISp-Miner Control Language.

LISp-Miner Control Language (LMCL) is a scripting language based on Lua language ([70]) with a simple and easily understandable syntax and good support through a wide community of users. Syntax highlighting is supported in common text editors for programming, so code is readable – see Fig. 5.5.

```
24    lm.log( "Initializing data tables");
25    lm.logIndentUp();
26
27    -- Prepare dataTableArray
28    local dataTableArray= lm.explore.prepareDataTableArray();
29
30    -- Iterate through all the dataTables
31    for i, dataTable in ipairs( dataTableArray) do
32
33        lm.log( "Initializing data table ".. dataTable.Name);
34        lm.logIndentUp();
35
36        dataTable.init();
37
38        if ( not dataTable.isPrimaryKeyDefined()) then
39            -- Primary key not set yet
40
41            -- Use the default ID column created during the text data import
42            dataTable.markPrimaryKey({
43                columnName= lm.data.IDColumnNameDefault
44                -- name of column to become the primary key
45            });
46
47        end;
48
49        -- Check the primary key being set properly and stop if not
50        assert( dataTable.checkPrimaryKey(),
51               "Error checking the primary key for table "..dataTable.Name);
52
53        -- Enable data caching to speed-up analytical task processing
54        dataTable.LocalDataCacheFlag= true;
55
56        lm.logIndentDown();
57
58    end;
```

Figure 5.5: Example of a LMCL script with syntax highlighted.

Syntax of scripting language incorporates all the common programming features – variables, expression evaluation, and execution controlling structures (conditions, cycles, and calling of subroutines).

Programmed scripts are executed many times faster than could be achieved manually using the standard user interface. It is useful mainly for time-consuming and repeated activities in the Data preparation phase or for laborious fine-tuning of data-mining task parameters to obtain a suitable number of interesting patterns.

The used Lua interpreter is really lightweight and proved to be fast, so far tested with medium-sized scripts (up to thousands of code-lines). Costs of script parsing and execution overhead are insignificant compared to data mining task solution times or to data transfers from a database. The performance of LMCL scripts therefore depends solely on the ability of the LISp-Miner system modules to solve data mining tasks. It has been proved already that the algorithms and optimizations techniques implemented in the LISp-Miner system lead to solution times linearly dependant on the number of rows (objects) in analysed data [127].

LMCL makes the internal objects and functions of the LISp-Miner system accessible and allows manipulations with them on a higher level of abstraction in a user-written script. There are means available for importing analysed data and its preparation, for data-mining tasks setup and finally means for browsing of found patterns. A partial data model is shown in Fig. 5.6.

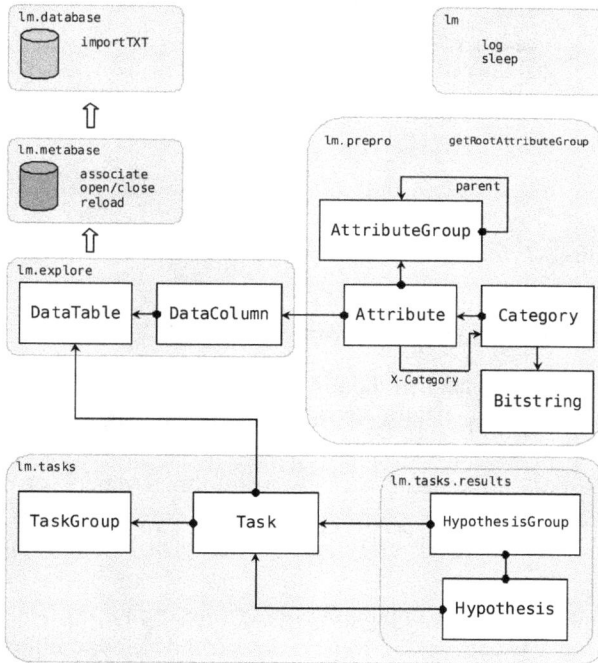

Figure 5.6: Namespaces and classes in LMCL language.

Every script action is logged for further analysis and simultaneously presented to the user so he is instantly informed about progress and decisions taken, which is useful mainly for long-running analyses.

Documentation for programmers writing LMCL scripts consists of a detailed reference description of object classes and functions, see Fig. 5.7 for an example of reference documentation for class Task.

Documentation follows a widely-used structure, as is common for example in JavaDoc. An important feature is that this documentation is self-generated from C++ implementation of the LM Exec module. So its completeness and up-to-dateness is guaranteed.

LISp-Miner Control Language Reference, version: 23.02.02 of 6 Jan 2014

Main Page

LM Exec
LMCL Language Basics
LMCL Diagrams
Demo Examples
EverMinerSimple Demo
Predefined Constants

Namespaces

lm
lm.data
lm.explore
lm.metabase
lm.prepro
lm.tasks
lm.tasks.results
lm.tasks.settings

Classes

Attribute
AttributeGroup
Category
DataColumn
DataTable
FTLiteral
FTLiteralSetting

Class Task

LISp-Miner data-mining analytical task of an unspecified type
inherits from LMWrapName
namespace: lm.tasks

Constructor

No constructor available. Objects of this class could not be instantiated from Lua scripts.

Properties List

DataTable	the DataTable, this task is based on
ID	Unique identifier (primary key) of the object
Name	Unique name of this object. Cannot be empty.
Note	Text description of this object
TaskGenerationStatus	Status code for task, see TaskGenerationStatus codes
TaskGroup	the TaskGroup, this task belongs to

Methods List

canDel ()	Returns true if the object could be deleted (is not used)
clone ()	Creates a clone (an exact copy) of this task
findFTWholeCedentSetting ()	Tries to look-up a FTWholeCedentSetting with a unique property given as parameter. Just one parameter has to be specified. Returns nil if FTWholeCedentSetting with this property doesn't exists.
findHypothesis ()	Tries to look-up a Hypothesis with a unique property given as parameter. Just one parameter has to be specified. Returns nil if

Figure 5.7: Auto-generated documentation reference for Task class.

5.4 EverMinerSimple Demo

There are several examples included in the LM Exec installation package. They range from the obligatory "Hello, world!" example to a more complex one called EverMinerSimple.

The EverMinerSimple demo is really a simplified version of the EverMiner (see [160]) concept, rather a prototype of it. Its only purpose is to prove that the LM Control Language is really able to automate a part of the data mining process.

This prototype solution implements only one iteration of the main phases of the data mining process with no new domain knowledge inferred yet. But it already incorporates the inner cycle of fine-tuning task parameters to obtain an acceptable number of patterns in the results. Only one type of pattern is used for now – 4ft-association rule.

There is a conceptual diagram of EverMinerSimple steps in Fig. 5.8. Few user-defined parameters provide all the necessary input to the whole process. The first group of parameters defines the text file with analysed data to import, destinations to store the created database with analysed data and the database with meta-data. Finally, it defines the ODBC DataSourceName to identify this data + meta-data pair within the operating system.

The second group of parameters provides a bit of domain knowledge – groups of attributes the analysed data columns should be grouped into. This information is important for analytical tasks construction, where all the possible combinations

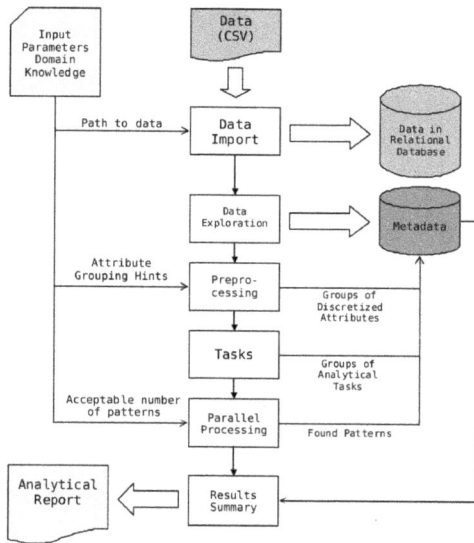

Figure 5.8: EverMinerSimple algorithm overview.

of groups of attributes in antecedents and succedents of patterns to be mined are created.

The most interesting input parameters are the minimal and maximal number of patterns to mine, regardless of the combination of groups this particular task is concerning. There are several ways to reduce (or enlarge) task search space to influence the number of found patterns, but they are out of the scope of this paper. Nevertheless, they are exploited in the *Parallel processing* phase to ensure the number of found patterns is within a given range. (A limit for a maximal number of iterations is implemented to avoid a never-ending cycle.)

An important feature is that no task settings are changed after it was processed. Every time a change is necessary to task settings, its exact clone is created first and the desired change is made to this cloned task setting instead. Therefore, a complete history of task settings evolution, together with a number and an exact form of found patterns is preserved in the meta-data database. It could be used later for investigations of steps and decisions taken during the automated data mining process – either for debug purposes or to help with a proper interpretation of found patterns.

The last input parameter specifies the path and name of file for the analytical report to be written into. In this example, an HTML report is created and opened in an operating system default web browser. A manually shortened version of it is shown in Fig. 5.9.

In this particular case, amount of found patterns for an analytical task *Calendar, Guest, Visit implying Weather type* was not reduced to the number requested

EverMinerSimple Demo Analytical Report

Input parameters

Data Exploration

Hotel

- Age: Long integer
- Nationality: String
- ...

Derived columns:

- VisitFrom.DayOfWeek: Long integer

Data Preprocessing

Guest

Age: <18;23), <23;31), <31;38), <38;43), <43;49), <49;54), <54;59), <59;65), <65;70), <70;82>
Nationality: AT, CZ, GE, PL, SK

...

Tasks

Calendar, Guest, Visit -> Weather (08)

Task finished but an acceptable number of patterns has not been achieved
Could not reduce the search space
Number of iterations: 8
Found patterns: 36
The most interesting ones:

- **Nationality**(*SK*) & **Persons**(*2*) & **VisitType**(*Tourist*) >÷< **Weather**(*Cloudy*)
- **Nights**(*7*) & **Persons**(*1*) & **VisitType**(*Tourist*) >÷< **Weather**(*Sunny*)

Figure 5.9: Example of an automatically created EverMinerSimple analytical report (shortened).

by input parameters (even after reaching an arbitrary limit of eight iterations of task settings changes). This failure is mentioned in the report and only two of found patterns are shown. Please remember also, that the *EverMinerSimple* example is just a prototype with no ambition to provide real results yet.

5.5 System Design and Implementation

Right from the beginning, an object-oriented approach was applied to the LISp-Miner system design. This approach greatly contributed to the maintainability of the quietly large system (ca. 1 million lines of source code) and also significantly simplified further development of the system and adding of new features. The object-oriented approach was based on the OMT methodology [143].

System architecture closely follows the phases of the data-mining process as defined by the CRISP-DM methodology. Sharing information among its modules is on the data level, made possible by the existence of the metabase and its relational structure.

The proposed architecture allows for an easy extension of the system by new modules. At the same time, the computational core of the system could be highly optimized, regardless of other existing or newly proposed modules. Standalone applications in sense of tailor-made interfaces for particular domain/analysis are possible due to modules for remote/distribute solution of data-mining tasks, without any compromises in task solution times (see Fig. 5.1).

5.5.1 Programming language and environment

There were three options of programming languages available back in the year 1996: C/C++, Pascal (Delphi), and a little bit later a brand new Java. For the sake of maximal speed and for maximal stability and time-proved robustness, we choose C/C++. As an integrated development environment we choose MS Visual C++, the environment LISp-Miner is developed up to now.

An important consequence of choosing C/C++ as a programming language is a more labor-intensive way of programming but more than counterweighted by an ability to develop highly compact and feature-rich applications with an above average speed of computation of data-mining tasks. The size of the system on disk is in units of megabytes and no other additional frameworks or libraries have to be installed. The LISp-Miner system installations process is therefore completed simply by extracting files from an installation ZIP file. No administration or elevated-user privileges are necessary during installation or running the system, so there are no problems to use the system in environments like computer labs, where individual user rights could be severely restricted. Thus, the requirement to use the LISp-Miner system for educational purposes is met.

5.5.2 Implementation layers

There are implementation layers of the LISp-Miner system depicted in Fig. 5.10. The system architecture is incrementally layered. That means that a rule is obeyed, in which each layer access functions and data from the nearest lower layer only. This requires a little bit more work while implementing new functions on lower levels, but preservers the consistency of each layer, so inner changes to each of the layers could be done with a low risk of messing things up and simplifies maintenance of the large source code.

There is a Database Layer at the lowest level accomplishing all the communication with the metabase and with analysed data (through the Microsoft ODBC database API). The basic business logic of the LISp-Miner system is defined in the LM Shared Layer. It includes a data-mining object model and basic functions used in all the modules of the system (for example, the fast computation of bitwise operations using Bitstring structure – see 5.5.3). General classes and functions of patterns generation and verification, common to all types of GUHA-procedures are defined in the CG Layer. Task-specific computation codes of pat-

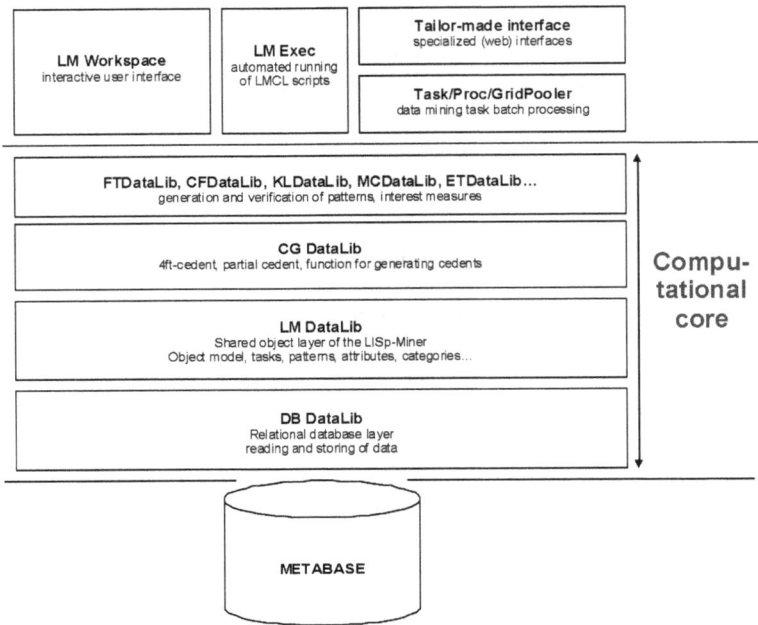

Figure 5.10: Implementation layers of the LISp-Miner systems.

terns generation and verification are defined in a corresponding library for each GUHA-procedure in the layer just above CG Layer. Finally, there is a layer of implementation of all the modules of the LISp-Miner system (LM Workspace, LM Exec...).

5.5.3 Bitstrings

Dealing with general Boolean attributes and with data with incomplete information, together with a necessity to quickly compute 4ft-tables and nine-fold tables, led to the employment of bitstrings that have been described in [101], see also [127, 138]. Each attribute A with categories $c_1,\ldots,c_u$ is represented by $u+1$ strings of n bits $\mathscr{C}(A(c_1)),\ldots,\mathscr{C}(A(c_u)),\mathscr{C}(A(X))$, where n is the number of rows of an analysed data matrix $\mathscr{M}$. There is 1 in a i-th bit of $\mathscr{C}(A(c_j))$ if and only if $A[o_i] = c_j$, where o_i is an i-th row of $\mathscr{M}$, $i = 1,\ldots,n$, $j = 1,\ldots,u$; analogously for missing information X. The string $\mathscr{C}(A(c_i))$ is called *a card of the category* c_i *of* A for $i = 1,\ldots,u$ and $A[X]$ is called an *X-card of* A. Examples of such cards for an attribute A_1 with categories $1,2,3$ in a data matrix $\mathscr{M}$ with missing information are in Fig. 5.11.

Cards of Boolean attributes φ and ψ are used to compute frequencies from 4ft-tables and nine-fold tables. The card of the Boolean attributes φ is denoted by $\mathscr{C}(\varphi)$. The card $\mathscr{C}(\varphi)$ is a string of bits that is analogous to a card

row	attributes of $\mathcal{M}$				cards of categories			X-card
of $\mathcal{M}$	A_1	A_2	...	A_K	$A_1[1]$	$A_1[2]$	$A_1[3]$	$A_1[X]$
o_1	1	7	...	2	1	0	0	0
o_2	3	4	...	1	0	0	1	0
o_3	X	6	...	3	0	0	0	1
⋮	⋮	⋮	⋱	⋮	⋮	⋮	⋮	⋮
o_{n-1}	2	2	...	8	0	1	0	0
o_n	1	1	...	6	1	0	0	0

Figure 5.11: Cards of categories and X-card of the attribute A_1.

of category. Each row of the data matrix corresponds to one bit of $\mathcal{C}(\varphi)$ and there is "1" in the i-th bit if and only if φ is true in row o_i. It is evident that $\mathcal{C}(\varphi \wedge \psi) = \mathcal{C}(\varphi) \wedge \mathcal{C}(\psi)$, $\mathcal{C}(\varphi \vee \psi) = \mathcal{C}(\varphi) \mathbin{\dot{\vee}} \mathcal{C}(\psi)$, $\mathcal{C}(\neg\varphi) = \mathbin{\dot{\neg}} \mathcal{C}(\varphi)$. Here, $\mathcal{C}(\varphi) \wedge \mathcal{C}(\psi)$ is a bit-wise conjunction of bitstrings $\mathcal{C}(\varphi)$ and $\mathcal{C}(\psi)$, analogously for $\dot{\vee}$ and $\dot{\neg}$. Moreover, it is $\mathcal{C}(A_1(1,2)) = A_1[1] \mathbin{\dot{\vee}} A_1[2]$ for the basic Boolean attribute $A_1(1,2)$, etc.

It is important that the bit-wise Boolean operations $\wedge$, $\dot{\vee}$ and $\dot{\neg}$ are carried out by very fast processor instructions. An optimised algorithm is also used to carry out a bitstring function $Count(\xi)$, returning the number of values "1" in the bitstring ξ. This function is used to compute frequencies from 4ft-table $4ft(\varphi, \psi, \mathcal{M}) = \langle a,b,c,d \rangle$. It is $a = Count(\mathcal{C}(\varphi) \wedge \mathcal{C}(\psi))$, $b = Count(\mathcal{C}[\varphi]) - a$, $c = Count(\mathcal{C}[\psi]) - a$, $d = n - a - b - c$, where n is the total number of rows in the data matrix $\mathcal{M}$.

The relevant partial cedents are generated in a depth-first way. This can be outlined by a toy example concerning a set of relevant partial cedents created from attributes A, B, each with categories $1, 2, 3$. We assume that basic Boolean attributes $A(\alpha)$, $B(\beta)$, where $\alpha, \beta \subsetneq \{1,2,3\}$ can only be used and that we are interested in conjunctions. Then the following partial cedents are generated: $A(1)$, $A(1) \wedge B(1)$, $A(1) \wedge B(1,2)$, $A(1) \wedge B(1,3)$, $A(1) \wedge B(2)$, $A(1) \wedge B(2,3)$, $A(1) \wedge B(3)$, $A(1,2)$, ..., $A(1,2) \wedge B(3)$, $A(1,3)$, ..., $A(1,3) \wedge B(3)$, $A(2)$, ..., $A(2) \wedge B(3)$, $A(2,3)$, ..., $A(2,3) \wedge B(3)$, $A(3)$, ..., $A(3) \wedge B(3)$, $B(1)$, $B(1,2)$, $B(1,3)$, $B(2)$, $B(2,3)$, $B(3)$.

A more detailed description of the 4ft-Miner algorithm and its implementation is out of the range of this chapter. The basic ideas come from [101], for more details see [127]. Let us note that there is an analogy to the important fact used in the apriori algorithm, i.e., if U, V are itemsets satisfying $supp(U) < minS$ and $U \subsetneq V$, then $supp(V) < minS$ also holds. The ASSOC procedure as well as the 4ft-Miner procedure use an analogous optimisation criterion. Let $Fr(\omega)$ denote the number of rows of a data matrix $\mathcal{M}$ satisfying a Boolean attribute ω. Then if $Fr(\varphi) < B$, then also $Fr(\varphi \wedge \psi) < B$, where φ and ψ are Boolean attributes.

APPLYING THE GUHA PROCEDURES

II

Section II is the core of the book. About forty examples of applications of the seven GUHA procedures introduced in Section I are presented. Examples show how particular configurations of parameters for each of the described procedures can be used to get useful insight from data. This encompasses, among others, various approaches to exception rules and subgroup discovery. Section II starts with a structured overview of all examples available in Chapter 6. Examples of applications of particular GUHA procedures in datasets are presented in Chapters 7–13. Chapter 14 is devoted to a relation of the GUHA procedures and business intelligence. First experiences with the implementation of the GUHA procedures in the Python language are presented in Chapter 15.

Chapter 6

Examples Overview

Examples of applications of particular GUHA procedures are presented in Chapters 7–13. The GUHA procedures of the LISp-Miner system are used in all examples. Each of Chapters 7–13 starts with a description of the parameters used in the presented examples. However, only selected important features of particular procedures are presented by examples. Various useful features of the GUHA procedures are not mentioned due to limited space. Overview of all presented examples is available in Sections 6.1–6.7.

The possibilities of applications of the GUHA procedures in Business Intelligence are discussed in Chapter 14. Examples of such applications are also presented. A short introduction of this discussion and examples is available in Section 6.8. Possibilities of transferring the GUHA procedures into currently most used data science languages are discussed in Chapter 15. An example of implementation of the GUHA procedure CF-Miner procedure in the Python language is also introduced. A short summary of Chapter 15 is presented in Section 6.9. A summary of all examples is available in Section 6.10. Several important notes are presented in Section 6.11.

6.1 Overview of 4ft-Miner Application Examples

Applications of the 4ft-Miner procedure concern three topics – comparison of the 4ft-Miner and arules package of the R-system (see Section 6.1.1), presentation of possibilities of the 4ft-Miner procedure related to important features of the GUHA association rules (see Section 6.1.2), and data mining with exception rules (see Section 6.1.3).

6.1.1 4ft-Miner and arules

The first goal of 4ft-Miner applications is to compare the 4ft-Miner procedure with arules package. The implementation of the apriori algorithm in arules package of the R-system is compared with both the slightly modified version of the apriori algorithm implemented into the 4ft-Miner procedure described in Section 7.2 and with the full version of the 4ft-Miner dealing with the secured approach to missing information. Dataset *Adult* described in Section 2.2 is used. Results can be summarized as follows.

■ Apriori implemented in the arules package is faster than apriori implemented in the 4ft-Miner, see Table 7.2 in Section 7.2.2. However, the time used by the 4ft-Miner is not much longer, and it is only a fraction of the whole time necessary to solve a real task.

■ Solution times for the secured completion are longer, see Table 7.2 in Section 7.2.2. This is because the GUHA approach of the depth-first walking is used and simultaneously, it is necessary to maintain additional information to compute nine-fold tables. The time consumed by the secured completion is still acceptable, and it is only a fraction of the time necessary to solve the whole analytical task concerning real data, which usually does contain missing information.

■ There are many extra rules produced by the apriori algorithm. Each of these extra rules has a problem – there is a completion of the analysed data matrix *Adult* in which the rule is not valid, see Section 7.2.3.

■ There is an obligatory use of minimal values of both support and confidence in the arules package. This can lead to the loss of rules, which is interesting because of the high value of the lift measure. This is presented in Section 7.2.4.

6.1.2 Applying important features of GUHA association rules

The second goal is to present possibilities of the 4ft-Miner procedure related to important features of the GUHA association rules. Dataset *Adult* described in Section 2.2 is again used. Four applications of the 4ft-Miner are described.

■ Possibilities of dealing with basic Boolean attributes $A(\alpha)$ where α is a coefficient are presented in Section 7.3.1. Two types of coefficients – *Sequences* and *Right cuts* are used to find segments of persons with a remarkable chance to have an extreme gain. This can be also understood as the subgroup discovery.

■ Usefulness of a possibility to deal with rules with a *conjunction* of basic Boolean attributes in consequent are presented in Section 7.3.2. Note that this is not available in arules [20]. A conjunction of extreme gain and large income is used to define extremely rich persons. Segments of such persons are identified by a 4ft-Miner application.

■ Since the chance to be extremely rich is not very high, a *disjunction* of extreme gain and large income is used to define rich persons. Such disjunctions in consequent are used to identify segments of rich persons, see Section 7.3.3.

■ Disjunction of basic Boolean attributes in combination with some 4ft-quantifiers open a possibility to use logical deduction to decrease the number of output rules. This is demonstrated in Section 7.3.4.

6.1.3 Mining for exception GUHA association rules

The third goal is to present possibilities of the 4ft-Miner procedure to mine exception rules. Dataset *Accidents* described in Section 2.3 is used. Three applications of the 4ft-Miner are described.

■ Mining for association rules pointing to segments of accidents with a significantly higher percentage of accidents with a particular degree of severity is presented in Section 7.4.1. This is illustrated in Fig. 6.1, see also Fig. 7.27.

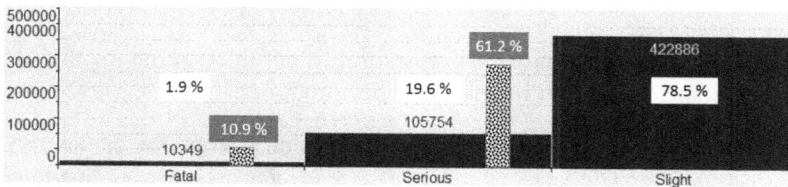

Figure 6.1: Percentages of degrees of severity with exceptions – increasing columns.

■ Mining for association rules pointing to segments of accidents with a significantly lower percentage of accidents with a particular degree of severity is presented in Section 7.4.2.

■ Rules resulting from applications described in Section 7.4.2 are used to formulate a new analytical question to demonstrate possibilities of mining for "exceptions from exceptions". A possibility to get higher confidence by extending the given antecedent is shown in Section 7.4.3.

6.2 Overview of CF-Miner Application Examples

Several examples concerning the subgroup discovery with the CF-Miner procedure are presented. All subgroups are defined by an interesting shape of a histogram. Datasets *Adult* and *Accidents* introduced in Sections 2.2 and 2.3 are
used. Applications of the CF-Miner to the *Adult* dataset are summarized in Section 6.2.1; applications to the *Accidents* dataset are summarized in Section 6.2.2.

6.2.1 Subgroup discovery in Adult dataset

Applications of the CF-Miner procedure to the *Adult* dataset are inspired by
histograms of attributes age_5 and age_exp presented in Fig. 6.2. Both histograms
are first increasing and then decreasing. Applications aim to find subgroups of
persons for which histograms of attributes have a different shape.

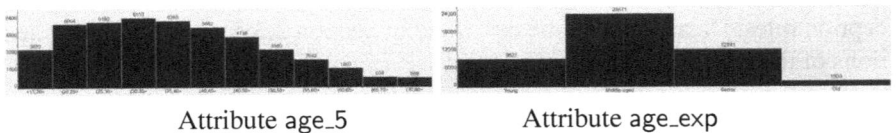

| Attribute age_5 | Attribute age_exp |

Figure 6.2: Histograms of attributes age_5, and age_exp.

- ■ Mining for subgroups of persons for which histograms of the age_exp
  attribute increase is presented in Section 8.2.1.

- ■ Mining for subgroups of persons for which histograms of the age_exp
  attribute decrease is described in Section 8.2.2.

- ■ Mining for subgroups of persons for which histograms of the age_5 attribute is first decreasing and then increasing is presented in Section 8.2.3.

6.2.2 Subgroup discovery in Accidents dataset

Four interesting analytical questions are formulated and solved for the *Accidents*
dataset. They are inspired by the decreasing trend of the numbers of accidents in
the period 2005–2015, see Fig. 6.3.

- ■ A question is solved, if there are large subgroups of accidents defined by
  a type of vehicle and the sex of a driver with a decreasing trend similar
  to the trend for all accidents. Several subgroups are found. The subgroup
  defined by the Boolean attribute Type(*Motor cycle over 500 cc*) is an
  example, see Section 8.3.1.

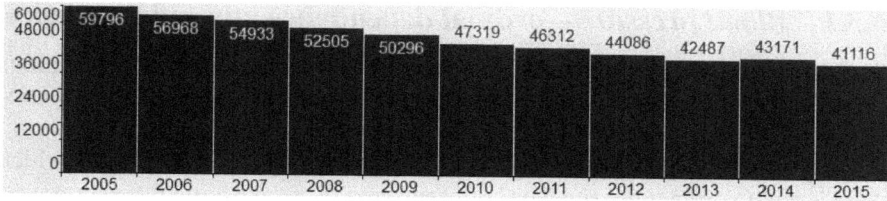

Figure 6.3: Decresing trend of all accidents in period 2005–2015.

- Thus, a question arises as to whether there are large enough subgroups of accidents with an increasing trend during the period 2005–2015. The subgroup defined by the Boolean attribute Type(*Car*) ∧ Speed_Limit(*20*) is an example of the CF-Miner output, see Section 8.3.2.

- An additional question is a question if there is a subgroup

$$\text{Type}(\textit{Motor cycle over 500 cc}) \wedge \omega$$

of the subgroup Type(*Motor cycle over 500 cc*) such that the trend of accidents in Type(*Motor cycle over 500 cc*) ∧ ω is increasing or at least non-decreasing even if a trend in the subgroup Type(*Motor cycle over 500 cc*) is decreasing. This question is solved in Section 8.3.3. A Boolean attribute

$$\omega = \text{Vehicle\_Age}(\textit{12, 13, 14, 15, 16} - \textit{20}) \wedge \text{Age}(\textit{46–55})$$

defines one of the found subgroups.

- The subgroup Type(*car*) ∧ Speed_limit(*20*) is the largest found subgroup with increasing trend – an exception to the generally decreasing trend. Thus a question arises, whether there is a local authority defined by one of the attributes District, Highway, Police which can be considered as an exception to this exception. This is answered in Section 8.3.4.

6.3 Overview of KL-Miner Application Examples

There are three applications of the KL-Miner procedure. Two applications concerning ordinal dependence and independence among blood pressure and additional medical characteristics are briefly described in Section 6.3.1. The third application introduces a subgroup discovery using a range of quantifiers. It is introduced in Section 6.3.2.

6.3.1 Blood pressure—ordinal dependence and independence

The dataset *STULONG* introduced in Section 2.4 is used in two applications concerning ordinal dependence and independence among categorical attributes. Kendall's coefficients τ_B is used as a measure of ordinal dependence.

A goal of the first application is to find subgroups of patients for which there is a high ordinal dependence between systolic and diastolic blood pressure. Boolean attributes derived from the groups Personal, Anamnesis, Risks, Measurement, Alcohol consumption, and Biochemical examination of attributes are used to describe subgroups. A contingency table for ordinal dependence corresponding to $\tau_B = 0.82$ is graphically represented by columns in the left part of Fig. 6.4. For details see Section 9.2.1.

A goal of the second application is to find subgroups of patients for which there is almost ordinal independence between Diastolic attribute and one of the attributes the Subscapularis, Triceps, Cholesterol, and Triglycerides. The same Boolean attributes are used to describe subgroups as in the previous application. A contingency table for ordinal independence corresponding to $\tau_B = 0$ is outlined in the right part of Fig. 6.4. For details see Section 9.2.2.

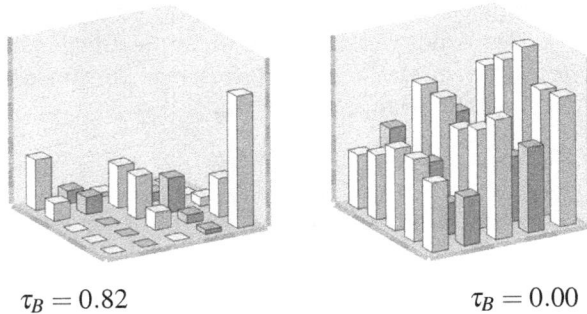

$$\tau_B = 0.82 \qquad\qquad\qquad \tau_B = 0.00$$

Figure 6.4: Contingency tables for $\tau_B = 0.82$ and $\tau_B = 0.00$.

6.3.2 Subgroup discovery using range of quantifiers

A range of KL-quantifiers is used in an application of the KL-Miner procedure to the *Hotel* dataset described in Section 2.5. A goal is to find subgroups of stays in the hotel such that at least 50 percent of stays of each subgroup satisfy both QSatisfaction(*average, high*) and DayOfWeek(*Mon, Tue, Wed*). Boolean attributes derived from the groups Guest, Domicile, Meteo, Stay, Price, and Check-in are used to to describe subgroups.

A range $\{n_{i,j}\}_{i=2,3}^{j=1,2,3}$ of a contingency table $\mathscr{T} = KL(\text{QSatisfaction}, \text{DayOfWeek}, HotelPlusExternal)$ defined by the quadruple $\langle 2, 3, 1, 3 \rangle$ is used to solve this task. This range is outlined in Fig. 6.5. For details see Section 9.3.1.

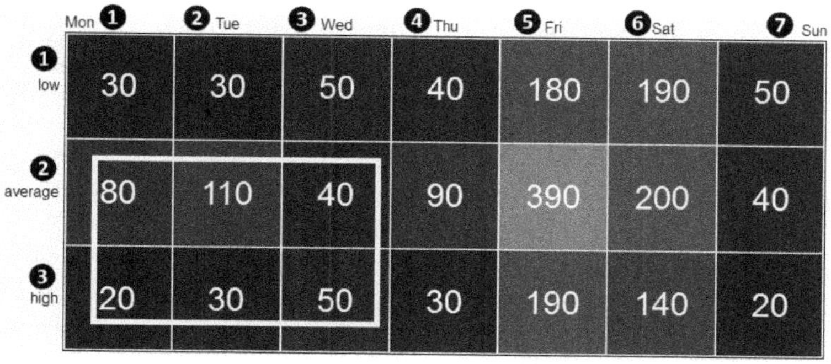

Figure 6.5: Range $\langle 2,3,1,3 \rangle$ of contingency table $\mathscr{T}$.

6.4 Overview of SD4ft-Miner Application Examples

All applications of the SD4ft-Miner procedure deal with the *Accidents* dataset introduced in Section 2.3. There are applications concerning differences between districts and applications concerning differences between female and male drivers. They are described in Sections 6.4.1 and 6.4.2.

6.4.1 Comparing districts

The first part of the SD4ft-Miner applications can be seen as inspired by a couple of conditional association rules:

$$\mathscr{R}_1 : \text{Sex}(\textit{Male}) \Rightarrow_{0.043;90} \text{Severity}(\textit{Fatal})/\text{District}(\textit{Highland}) \qquad (6.1)$$

$$\mathscr{R}_2 : \text{Sex}(\textit{Male}) \Rightarrow_{0.011;62} \text{Severity}(\textit{Fatal})/\text{District}(\textit{Westminster}) . \qquad (6.2)$$

It holds that a confidence $Conf(\mathscr{R}_1)$ of the rule $\mathscr{R}_1$ is 0.043, a confidence $Conf(\mathscr{R}_2)$ of the rule $\mathscr{R}_2$ is 0.011 and $\frac{Conf(\mathscr{R}_1)}{Conf(\mathscr{R}_2)} = 3.74$.

This means that a relative frequency of fatal accidents in a segment of accidents given by Sex(*Male*) in district Highland, is 3.74 times higher than in district Westminster. This can be expressed by an SD4ft-pattern

$$\textit{Highland} \bowtie \textit{Westminster} : \text{Sex}(\textit{Male}) \rightarrow_{*3.74} \text{Severity}(\textit{Fatal})$$

where we write "*Highland*" instead of "District(*Highland*)" and similarly for District(*Westminster*). Symbol $\rightarrow_{*3.74}$ is a suitable SD4ft-quantifier. These considerations lead to the following analytical questions:

■ "*Which segments S of accidents and which couples of districts $\langle D_1; D_2 \rangle$ satisfy that relative frequency of fatal and/or serious accidents in segment S for district D_1 is at least 2 times higher than for D_2?*" This can be formalized such that we are interested in true SD4ft-patterns

$$\text{District}(D_1) \bowtie \text{District}(D_2) : S \approx \text{Severity}(\omega) , \qquad (6.3)$$

where ≈ is a suitable SD4ft-quantifier. This task is solved in Section 10.2.1. The task can be also considered as a task of the subgroup discovery. The found segments of accidents can be considered as interesting subgroups.

■ *"Which segments S of accidents and which districts D satisfy that relative frequency of fatal accidents in segment S for district D is at least three times higher than for the whole dataset?"* This can be formalized such that we are interested in true SD4ft-patterns

$$\text{District}(D) \bowtie True :\approx \text{Severity}(Fatal) \,,$$

where ≈ is a suitable SD4ft-quantifier. This task is solved in Section 10.2.2. The task can be also considered as a task of exception rules discovery.

■ *"Which segments S of accidents and which districts D satisfy that relative frequency of fatal accidents in segment S for district D is at least 2.5 times lower than for the whole dataset?"* This can be formalized such that we are interested in true SD4ft-patterns

$$\text{District}(D) \bowtie True : S \approx \text{Severity}(Fatal) \,,$$

where ≈ is a suitable SD4ft-quantifier. This task is solved in Section 10.2.2. The task can be also considered as a task of exception rules discovery.

6.4.2 Comparing female and male drivers

There are three similar analytical questions related to an SD4ft-pattern

$$\text{Sex}(Male) \bowtie \text{Sex}(Female) : S \approx \text{Severity}(\omega)$$

where S belongs to a suitably defined set of relevant segments, Severity(ω) belongs to a suitably defined set of Boolean attributes derived from the Severity attribute and ≈ is a suitable SD4ft-quantifier:

■ *"For which segments S of accidents, is a relative frequency of fatal or serious accidents remarkably higher for male than for female drivers?"* This task is solved in Section 10.2.4.

■ *"For which segments S of accidents, is a relative frequency of fatal or serious accidents remarkably higher for female than for male drivers?"* This task is solved in Section 10.2.5.

■ *"For which segments S of accidents is a relative frequency of fatal or serious accidents for female and for male drivers approximately the same?"* This task is solved in Section 10.2.6.

All three tasks can be also considered as tasks of the subgroup discovery. The found segments of accidents can be considered as interesting subgroups.

6.5 Overview of SDCF-Miner Application Examples

Several analytical tasks concerning the subgroup discovery and histograms are described in Chapter 8. Their summary is in Section 6.2. All these tasks are solved by the CF-Miner procedure. The SDCF-Miner procedure brings additional possibilities to solve tasks concerning the subgroup discovery and histograms.

Two such analytical questions concerning the *Accidents* dataset are solved in Chapter 11. The first one concerns searching for authorities with exceptional trends of accidents, see Section 6.5.1. The second one concerns different trends of the number of accidents among police forces, see Section 6.5.2.

6.5.1 Exceptional histograms and authorities

We are interested in authorities with exceptional trends of accidents. It is known that the trend of accidents in the period 2005–2015 is decreasing, see Fig. 6.3. We consider an authority $\mathscr{A}$ as authority with an exceptional trend of accidents if

- there is a segment of accidents defined by a Boolean attribute $\mathscr{S}$ such that a trend of accidents for this segment is non-decreasing

- trend of accidents for a segment defined by a Boolean attribute $\mathscr{S} \wedge \mathscr{A}$ is at least partly decreasing.

This analytical question can be seen as a task of searching for two related subgroups of accidents defined by the Boolean attributes $\mathscr{S}$ and $\mathscr{A}$. Note that mode $\alpha \bowtie \alpha \wedge \beta$ was used to compare segments $\mathscr{S}$ and $\mathscr{S} \wedge \mathscr{A}$ whereas mode $\alpha \bowtie \beta$ was used to compare districts, see equation (6.3). A run of the SDCF-Miner procedure was used to get a conclusion that there are not too many interesting results. The most interesting result is shown in Fig. 6.6. It holds

$$\mathscr{S}_{Exm} = \mathsf{Vehicle\_Age}(\mathit{14,15}) \wedge \mathsf{Age}(\mathit{46\text{-}55}) \wedge \mathsf{Sex}(\mathit{Male}) \,.$$

For details see Section 11.2.1.

6.5.2 Trends of the number of accidents and police forces

The analytical task introduced in the previous section was modified such that we are interested only in 2011–2015 instead of the entire period 2005–2015. A simplified task to find couples of Polices $\langle \mathsf{Police}(P_1), \mathsf{Police}(P_2) \rangle$ and types of vehicles $\mathsf{Type}(T)$ such that

Year/ $\mathscr{S}_{Exm}$

Year/ $\mathscr{S}_{Exm} \wedge$ Police(*Sussex*)

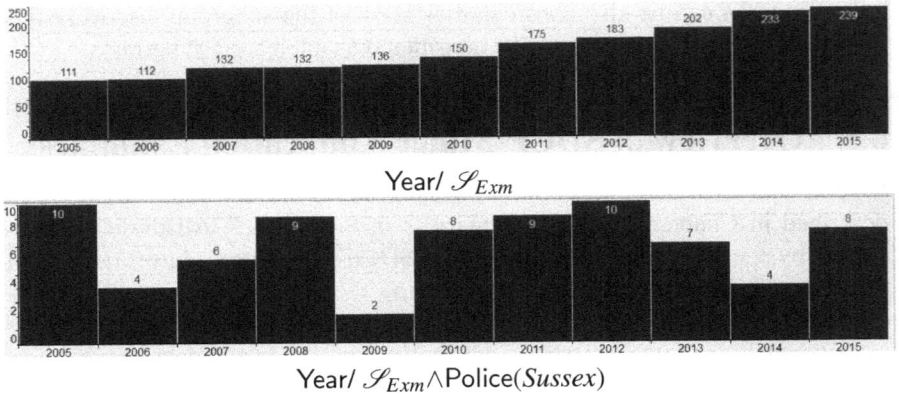

Figure 6.6: Histograms Year/ $\mathscr{S}$ and Year/$\mathscr{S} \wedge$ Police(*Sussex*).

■ a trend of accidents of vehicles of type Type(T) belonging to Police(P_1) in period 2011–2015 is non-decreasing.

■ a trend of accidents of vehicles of type Type(T) belonging to Police(P_2) in period 2011–2015 is non-increasing.

The mode $\alpha \bowtie \beta$ was used to solve this task. An example of result is shown in Fig. 6.7.

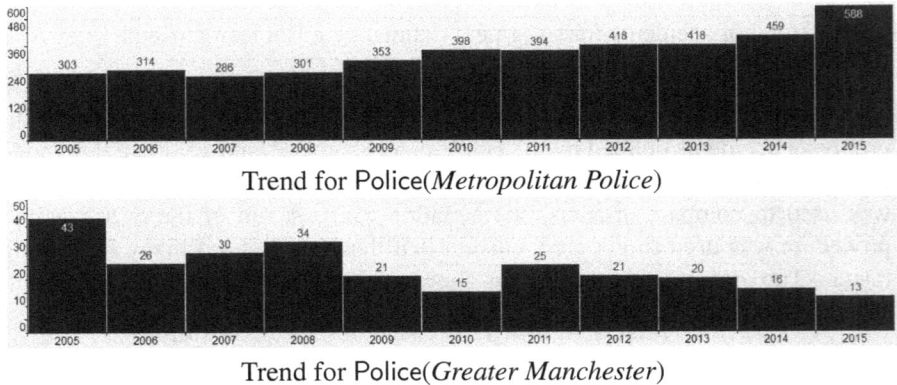

Trend for Police(*Metropolitan Police*)

Trend for Police(*Greater Manchester*)

Figure 6.7: Histograms for Type(*Motorcycle 125cc and under*).

This analytical question can be seen as a task of searching for subgroups of accidents defined by a type of vehicle and two histograms with given properties. Note that mode $\alpha \bowtie \beta$ was used to compare Police(P_1) and Police(P_2). For details see Section 11.2.2.

6.6 Overview of SDKL-Miner Applications

Three analytical tasks concerning subgroup discovery in the *STULONG* dataset described in Section 2.4 are solved. The first one is a task of searching for subgroups defined by Boolean attributes χ derived from attributes Education, MaritalStatus, Responsibility, and Age.

We search for a subgroup χ and categories c_1 and c_2 of attribute Liquors corresponding to different levels of drinking liquors with the highest difference what concerns ordinal dependencies between two attributes R and C chosen from attributes BMI, Subscapularis, Triceps, Diastolic, Systolic, Cholesterol, Triglycerides, Vine, and Beer.

In other words, we search for two sub-matrices $\mathcal{M}_1$ and $\mathcal{M}_2$ given by Boolean attributes $\chi \wedge \text{Liquors}(c_1)$ and $\chi \wedge \text{Liquors}(c_2)$ and for attributes R and C such that there is the biggest difference between ordinal dependence R and C at $\mathcal{M}_1$ and $\mathcal{M}_2$. The intensity of ordinal dependence R and C at a general data matrix $\mathcal{M}$ is measured by Kendall's coefficients τ_B. Kendall's coefficients τ_B for R and C at data matrix $\mathcal{M}$ is computed from a contingency table $KL(R,C,\mathcal{M})$ of R and C at $\mathcal{M}$. This task is solved in Section 12.2.1.

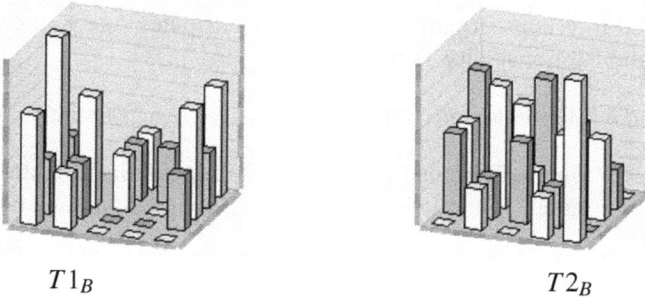

$T1_B$ $T2_B$

Figure 6.8: Contingency tables $T1_B$ and $T2_B$.

Analogous tasks for attributes Wine and Beer are solved in Sections 12.2.2 and 12.2.3. A result for attribute Beer is partly presented in Fig. 6.8 where contingency tables

$$T1_B = KL(\text{Diastolic}, \text{Cholesterol}, Entry/\chi_B \wedge \text{Beer}(no))$$

and

$$T2_B = KL(\text{Diastolic}, \text{Cholesterol}, Entry/\chi_B \wedge \text{Beer}(up\ to\ 1\ litre\ /\ day))$$

are outlined. Here *Entry* is a data matrix corresponding to the *STULONG* dataset, and it holds

$$\chi_B = \text{Age}(\leq 45) \wedge partly\ independent \wedge married \wedge secondary\ .$$

We use "*partly independent*" instead of "Responsibility(*partly independent*)", "*married*" instead of "MaritalStatus(*married*)" and "*secondary*" instead of "Education(*secondary*)".

6.7   Overview of Ac4ft-Miner Application Examples

There are four applications of the Ac4ft-Miner procedure described in this book. Two of them concern blood pressure of patients. They are introduced in Section 6.7.1. Two additional applications deal with guests satisfaction, see Section 6.7.2.

6.7.1   *Action rules and blood pressure*

The applications dealing with blood pressure concern the *STULONG* dataset described in Section 2.4. They have common features described in Section 13.2.1. Note that a goal is to demonstrate possibilities of the Ac4ft-Miner procedure, not to get new medical knowledge.

The first application mines subgroups of patients and changes of the attribute BMI leading to at least two times lower probability of high blood pressure. We are interested in groups of patients defined by attributes Age, Education, MaritalStatus, and Responsibility. Action rule

$$married \wedge \mathsf{BMI}(obese\ class\ I \mapsto normal) \downarrow_{3.137,47,61} \mathsf{Systolic}(very\ high)$$

where we use "*married*" instead of "MaritalStatus(*married*)" is an example of resulting action rules. This rule is equivalent to a couple of association rules

$$married \wedge \mathsf{BMI}(obese\ class\ I) \rightarrow_{0.423,47} \mathsf{Systolic}(very\ high)$$

and

$$married \wedge \mathsf{BMI}(normal) \rightarrow_{0.135,61} \mathsf{Systolic}(very\ high).$$

The action rule means that if married patients change a level of their BMI from *obese class I* to *normal*, then probability of having very high systolic pressure decreases from 0.423 to 0.135, i.e., 3.137 times. The first application is described in Section 13.2.2.

The second application is similar. It mines for subgroups of patients and changes of BMI leading to increasing probability of average blood pressure. This application is introduced in Section 13.2.3.

6.7.2   *Action rules and guest satisfaction*

Action rules dealing with guests' satisfaction concern the *Hotel* dataset described in Section 2.5. The first application of the Ac4ft-Miner procedure searches for possibilities to increase the relative frequency of high overall satisfaction of guests. Subgroups of stays in hotel defined by attributes Age and Sex from the group Guest of attributes, Sky and Temperature from the group Meteo of attributes, and by attributes DayOfWeek, WeekEnd, Month, and Season from the group Check-in are taken into considerations as well as possibilities to increase partial evaluations given by attributes QRoom, QFood, QService, and QCulture. This task is solved in Section 13.3.1.

The second analytical question task is similar. However, we are also interested in the consequences of increasing partial evaluations. This is formalised by an application of changes of succedents. Details are available in Section 13.3.2.

6.8 GUHA and Business Intelligence—Overview

Chapter 14 shows that business intelligence can benefit from the GUHA method. Business intelligence together with related self service business intelligence is presented in Section 14.1. Several definitions of business intelligence are introduced. The process of analysis using self service business intelligence tool and GUHA procedures, its common parts and differences are presented in Section 14.2.

Two scenarios of possible complementary usage of self service business intelligence and the ability to generate and verify a large set of relevant patterns by GUHA procedures are introduced in Section 14.3 Usage of these scenarios is demonstrated on practical examples presented in Section 14.4. The procedures 4ft-Miner and CF-Miner are applied to the dataset *Accidents* described in Section 2.3.

6.9 GUHA and Python—CleverMiner Project

All the examples of the GUHA procedures applications concern the LISp-Miner system described in Chapter 5. Note that additional implementations of the GUHA method are introduced in [44]. However, many data scientists use languages Python and R. It is crucial that all important operations with data must be available in these languages. Also, tools for data visualization are easily available. Thus, there is a natural requirement to make the GUHA procedures available also in at least one of these languages. This is a goal of the CleverMiner project described in Chapter 15.

The chapter starts with a discussion about which of the mentioned languages is more suitable to be used for the project. Arguments for a decision to use the Python language are discussed. The goals and main features of the CleverMiner project are then introduced. The current state of the project involving the 4ft-Miner and CF-Miner procedures is described. An example of the application of the CF-Miner procedure implemented in the CleverMiner project is presented. Future plans with the CleverMiner project are listed at the end of Chapter 15.

6.10   Examples Summary

All the presented GUHA procedures have very fine possibilities to tune a definition of a set of relevant Boolean attributes to be generated and used in verified patterns. Particular tools to define sets of relevant Boolean attributes are described in Section 4.4. This way, both conjunction and disjunction of basic Boolean attributes $A(\alpha)$ can be generated instead of conjunctions of basic Boolean attributes $A(a)$ used in very well known apriori algorithm. Let us emphasize that α is a subset of all possible values of A and a is a particular possible value of A.

Basic Boolean attributes $A(\alpha)$ are applied through coefficients, see Section 6.10.1. Conjunctions and disjunctions of basic Boolean attributes are applied through partial cedents, see Section 6.10.2.

6.10.1   Applying coefficients

Recall that a literal is a basic Boolean attribute $A(\alpha)$ or its negation $\neg A(\alpha)$; α is a coefficient of the basic Boolean attribute $A(\alpha)$. A set of relevant literals to be generated from an attribute A is defined by an $\mathscr{R}\mathscr{L}$-expression $\mathscr{R}\mathscr{L}(A)$ in a form $\langle A; Coef, Min, Max, Gace, B/R, Class \rangle$ see Section 4.4.2. Here $Coef$ is a type of a coefficient, Min and Max mean the minimal and the maximal length of a coefficient (i.e., the number of categories).

There are seven types of coefficients – *Subsets, Sequences, Cyclical sequences, Left cuts, Right cuts, Cuts* (i.e., *Left cuts* or *Right cuts*), and *One category*. Our goal is to point out to examples of applications of GUHA procedures using particular types of coefficients.

Subsets are used in many applications, see, e.g., Sections 7.3 and 7.2.1. Let us note that the coefficients of the type *Subsets* are usually used with minimal and maximal lengths 1. This is because of the danger of the fast growing number of such coefficients. However, in some cases it makes sense to use coefficients – subsets of length 1–2 or even longer. This may be used with the attribute State which has five categories, see Section 2.5.3. Subsets of length 1–2 lead to 15 basic Boolean attributes State(α). However subsets of length 1–2 for attribute City with 28 categories lead already to 406 basic Boolean attributes City(α).

Sequences are used in many applications, let us point to Sections 7.3.1 and 8.3.2. *Cyclical sequences* are used in Sections 9.3.1 and 10.2.4.

Left cuts are used in Section 10.2.1. *Right cuts* are used in Sections 7.3 and 13.2. *Cuts* are not used.

Type *One category* is used in Sections 7.3.2, 7.3.3, 7.4.3, 8.3.3, 8.3.4 and additional.

The remaining parameters of an $\mathscr{RL}$-expression are *Gace*, *B/R*, *Class*. The value *Pos* is the only value used for parameter *Gace* in all applications described in this book. Parameters *B/R* and *Class* are related to partial cedents, they are explained in Sections 4.4.3 and 4.4.4. Their use is described in the next section.

6.10.2 Applying partial cedents

Each partial cedent is a conjunction or a disjunction of literals. A set of relevant partial cedents is given by an $\mathscr{RP}$-expression

$$\langle Name_P, Type_P, Min_P, Max_P; \mathscr{RL}(A_1), \ldots, \mathscr{RL}(A_u) \rangle \, ,$$

which is explained in Sections 4.4.3 and 4.4.4. $\mathscr{RL}(A_1), \ldots, \mathscr{RL}(A_u)$ are $\mathscr{RL}$-expression in which parameters *B/R* and *Class* for particular attributes are used.

$Type_P$ is a type of partial cedents with possible values *Conjunction* and *Disjunction*. Most of applications of the GUHA procedures described in this book uses $Type_P = Conjunction$, see, e.g., Section 7.3. A value $Type_P = Disjunction$ is used in Section 7.3.3.

Parameter *B/R* has possible values *B* and *R*. The value *B* is used in most applications. The value *R* is used in Section 11.2.1, see Fig. 11.4. An example of application of the parameter *Class* is available in Fig. 13.11, Section 13.3.1.

6.11 Important Notes

Let us again emphasize that only selected important features of particular procedures are presented by examples. Various useful possibilities of the GUHA procedures are not mentioned due to limited space.

In addition, we do not describe the whole CRISP-DM process. Only the final application of a chosen procedure is introduced in most cases. Values of used parameters usually resulted from several not described experiments.

Also, let us emphasize that secured completion of missing values described in Section 4.5.2 is used when dealing with Boolean attributes. Secured completion of missing information is also used when applying the 4ft-Miner procedure in Sections 7.2 and 7.3. Otherwise, a deleting approach is used. Details are introduced in particular chapters where contingency tables used to verify particular patterns are described.

Finally, let us note that a PC with 8GB RAM and Intel(R) Core(TM) i5-5200U, processor at 2.2 GHz running OS Windows 7 was used for all runs of the GUHA procedures described in Chapters 7–13.

Chapter 7

4ft-Miner—GUHA Association Rules

There are many analytical tools producing association rules. Most of them are based on the apriori algorithm [2]. The 4ft-Miner procedure deals with more general rules introduced in connection with the GUHA method [41]. Thus, if there is a need to emphasize that we deal with these more general association rules, we call such rules *GUHA association rules*.

The first goal is to introduce the GUHA association rules and to present their specific features. The specific features can be summarized as follows:

- GUHA association rules are in the form $\varphi \approx \psi$, where φ and ψ are general Boolean attributes. Conditional association rules $\varphi \approx \psi/\chi$ are also considered, here χ is a general Boolean attribute

- The secured approach to missing information is applied not only to Boolean attributes, but also to the association rules.

An additional goal is to compare the 4ft-Miner with a typical implementation of the apriori algorithm.

The GUHA association rules are introduced in Section 7.1 together with the 4ft-Miner procedure. A detailed comparison of the 4ft-Miner with the implementation of the apriori algorithm in the popular R system is presented in Section 7.2. Applications of the 4ft-Miner presenting important features of the GUHA association rules are described in Sections 7.3 and 7.4. The datasets *Adult* and *Accidents* are used. Additional examples of applications of the 4ft-Miner procedure are available in [121, 124, 125, 126, 127, 138, 155, 156].

7.1 GUHA Association Rules and 4ft-Miner Procedure

GUHA association rules and related notions are introduced in Section 7.1.1. The 4ft-Miner can work in two modes. In a *classical mode*, general association rules are generated and verified; 4ft-quantifiers described in Section 7.1.2 can be used. In a *histogram mode*, association rules – potential exceptions to a histogram are generated and verified; 4ft-quantifiers introduced in Section 7.1.3 are applied.

Principles of dealing with missing information in the 4ft-Miner procedure are introduced in Section 7.1.4. Secured completion of missing information is described in Section 7.1.5. Another approach to missing information introduced in Section 7.1.6 is used in the R system. This approach is also implemented in the 4ft-Miner procedure. Prime association rules introduced in Section 7.1.7 can be applied to reduce the output of the 4ft-Miner procedure. Input and output of the 4ft-Miner procedure are summarized in Section 7.1.8.

7.1.1 GUHA association rules and related notions

The 4ft-Miner procedure deals with association rules $\varphi \approx \psi$ and conditional association rules $\varphi \approx \psi / \chi$. Here φ, ψ, and χ are Boolean attributes defined in Section 1.4.3. A Boolean attribute φ is called an *antecedent*, a Boolean attribute ψ is called a *succedent* or *consequent* and a Boolean attribute χ is called a *condition*. Symbol $\approx$ is a 4ft-quantifier.

An association rule $\varphi \approx \psi$ is true in a data matrix $\mathcal{M}$ if a condition related to a 4ft-quantifier $\approx$ is satisfied for a 4ft-table $4ft(\varphi, \psi, \mathcal{M})$ of φ and ψ in $\mathcal{M}$. $4ft(\varphi, \psi, \mathcal{M})$ is introduced in Fig. 7.1. Here a is the number of rows of $\mathcal{M}$ satisfying both φ and ψ, b is the number of rows satisfying φ and not satisfying ψ, etc., see also Section 4.2. A conditional association rule $\varphi \approx \psi / \chi$ is true in a data matrix $\mathcal{M}$ if and only if the rule $\varphi \approx \psi$ is true in a data matrix $\mathcal{M}/\chi$.

Each 4ft-table $4ft(\varphi, \psi, \mathcal{M})$ is a quadruple $\langle a,b,c,d \rangle$ of non-negative integers. Thus, we consider a 4ft-quantifier $\approx$ as a $\{0,1\}$-valued function defined for all quadruples of non-negative integers. This makes it possible to define the truthfulness of association rules in a more precise way: We say that an *association rule $\varphi \approx \psi$ is true in a data matrix $\mathcal{M}$* if $\approx (a,b,c,d) = 1$ where $\langle a,b,c,d \rangle = 4ft(\varphi, \psi, \mathcal{M})$. If $\approx (a,b,c,d) = 0$, then we say that the *association rule $\varphi \approx \psi$ is false in a data matrix $\mathcal{M}$*. This means that a conditional association rule $\varphi \approx \psi / \chi$ is true in a data matrix $\mathcal{M}$ if $\approx (a,b,c,d) = 1$ where $\langle a,b,c,d \rangle = 4ft(\varphi, \psi, \mathcal{M}/\chi)$, otherwise it is false.

$\mathcal{M}$	ψ	$\neg\psi$
φ	a	b
$\neg\varphi$	c	d

Figure 7.1: 4ft-table of φ and ψ in a data matrix $\mathcal{M}$.

Let us note that here we use the 4ft-quantifier $\approx$ both as a symbol – a part of an association rule $\varphi \approx \psi$ and as a function $\{0,1\}$-valued function. A more precise approach is presented in Chapter 18.

7.1.2   4ft-quantifiers for classical mode of 4ft-Miner

There are many 4ft-quantifiers for the classical mode of the 4ft-Miner. They are based on statistical hypothesis tests – Fisher's test, χ^2 test, etc., see [121]. We use only four 4ft-quantifiers defined by measures of interestingness of association rules and introduced in Table 7.1. The parameters – real numbers p, s, q satisfying $0 < p \leq 1, 0 < s \leq 1, q > 0$ and integer number $Base > 0$ are used. The condition $a + b > 0$ in the row 1 of Table 7.1 is used to ensure that the fraction $\frac{a}{a+b}$ is defined. There are analogous conditions in the additional rows of Table 7.1.

Table 7.1: 4ft-quantifiers defined by measures of interestingness of association rules.

#	Name	Symbol $\approx$	$\approx (a,b,c,d) = 1$ if and only if
1	confidence	$\rightarrow_p$	$\frac{a}{a+b} \geq p \wedge a+b > 0$
2	lift	$\sim_q$	$\frac{a(a+b+c+d)}{(a+b)(a+c)} \geq q \wedge (a+b)(a+c) > 0$
3	support	$\odot_s$	$\frac{a}{a+b+c+d} \geq s \wedge a+b+c+d > 0$
4	Base	$\oplus_{Base}$	$a \geq Base$

There are 4ft-quantifiers abbreviating conjunctions of couples of some of 4ft-quantifiers introduced in Table 7.1:

■ 4ft-quantifier $\Rightarrow_{p,Base}$ called *founded confidence* abbreviates a condition $\frac{a}{a+b} \geq p \wedge a \geq Base$ corresponding to a conjunction $\rightarrow_p$ and $\oplus_{Base}$.

■ 4ft-quantifier $\rightarrow_{p,s}$ called *confidence - support* abbreviates a condition $\frac{a}{a+b} \geq p \wedge \frac{a}{a+b+c+d} \geq s$ corresponding to a conjunction of $\rightarrow_p$ and $\odot_s$.

■ 4ft-quantifier $\approx_{q,Base}$ called *founded lift* abbreviates a condition $\frac{a(a+b+c+d)}{(a+b)(a+c)} \geq q \wedge a \geq Base$ corresponding to a conjunction of $\sim_q$ and $\oplus_{Base}$.

■ 4ft-quantifier $\sim_{q,s}$ called *supported lift* abbreviates a condition $\frac{a(a+b+c+d)}{(a+b)(a+c)} \geq q \wedge \frac{a}{a+b+c+d} \geq s$ corresponding to a conjunction of $\sim_q^+$ and $\odot_s$.

7.1.3   4ft-quantifiers for histogram mode of 4ft-Miner

The histogram mode concerns association rules $\varphi \approx A(a_i)$ where $\approx$ is a 4ft-quantifier for the histogram mode and a_i is one of categories $a_1, \ldots, a_t$ of an

$\mathcal{M}$	a_1	$\cdots$	a_i	$\cdots$	a_t	Σ
φ	n_1	$\cdots$	n_i	$\cdots$	n_t	n
$True$	m_1	$\cdots$	m_i	$\cdots$	m_t	m

$\mathcal{M}$	$A(a_i)$	$\neg A(a_i)$
φ	a	b
$\neg\varphi$	c	d

Figure 7.2: Frequency table for definition of 4ft-quantifiers for the histogram mode.

attribute A. Our motivation is to find rules which can be considered as exceptions to a histogram of attribute A at a data matrix $\mathcal{M}$.

Recall a CF-table $CF(A, \mathcal{M}/\varphi)$ introduced in Section 4.2 and shown in the left part of Fig. 7.2. Here, n_i denotes the number of rows o of a data matrix $\mathcal{M}/\varphi$ for which $A(o) = a_i$ and m_i denotes the number of rows o of a data matrix $\mathcal{M}$ for which $A(o) = a_i$. We assume $i = 1, \ldots, t$. In addition, $n = n_1 + \cdots + n_t$ and $m = m_1 + \cdots + m_t$.

Our goal is to express that the rule $\varphi \approx A(a_i)$ is an exception to the histogram of attribute A at a data matrix $\mathcal{M}$. We deal with histograms showing percentage of frequencies of particular categories from the number of rows of $\mathcal{M}$. A percentage of frequencies of a category a_i from the number of rows of $\mathcal{M}$ is equal to $100 * \frac{m_i}{m}$.

The rule $\varphi \approx A(a_i)$ concerns a column corresponding to a category a_i of a histogram of attribute A in a data matrix $\mathcal{M}/\varphi$. Let us assume that $4ft(\varphi, A(a_i), \mathcal{M}) = \langle a, b, c, d \rangle$ is a 4ft-table of $\varphi, A(a_i)$ in $\mathcal{M}$, see the right part of Fig. 7.2. If we consider a histogram showing percentage of frequencies of particular categories from the number of rows of $\mathcal{M}/\varphi$, then $100 * \frac{a}{a+b}$ is a height of a column corresponding to the category a_i. Note that $a = n_i$, $a+b = n$, $m_i = a+c$, and $a+b+c+d = m$.

If $\frac{a}{a+b} / \frac{a+c}{a+b+c+d} \geq 2$ then we can conclude that the association rule $\varphi \approx A(a_i)$ indicates an exception to a histogram of attribute A at a data matrix $\mathcal{M}$ such that height of a column corresponding to the category a_i is at least two times higher for the data matrix $\mathcal{M}/\varphi$ than for $\mathcal{M}$. If $\frac{a}{a+b} / \frac{a+c}{a+b+c+d} \leq 1/2$ then we can conclude that the association rule $\varphi \approx A(a_i)$ indicates an exception to a histogram of attribute A at a data matrix $\mathcal{M}$ such that height of a column corresponding to the category a_i is at least two times lower for the data matrix $\mathcal{M}/\varphi$ than for $\mathcal{M}$.

Recall that $\frac{a}{a+b} / \frac{a+c}{a+b+c+d} = \frac{a(a+b+c+d)}{(a+b)(a+c)}$ and this fraction defines a very known interestingness measure lift. We already have a 4ft-quantifier $\sim_q$ defined by a condition $\frac{a(a+b+c+d)}{(a+b)(a+c)} \geq q \wedge a > 0$, see row 2 in Table 7.1. Thus, we only need to enhance the 4ft-quantifier $\sim_q$ by possibilities to use additional relations. We define an enhanced lift quantifier $\sim_{\circledast q}$ where $\circledast$ is one symbols $<, \leq, \geq, >$ and q is a real positive number. The enhanced lift quantifier $\sim_{\circledast q}$ is defined such that

$$\sim_{\circledast q}(a, b, c, d) = 1 \text{ if and only if } \frac{a(a+b+c+d)}{(a+b)(a+c)} \circledast q \wedge a > 0. \quad (7.1)$$

Examples of applications of 4ft-quantifiers for the histogram mode of 4ft-Miner are available in Sections 7.4.1 and 7.4.2. Let us note that there are additional 4ft-quantifiers for the histogram mode available.

7.1.4 Association rules and missing information

Principles of dealing with missing information are introduced in Section 4.5. Let us recall that we start with data matrices with missing information introduced in Section 4.5.1. This leads to *Boolean attributes with missing values* also called *three-valued attributes*. A three-valued attribute has three possible values: *true*, *false* and *X* meaning *unknown*.

The principle of *secured completion of missing values* for Boolean attributes is defined in Section 4.5.2. A value of a Boolean attribute φ for a row o of a data matrix $\mathcal{M}^X$ with missing values is true (false) if it is true (false) in each completion of $\mathcal{M}^X$ by regular values. If there are completions $\mathcal{M}_1$ and $\mathcal{M}_2$ such that φ is true for a row o in $\mathcal{M}_1$ and false in $\mathcal{M}_2$, then a value of φ for a row o in $\mathcal{M}^X$ is X.

This way, we get a *nine-fold table* $9ft(\varphi, \psi, \mathcal{M}^X)$ of three-valued attributes φ and ψ in a data matrix $\mathcal{M}^X$ with missing information. It is shown in the left part of Fig. 7.3.

Here, $f_{1,1}$ is the number of rows o of $\mathcal{M}^X$ satisfying both $\varphi(o, \mathcal{M}^X) = 1$ and $\psi(o, \mathcal{M}^X) = 1$, $f_{1,X}$ is the number of rows satisfying both $\varphi(o, \mathcal{M}^X) = 1$ and $\psi(o, \mathcal{M}^X) = X$, etc. We write

$$9ft(\varphi, \psi, \mathcal{M}^X) = \langle f_{1,1}, f_{1,X}, f_{1,0}, f_{X,1}, f_{X,X}, f_{X,0}, f_{0,1}, f_{0,X}, f_{0,0} \rangle . \qquad (7.2)$$

There are four ways of dealing with missing information in the 4ft-Miner procedure. *Deleting missing information* means that all frequencies concerning missing values are deleted before verification of an association rule. This means that a 4ft-table $4ft(\varphi, \psi, \mathcal{M}) = \langle f_{1,1}, f_{1,0}, f_{0,1}, f_{0,} \rangle$ shown in the right part of Fig. 7.3 is used to verify an association rule $\varphi \approx \psi$. Here $\mathcal{M}$ is a suitable sub-matrix of $\mathcal{M}^X$.

$\mathcal{M}^X$	ψ	ψ_X	$\neg\psi$
φ	$f_{1,1}$	$f_{1,X}$	$f_{1,0}$
φ_X	$f_{X,1}$	$f_{X,X}$	$f_{X,0}$
$\neg\varphi$	$f_{0,1}$	$f_{0,X}$	$f_{0,0}$

$\mathcal{M}$	ψ	$\neg\psi$
φ	$f_{1,1}$	$f_{1,0}$
$\neg\varphi$	$f_{0,1}$	$f_{0,0}$

Figure 7.3: $9ft(\varphi, \psi, \mathcal{M}^X)$ and $4ft(\varphi, \psi, \mathcal{M})$ – deleting missing information.

Secured handling of missing information can also be applied to association rules, see Section 7.1.5. It is also possible to use ignoring missing information, which is used in the R system and introduced in Section 7.1.6. An *optimistic handling with missing values* [121] is also implemented in the 4ft-Miner. However, it is not used in the examples described below.

7.1.5   Secured completion and association rules

Let Ω denote an association rule $\varphi \approx \psi$ or a conditional association rule $\varphi \approx \psi / \chi$. If $\mathcal{M}^X$ is a data matrix with missing information, then:

- Ω *is true in* $\mathcal{M}^X$ if Ω is true in each completion $\mathcal{M}$ of $\mathcal{M}^X$.

- Ω *is false in* $\mathcal{M}^X$ if Ω is false in each completion $\mathcal{M}$ of $\mathcal{M}^X$.

- If there are completions $\mathcal{M}_1$ and $\mathcal{M}_2$ of $\mathcal{M}_X$ such that Ω is true in $\mathcal{M}_1$ and Ω is false in $\mathcal{M}_2$ then we say that a value of Ω in $\mathcal{M}^X$ is X or that a value of Ω in $\mathcal{M}^X$ is unknown.

If we evaluate an association rule $\varphi \approx \psi$ in a data matrix $\mathcal{M}^X$ with missing information, we deal with nine-fold tables instead of four-fold tables. Let us have nine-fold table $9ft(\varphi, \psi, \mathcal{M}^X)$ shown in the left part of Fig. 7.4.

$\mathcal{M}^X$	ψ	ψ_X	$\neg\psi$
φ	$f_{1,1}$	$f_{1,X}$	$f_{1,0}$
φ_X	$f_{X,1}$	$f_{X,X}$	$f_{X,0}$
$\neg\varphi$	$f_{0,1}$	$f_{0,X}$	$f_{0,0}$

$\mathcal{M}$	ψ	$\neg\psi$
φ	$f_{1,1}+f_{1,X,a}+$ $+f_{X,1,a}+f_{X,X,a}$	$f_{1,0}+f_{1,X,b}+$ $+f_{X,0,b}+f_{X,X,b}$
$\neg\varphi$	$f_{0,1}+f_{X,1,c}+$ $+f_{0,X,c}+f_{X,X,c}$	$f_{0,0}+f_{X,0,d}+$ $+f_{0,X,d}+f_{X,X,d}$

Figure 7.4: Nine-fold table $9ft(\varphi, \psi, \mathcal{M}^X)$ ands its completions $4ft(\varphi, \psi, \mathcal{M})$.

If $\mathcal{M}$ is a completion of $\mathcal{M}^X$, then a 4ft-table $4ft(\varphi, \psi, \mathcal{M})$ can be derived from $9ft(\varphi, \psi, \mathcal{M}^X)$ in a way outlined in the right part of Fig. 7.4. Here, $f_{1,X,a}, f_{1,X,b}, f_{X,1,a}, f_{X,1,c}, f_{X,0,b}, f_{X,0,d}, f_{0,X,c}, f_{0,X,d}, f_{X,X,a}, f_{X,X,b}, f_{X,X,c}, f_{X,X,d}$ are non-negative integers satisfying $f_{1,X,a}+f_{1,X,b}=f_{1,X}$, $f_{X,1,a}+f_{X,1,c}=f_{X,1}$, $f_{X,0,b}+f_{X,0,d}=f_{X,0}$, $f_{0,x,c}+f_{0,X,d}=f_{0,X}$, $f_{X,X,a}+f_{X,X,b}+f_{X,X,c}+f_{X,X,d}=f_{X,X}$. Moreover, $f_{1,X,a}$ denotes the number of rows o, which satisfy $\varphi(o, \mathcal{M}^X)=1$ and $\psi(o, \mathcal{M}^X)=X$ as well as $\varphi(o, \mathcal{M})=1$ and $\psi(o, \mathcal{M})=1$. In other words, $f_{1,X,a}$ is the number of rows o which "*move from* $f_{1,X}$ *to* $f_{1,1}$" when completing missing information. (Recall that $f_{1,1}$ correspond to a-frequency in a 4ft-table $\langle a, b, c, d \rangle$.) This is analogous for additional frequencies $f_{X,1,a}, f_{X,1,c}, \ldots, f_{X,X,d}$. Thus, a 4ft-table $4ft(\varphi, \psi, \mathcal{M})$ of each of completions $\mathcal{M}$ of a data matrix $\mathcal{M}^X$ with missing information can be written in a form shown in the right part of Fig. 7.4.

We are interested if a given association rule $\varphi \approx \psi$ is true in a given data matrix $\mathcal{M}^X$ with missing information. This means that we have to test if $\approx (a, b, c, d) = 1$ holds for each 4ft-table $\langle a, b, c, d \rangle = 4ft(\varphi, \psi, \mathcal{M})$, where $\mathcal{M}$ is a completion of $\mathcal{M}^X$.

There is a 4ft-table $\langle a_s, b_s, c_s, d_s \rangle$ such that $\varphi \approx \psi$ is true in $\mathcal{M}^X$ if and only if $\approx (a_s, b_s, c_s, d_s) = 1$ for many important 4ft-quantifiers [121]. Such 4ft-table is defined using $9ft(\varphi, \psi, \mathcal{M}^X)$ and it is called a *secured completion of*

$9ft(\varphi,\psi,\mathscr{M}^X)$. Secured completions for 4ft-quantifiers introduced in Section 7.1.2 are defined in the following ways:

- It holds for $\rightarrow_p$, $\odot_s$, $\oplus_{Base}$, $\rightarrow_{p,s}$ and $\Rightarrow_{p,Base}$: $a_s = f_{1,1}$,
  $b_s = f_{1,0} + f_{1,X} + f_{X,X} + f_{X,0}$, $c_s = f_{0,1} + f_{X,1} + f_{0,X}$, $d_s = f_{0,0}$.

- It holds for $\sim_q$, $\sim_{q,s}$, $\approx_{q,Base}$, and $\sim_{q,Base}^+$: $a_s = f_{1,1}$,
  $b_s = f_{1,0} + f_{1,X} + f_{X,X,b} + f_{X,0}$, $c_s = f_{0,1} + f_{X,1} + f_{X,X,c} + f_{0,X}$, $d_s = f_{0,0}$.
  Here $f_{X,X,b} \geq 0$, $f_{X,X,c} \geq 0$, $f_{X,X,b} + f_{X,X,c} = f_{X,X}$ and $f_{X,X,b}, f_{X,X,c}$ are chosen such that $|b_s - c_s|$ is minimal.

Note that there are more possibilities on how to define the secured completion $\langle a_s, b_s, c_s, d_s \rangle$ for $\rightarrow_p$, $\odot_s$, $\oplus_{Base}$, $\rightarrow_{p,s}$ and $\Rightarrow_{p,Base}$.

The idea behind the definition of the secured completion is to minimize a value of a corresponding measure of interest; for details see [121]. The same idea is used to define secured completions for conditional association rules.

7.1.6 *Ignoring missing information*

One of our goals is to compare the 4ft-Miner procedure with arules – an implementation of the apriori algorithm in the R system [20]. Thus, we have also to compare dealing with missing information. Let us emphasize that only attribute-value pairs $A(a)$ are used in the apriori algorithm. Basic Boolean attribute $A(\alpha)$ where α is a general own subset of a set of all categories of A are not used.

In [20], dealing with missing information is described as: *By default it is assumed that missing values do not carry information and thus all of the corresponding dummy items are set to zero. If the fact that the value of a specific attribute is missing provides information (e.g., a respondent in an interview refuses to answer a specific question), the domain expert can create for the attribute a category for missing values which then will be included in the transactions as its own dummy item.*

This can be generalized for basic Boolean attributes $A(\alpha)$ in the following way. Let $\mathscr{M}^X$ be a data matrix with missing information, let A be an attribute of $\mathscr{M}^X$ and let $A(\alpha)$ be a basic Boolean attribute. If o is a row of $\mathscr{M}^X$, then it holds:

- if $A(o, \mathscr{M}^X) \in \alpha$ then $A(\alpha)$ is true for a row o

- if $A(o, \mathscr{M}^X) \notin \alpha$ then $A(\alpha)$ is false for a row o

- if $A(o, \mathscr{M}^X) = X$ then $A(\alpha)$ is false for a row o.

We call this approach *ignoring missing information*. It is also implemented in the 4ft-Miner procedure. A comparison of this approach with the secured approach is available in Section 7.2.

7.1.7 Prime association rules

We say that an *association rule* $\varphi \approx \psi$ *logically follows from a rule* $\kappa \approx \lambda$ if it holds for each data matrix $\mathscr{M}$: if $\kappa \approx \lambda$ is true in $\mathscr{M}$, then $\varphi \approx \psi$ is also true in $\mathscr{M}$. This is the same as to say that a deduction rule $\frac{\kappa \approx \lambda}{\varphi \approx \psi}$ is correct.

Prime association rules are defined this way: Let $\mathscr{M}$ be a data matrix. We say, that an *association rule* $\varphi \approx \psi$ *is prime according to a run of the 4ft-Miner* if both Pr1 and Pr2 hold:

Pr1: $\varphi \approx \psi$ belongs to output of the 4ft-Miner

Pr2: there is no association rule $\kappa \approx \lambda$ belonging to the output such that $\varphi \approx \psi$ logically follows from $\kappa \approx \lambda$.

Application of prime association rules is practically important when dealing with 4ft-quantifiers $\rightarrow_{p,s}$ and $\Rightarrow_{p,Base}$ introduced in Section 7.1.2. We use the following correct deduction rules:

$$\frac{\kappa \approx A(\alpha) \wedge \lambda}{\kappa \approx A(\alpha') \wedge \lambda}, \qquad \frac{\kappa \approx A(\alpha) \vee \lambda}{\kappa \approx A(\alpha') \vee \lambda}, \qquad \frac{\kappa \approx \lambda}{\kappa \approx \lambda \vee \omega}. \qquad (7.3)$$

We assume $\alpha \subsetneq \alpha'$. For additional information see Chapter 18.

If we ask the 4ft-Miner procedure to produce only prime association rules, we can remarkably decrease the number of output rules. Output of all practically redundant rules which can be easily deduced from the resulting association rules will be denied. An example of the application of prime association rules is available in Section 7.3.4.

7.1.8 4ft-Miner input and output

Input of the 4ft-Miner procedure consists of:

- an $\mathscr{RC}$-expression *Ant* defining a set *RelAnt* of relevant antecedents φ

- an $\mathscr{RC}$-expression *Succ* defining a set *RelSucc* of relevant antecedents ψ

- an $\mathscr{RC}$-expression *Cond* defining a set *RelCond* of relevant conditions χ, this $\mathscr{RC}$-expression can be missing

- a list *List4ftQ* of basic 4ft-quantifiers; we will use 4ft-quantifier introduced in Sections 7.1.2 and 7.1.3

- parameter *Missing* with possible values *Deleting*, *Ignore* and *Secured*

- parameter *Prime* with possible values *Yes*, *No*; a default value is *No*

- a data matrix $\mathscr{M}$.

A set of relevant association rules is defined as a set of all rules $\varphi \approx \psi$ where $\varphi \in RelAnt$, $\psi \in RelSucc$ and φ and ψ have no common attributes. If an $\mathscr{RC}$-expression *Cond* is used, then also all conditional association rules $\varphi \approx \psi/\chi$ where $\varphi \in RelAnt$, $\psi \in RelSucc$, $\chi \in RelCond$ and φ, ψ and χ have no common attributes are considered as relevant association rules. 4ft-quantifier $\approx$ is defined as a conjunction of all basic 4ft-quantifiers introduced in the list *List4ftQ*.

The 4ft-Miner procedure generates all relevant association rules and verifies them in a given data matrix $\mathscr{M}$. If there is a missing information in the $\mathscr{M}$ data matrix, then the approach given by the parameter *Missing* is applied, see Sections 7.1.4–7.1.6.

If a value of the parameter *Prime* is *No*, then all true relevant association rules are listed in the output. If a value of the parameter *Prime* is *Yes*, then only all prime association rules are listed in the output, see Section 7.1.7.

Note that each $\mathscr{RC}$-expression is a list of $\mathscr{RP}$-expressions, see Section 4.4.5. If this list consists of only one $\mathscr{RP}$-expression, then this one $\mathscr{RP}$-expression can be used instead of $\mathscr{RC}$-expression. A minimal length of relevant antecedents can be 0. Relevant antecedent of length 0 corresponds to identically true Boolean attribute *True*, see Section 1.4.3.

7.2 Comparing 4ft-Miner and Arules

The goal of this section is to introduce a comparison of the 4ft-Miner procedure of the LISp-Miner system to an implementation of the apriori algorithm in arules package of the R-system. The arules package is described in [20]. A detailed comparison is published in [138]. We present here its shortened version. Principles of comparison are summarised in Section 7.2.1. A comparison of the performance of 4ft-Miner and arules is available in Section 7.2.2.

The ignoring approach to missing information described in Section 7.1.6 is implemented both in the arules package and in the 4ft-Miner procedure. The secured approach described in Section 7.1.5 is implemented in the 4ft-Miner only. A comparison of these approaches is presented in Section 7.2.3.

A possibility of a loss of some interesting rules related to the obligatory use of minimal values of both support and confidence in the arules package is studied in Section 7.2.4. The influence of using general basic Boolean attributes and additional features of definitions of a set of relevant rules to possibilities of solving various analytical tasks are shortly discussed in Section 7.2.5. Results of the comparison are summarized in Section 7.2.6.

7.2.1 Principles of comparison

The *Adult* dataset is used in one of the examples in [20] to describe the arules package. Therefore, we use the *Adult* dataset to compare the arules and the

4ft-Miner procedure. We use also the same definitions of categories of particular attributes. We compare the arules with both the slightly modified version of the apriori algorithm implemented into the 4ft-Miner procedure described in Section 7.1.6 and with the full version of the 4ft-Miner dealing with the secured approach to missing information.

The arules mines for association rules $\varphi \rightarrow_{p,s} \psi$ where φ is a conjunction of basic Boolean attributes $A(a)$ (attribute-value pairs), ψ is exactly one basic Boolean attribute and $\rightarrow_{p,s}$ is a 4ft-quantifier defined by the condition $\frac{a}{a+b} \geq p \wedge \frac{a}{a+b+c+d} \geq s$. This means that confidence is at least p and support is at least s, see Section 7.1.2. The input parameters for the 4ft-Miner procedure can be summarized this way: A set of relevant antecedents is given by a list of all 13 used attributes: education, income, sex, marital_status, relationship, workclass, occupation, race, native_country, age_exp, hours_per_week, capital_gain, and capital_loss. introduced in Table 2.1 and in Section 2.2.2. All possible attribute-value pairs and all possible conjunctions of 1–10 attribute-value pairs are considered as relevant antecedents. The empty antecedent considered as *True* is also relevant. We use conjunctions of length of 10 attributes maximally since there are no longer valid rules available in the analysed data.

We can say, that a set of relevant antecedents is given by the $\mathscr{RP}$-expression Antecedent_arules available in Fig. 7.5. Note that $\mathscr{RP}$-expression Antecedent_arules represents an $\mathscr{RC}$-expression – a list of $\mathscr{RP}$-expression consisting of one item.

A set of relevant succedents is given by the $\mathscr{RP}$-expression Succedent_arules, which is almost the same as the $\mathscr{RP}$-expression Antecedent_arules. The only difference is that *Length:* $1 - 1$ is used instead of *Length:* $0 - 10$. This

Name: Antecedent_arules		Type: *Conjunction*		Length: $0 - 10$	
Attribute	Coefficients	Length	Gace	B/R	Class
education	*Subsets*	$1 - 1$	*Pos*	*B*	—
income	*Subsets*	$1 - 1$	*Pos*	*B*	—
sex	*Subsets*	$1 - 1$	*Pos*	*B*	—
marital_status	*Subsets*	$1 - 1$	*Pos*	*B*	—
relationship	*Subsets*	$1 - 1$	*Pos*	*B*	—
workclass	*Subsets*	$1 - 1$	*Pos*	*B*	—
occupation	*Subsets*	$1 - 1$	*Pos*	*B*	—
race	*Subsets*	$1 - 1$	*Pos*	*B*	—
native_country	*Subsets*	$1 - 1$	*Pos*	*B*	—
age_exp	*Subsets*	$1 - 3$	*Pos*	*B*	—
hours_per_week	*Subsets*	$1 - 1$	*Pos*	*B*	—
capital_gain	*Subsets*	$1 - 1$	*Pos*	*B*	—
capital_loss	*Subsets*	$1 - 1$	*Pos*	*B*	—

Figure 7.5: $\mathscr{RP}$-expression Antecedent_arules.

means that all possible attribute-value pairs are considered as relevant succedents.

Finally 4ft-quantifiers $\rightarrow_{minC,minS}$ are used, *minC* means minimal confidence and *minS* means minimal support, *Missing = Ignore* and *Prime = No*.

To obtain meaningful results, we need to compare only the algorithm cores, excluding unrelated steps (i.e., data preparations steps or user-interface related steps like progress-reporting) and excluding times spent in communications with other systems (e.g., DBMS). Thus, the data preprocessing phase was completed in advance both in the R console and in the LISp-Miner environment. Both compared systems support logging with timestamps to get the most precise times. Both 32-bit and 64-bit versions are available in both systems. We compare the 32-bit versions.

We compare solutions times of the apriori function in the R console to solution times of command-line modules in LISp-Miner with all the progress-reporting and other user-interface features disabled. LISp-Miner offers both single-thread processing of tasks (by the LM TaskPooler module) and multi-thread parallel processing (by the LM SamePooler module). We provide both solution times.

Both systems have options to store results into a text file. Furthermore, LISp-Miner could store them into any DBMS accessible through ODBC. Anyway, storing of results was excluded from solution times because its duration is a pure function of HDD speed (in the first case) or the speed of the chosen DBMS and its ODBC database driver (in the second case).

The testing platform was the HP ProDesk with Intel i5-4590S (4 cores at 3GHz), 4 GB RAM and HDD Toshiba 500GB DT01ACA050 running Windows 7 Professional x64 with SP1. We have used the R system version of 3.2.3 and LISp-Miner system version of 27.00.01.

7.2.2 Performance

Results of the comparison are available in Table 7.2. Solutions times and the numbers of found association rules are presented for the apriori approach implemented in the arules package, for the apriori implemented in the 4ft-Miner and for the GUHA approach with secured completion of missing information implemented in the 4ft-Miner, see Section 7.1.5. There are two columns for the 4ft-Miner implementations–the first one for a single-thread solution using one processor marked "ST" and the second one for a multi-thread parallel solution using all available processor cores, which is marked "MT".

There are 17 variants of minimal support *minS* and minimal confidence *minC* used, see columns A and B. The rest of Table 7.2 presents solution times in seconds and the number of found rules for each combination of *minS* and *minC* and of a particular algorithm used. Solution times for apriori implementation in the arules package for R are in column C; solution times for 4ft-Miner apriori

Table 7.2: Comparing R-arules and 4ft-Miner in the *Adult* data matrix.

Parameters		apriori				secured		
		R	4ft-Miner		number	number	4ft-Miner	
minS	*minC*	arules	ST	MT	of rules	of rules	ST	MT
A	B	C	D	E	F	G	H	I
0.96	0.9	0.2	0.2	0.7	0	0	0.2	0.7
0.95	0.9	0.2	0.2	0.7	1	1	0.3	0.7
0.9	0.9	0.2	0.2	0.7	2	2	0.3	0.7
0.8	0.9	0.2	0.2	0.7	7	7	0.2	0.7
0.7	0.9	0.2	0.2	0.7	17	17	0.2	0.7
0.6	0.9	0.2	0.2	0.7	26	26	0.2	0.7
0.5	0.9	0.2	0.2	0.7	52	50	0.2	0.7
0.4	0.9	0.2	0.3	0.7	103	98	0.3	0.7
0.3	0.9	0.2	0.3	0.7	326	309	0.3	0.7
0.2	0.9	0.2	0.3	0.7	845	788	0.6	1.7
0.1	0.9	0.3	1.0	0.9	4 122	3 726	3.5	1.5
0.05	0.9	0.4	3.0	1.6	14 012	11 725	12.1	4.5
0.04	0.9	0.5	4.8	2.5	20 344	16 688	20.4	7.0
0.03	0.9	0.5	8.6	3.5	31 456	25 287	39.1	13.2
0.02	0.9	0.6	16.4	6.3	57 907	45 358	85.6	27.3
0.01	0.9	0.8	42.6	15.9	143 535	107 317	271.0	81.9
0.01	0.6	0.9	42.6	17.2	276 443	240 845	274.0	83.3

implementation are in column D (single-thread task solving) and in column E (multi-thread parallel task solving). The number of found rules is in column F (this number is the same for both apriori implementations).

The number of found rules with the secured completion used is in column G. It is significantly different from column F. Solution times for the GUHA approach with secured completion are in column H (single-thread task solving) and in column I (multi-thread parallel task solving).

We can see that apriori implemented in the arules package (column C) is faster than apriori implemented in the 4ft-Miner (columns D and E). The higher number of output rules, the higher difference between both implementations. However, the time used by the 4ft-Miner is not much too long, and it is only a fraction of the whole time necessary to solve a real task.

Regarding solutions times for the apriori algorithm in columns D and E, we have to stress that the implementation in the LISp-Miner system is meant as an *add-on* functionality provided just for a few cases when a task description is suitable for apriori (i.e., really simple, without missing information and does not use any of the rich-syntax possibilities the GUHA approach offers). It was implemented on the partial cedents level only, so no existing functionality of the LISp-Miner system is lost. Therefore, some optimisations regarding the whole association rule could not be used and some time is spent on preparing additional memory structures. It results in time overhead compared to the apriori implementation in the arules package.

Solution times for the secured completion are longer because the GUHA approach of the depth-first walking has to be used and simultaneously, it is necessary to maintain additional information to compute nine-fold tables. Thus, solution times in columns H and I are remarkably higher than in columns C, D and E. However, we state that arules results are misleading in the case of ignoring missing information, see Section 7.2.3. The time consumed by the secured completion is still acceptable, and it is only a fraction of the time necessary to solve the whole analytical task concerning real data, which usually does contain missing information.

Moreover, the GUHA approach makes it possible to use not only the secured completion but any of the available different types of handling of missing information (see Section 7.1.4), as well as dealing with general Boolean attributes and with conditional association rules. These additional features can be only hardly realised by apriori, see Section 7.2.5.

7.2.3 Comparing ignoring missings and secured completion

We can see a big difference between apriori (i.e., *Ignore*) and secured completion of missing information in columns F and G of Table 7.2. There are many extra rules produced by the apriori algorithm. Each of these extra rules has a problem – there is a completion of the analysed data matrix *Adult* in which the rule is not valid. Let us consider *minsup* = 0.5 and *minconf* = 0.9. One of the problematic rules is the rule

$$\mathsf{Capital\_loss}(\textit{None}) \wedge \mathsf{Native\_country}(\textit{United-States}) \wedge \mathsf{Sex}(\textit{Male})$$
$$\rightarrow \mathsf{Race}(\textit{White}).$$

We denote the antecedent

$$\mathsf{Capital\_loss}(\textit{None}) \wedge \mathsf{Native\_country}(\textit{United-States}) \wedge \mathsf{Sex}(\textit{Male})$$

as *Ant* and the succedent Race(*White*) as *Suc*. A nine-fold table $9ft(Ant,Suc,Adult)$ for the rule in question as well as a 4ft-table $4ft\_Ignoring$ resulting from the arules approach are available in Fig. 7.6. Both tables are included in a comprehensive protocol produced by the 4ft-Miner procedure for each output rule.

A secured four-fold table $4ft\_Secured$ for the rule in question is defined as $4ft\_Secured = \langle 24976, 2675 + 181, 16390 + 396, 4224 \rangle$, see Section 7.1.5. Let us note that the frequency 181 in the last column of the $9ft(Ant,Suc,Adult)$ in Fig. 7.6 can generate 182 4ft-tables $\langle 24976, 2675 + i, 16786, 4224 + 181 - i \rangle$, where $0 \le i \le 181$. These tables belong to at least 182 mutually distinct completions of the *Adult* data matrix. We need $\frac{24976}{48842} \ge 0.5 \wedge \frac{24976}{24976+2675+i} \ge 0.9$ which requires $0 \le i \le 100$. Thus, there are at least 81 mutually distinct completions of the *Adult* data matrix in which the rule in question is false. In addition, let us note

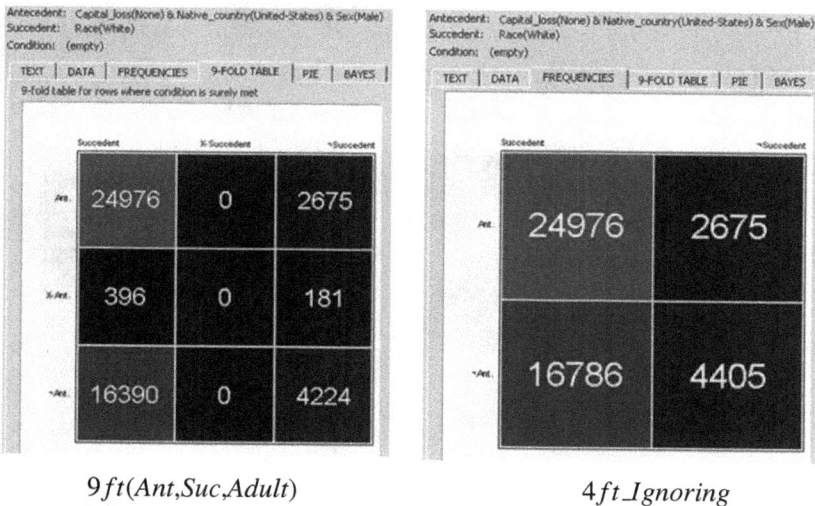

9*ft*(*Ant,Suc,Adult*)   4*ft_Ignoring*

Figure 7.6: Tables 9*ft*(*Ant,Suc,Adult*) and 4*ft_Ignoring*.

that the confidence of the rule

$$\text{Capital\_loss}(\textit{None}) \wedge \text{Native\_country}(\textit{United-States}) \wedge \text{Sex}(\textit{Male})$$
$$\rightarrow \text{Race}(\textit{White})$$

in the secured completion is $\frac{24976}{24976+2675+181} = 0.897 < 0.9$. Thus, this rule is not in output of any run of the 4ft-Miner with secured completion introduced in Table 7.2.

In Table 7.2, we deal with the 4ft-quantifier $\rightarrow_{p,s}$ – confidence-support introduced in Section 7.1.2. However, if we deal with the lift measure, the problem with missing values is even worse. Let us use the arules instructions

```
>rules<-apriori(Adult,parameter=list(support=0.01,confidence=0.6))
>Lift4 <- subset(rules, subset = rhs %in%''income=large"&lift>=4).
```
This results in a set `Lift4` of 88 rules with lift in the range $\langle 4.117980; 4.266398 \rangle$ and the instruction
```
> inspect(head(sort(Lift4, by = ''lift"), n = 1))
```
outputs the rule according to Fig. 7.7

```
> inspect(head(sort(Lift4, by = "lift"), n = 1))
  lhs                              rhs                  support confidence      lift
1 {marital-status=Married-civ-spouse,
   capital-gain=High,
   native-country=United-States}   => {income=large} 0.0156218  0.6849192  4.266398
```

Figure 7.7: The rule with the highest lift in the set Lift4 of rules.

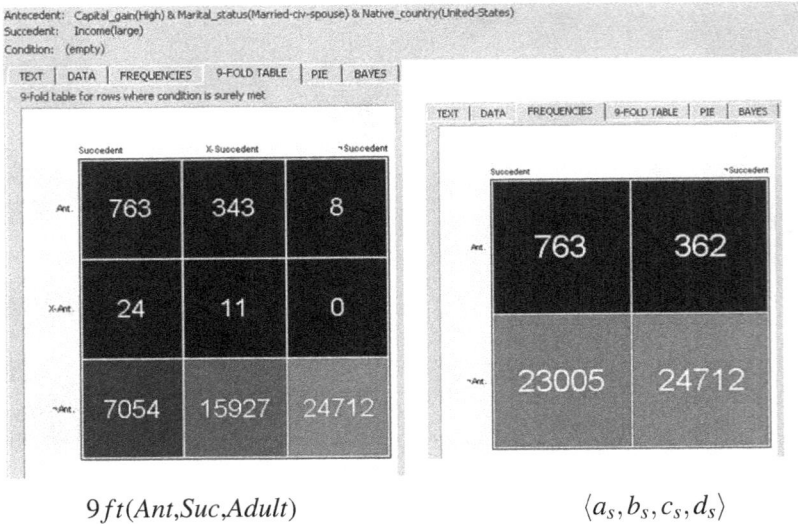

9 *ft(Ant,Suc,Adult)* $\langle a_s, b_s, c_s, d_s \rangle$

Figure 7.8: Tables $9ft(AntL,SucL,Adult)$ and $\langle a_s, b_s, c_s, d_s \rangle$.

However, if we use the secured approach for the lift, there is no rule with lift ≥ 4 and the lift is in the range $\langle 1.343493; 1.396369 \rangle$. Let us denote the rule in Fig. 7.7 as $AntL \rightarrow SucL$. Using the 4ft-Miner, we can get the nine-fold table $9ft(AntL,SucL,Adult)$ and the secured 4ft-table $\langle a_s, b_s, c_s, d_s \rangle$ for $9ft(AntL,SucL,Adult)$ and the 4ft-quantifier $\approx_{q,s}$ of supported lift, see Fig. 7.8.

It holds

$$\langle a_s, b_s, c_s, d_s \rangle = \langle 763, \ 8 + 343 + 11, \ 7054 + 24 + 15927, \ 24712 \rangle \, ,$$

see Section 7.1.5; i.e., $\langle a_s, b_s, c_s, d_s \rangle = \langle 763, 362, 23005, 24712 \rangle$. This means that for each completion of the *Adult* data matrix and the *lift* of the rule $AntL \rightarrow SucL$, we have $lift \geq \frac{763(763+362+23005+24712)}{(763+362)(763+23005)} = 1.393711$. It is easy to show that the maximal value of lift among all completions of the *Adult* data matrix is equal to $\frac{(763+343+24+11)(763+343+8+24+11+7054+15927+24712)}{(763+343+24+11+8)(763+343+24+11+7054)} = 5.918479$. This means that the lift of the rule $AntL \rightarrow SucL$ ranges from 1.393711 to 5.918479 if we consider all possible completions of the *Adult* data matrix. Thus, the value 4.266396 produced as a lift of the rule in question can be considered confusing.

We can conclude that:

■ the association rule mining process usually suffers from a problem of a large number of output rules

■ the ignoring of missing information approach used in the arules package produces rules, which do not satisfy the constraints (i.e., are false) in some completions of the analysed data matrix; the number of such rules can be relatively large

■ a value of lift of a rule produced by the arules package for a data matrix with missing information can be confusing

■ the secured completion produces a lower number of rules, which surely satisfy the constraints (i.e., are true) in all possible completions of the analysed data matrix. In the last three rows of Table 7.2, we can see that the number of rules decreases by 13–25 percent.

7.2.4 Loss of some interesting rules

We present an example of loss of rules, which is interesting because of the high value of the lift measure. The loss is related to the obligatory use of minimal values of both support and confidence in the arules package. The following instructions for R

```
>rules <- apriori(Adult,parameter = list(support=0.05, confidence=0.9))
>inspect(head(sort(rules, by = ''lift"), n = 1))
```

result to 14 012 rules, the highest lift is 2.956264, and it belongs to the rule

```
age=Young, relationship=Own-child, sex=Male, capital-loss=none,
native-country=United-States => marital-status=Never-married,
```

see Fig. 7.9. However, if we apply the 4ft-Miner procedure to search all rules satisfying $support \geq 0.05 \wedge lift \geq 2.96$ with even secured completion to missing information, we get 261 rules. Details concerning the rule with the highest lift are available in Fig. 7.10.

Confidence of this rule is $\frac{2449}{2449+477} = 0.84$. Overview of all rules with $lift \geq 2.96$ not produced by the instruction

```
>rules <- apriori(Adult,parameter = list(support=0.05, confidence=0.9))
```

is in Table 7.3. A workaround in arules would be to set *confidence* to a very low artificial number. But it would result in an explosion of rules mined and problems in processing such a large set later.

We can conclude that the obligatory use of minimal values of both support and confidence applied in the arules package can lead to remarkable loss of rules interesting for their high values of lift. It is practically not possible to estimate the degree of such loss. This is the same for additional measures of interestingness. This problem does not occur in the 4ft-Miner procedure, where conditions concerning various measures of interestingness can be freely combined.

```
> inspect(head(sort(rules, by = "lift"), n = 1))
    lhs                              rhs                                    support confidence    lift
1 {age=Young,
   relationship=Own-child,
   sex=Male,
   capital-loss=none,
   native-country=United-States} => {marital-status=Never-married} 0.0513902   0.975515 2.956264
```

Figure 7.9: The highest lift for apriori; support=0.05, confidence=0.9.

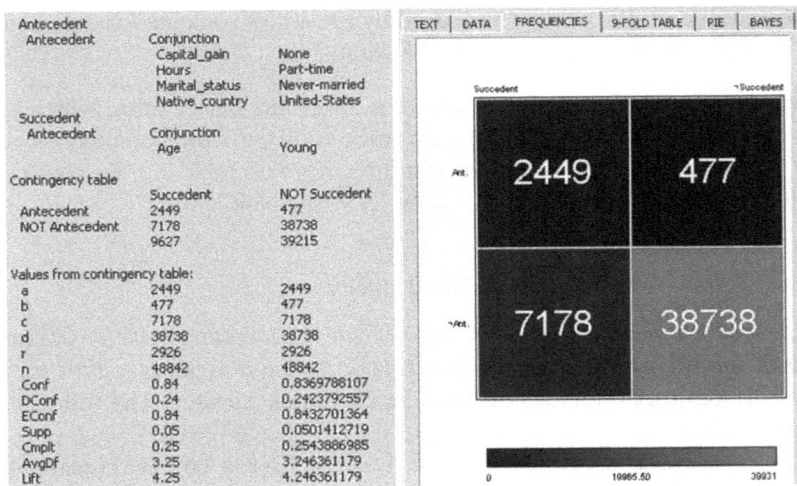

Figure 7.10: The rule with the highest *lift* for *support* ≥ 0.05.

Table 7.3: Rules with *support* ≥ 0.05, *lift* ≥ 2.96 and *confidence* < 0.9.

Confidence	Number of rules	Lift	
		min	max
(0.85, 0.90)	0	-	-
(0.80, 0.85)	7	4.189	4.246
(0.75, 0.80)	40	3.814	3.955
(0.70, 0.75)	44	3.555	3.724
(0.65, 0.70)	41	3.388	3.553
(0.60, 0.65)	49	3.109	4.175
(0.55, 0.60)	37	3.599	3.852
(0.50, 0.55)	34	3.236	3.542
(0.45, 0.50)	9	2.963	3.168

7.2.5 Applying GUHA Features

Let us recall that a set of relevant association rules to be generated by the 4ft-Miner procedure is defined by two $\mathscr{RC}$-expressions. This way, various tools to tune a set of relevant rules in a fine way are available. These tools include basic Boolean attributes $A(\alpha)$ with seven types of coefficients α, partial cedents - conjunctions or disjunctions of basic Boolean attributes with a possibility to use classes of equivalence of attributes and cedents - conjunctions of partial cedents. Additional possibilities are related to dealing with conditional association rules.

This brings possibilities, which can be only very hardly realized by the apriori. This is discussed in detail in [138]. An example of application of the 4ft-Miner to mine for segments of clients with extreme values of gain is used to show that solving of the same tasks by the arules will require several hundreds applications and many data transformations before applying the arules. The core of the problem is that coefficients – sequences and right cuts are used in the 4ft-Miner application. A similar task is solved in Section 7.3.1.

Various additional applications of the 4ft-Miner procedure are presented in this Chapter 7 below. Specific features related to the GUHA method are applied in all these applications. This means that the presented tasks are hard to be solved by the apriori algorithm. Note that the same possibilities to define sets of relevant Boolean attributes are available also in the additional GUHA procedures described in this book.

7.2.6 Summary of comparison

The comparison of the arules implementation of the apriori and the 4ft-Miner procedure can be summarized this way:

The apriori implemented in the rules package is faster than apriori implemented in the 4ft-Miner. The higher number of output rules, the higher difference between both implementations. However, the time used by the 4ft-Miner is not much too long, and it is only a fraction of the whole time necessary to solve a real task. Solution times for the secured completion are longer because the GUHA approach of the depth-first walking has to be used and simultaneously, it is necessary to maintain additional information to compute nine-fold tables.

The association rule mining process usually suffers from a problem of a large number of output rules. The ignoring of the missing information approach used in the arules package produces rules, which are false in some completions of the analysed data matrix. The number of such rules can be relatively large. A value of lift of a rule produced by the arules package for a data matrix with missing information can be confusing. The secured completion produces a lower number of rules, which are surely true in all possible completions of the analysed data matrix.

The obligatory use of minimal values of both support and confidence applied in the arules package can lead to remarkable loss of rules interesting for their high values of lift. This is the same for additional measures of interestingness. This problem does not occur in the 4ft-Miner procedure, where conditions concerning various measures of interestingness can be freely combined.

There are tools to tune in a fine way a set of relevant rules to be verified by the 4ft-Miner procedure. Additional possibilities are related to dealing with conditional association rules. This brings possibilities that can be only very hardly realized by the apriori. Note that an additional implementation of a GUHA procedure dealing with more general association rules than apriori algorithm is introduced in [99].

7.3 Applying 4ft-Miner in *Adult* Dataset

The goal of this section is to present possibilities of the 4ft-Miner procedure related to important features of the GUHA association rules. All presented examples are inspired by the examples of 4ft-applications presented in [138].

First, we present possibilities of dealing with basic Boolean attributes of the form $A(\alpha)$ where α is a coefficient – subset of a set of all categories of attribute A. We use two types of coefficients – *Sequences* and *Right cuts* to define suitable Boolean characteristics of education and extreme gain. We use them to solve a task to find segments of persons with remarkable chances to have an extreme gain. This is described in Section 7.3.1.

Then we present the usefulness of a possibility to deal with rules with a conjunction of basic Boolean attributes in consequent, which is not available in arules [20]. We use a conjunction of extreme gain and large income to define extremely rich persons, and we will mine for segments of such persons, see Section 7.3.2. Since the chance to be extremely rich is not too much high, we use a disjunction of extreme gain and large income to define rich persons. Then we will mine for segments of rich persons, see Section 7.3.3.

Disjunction of basic Boolean attributes in combination with some 4ft-quantifiers opens a possibility to use logical deduction to decrease the number of output rules. This is achieved by skipping all rules which can be considered as logical consequences of already output rules as introduced in Section 7.1.7. This is demonstrated in Section 7.3.4.

7.3.1 Applying sequences and right cuts—extreme gain

The goal of this section is to present possibilities of applications of two types of coefficients of basic Boolean attributes introduced in Section 4.4.2. We use sequences and cuts to solve a task to find segments of persons with a remarkable chance to have an extreme gain.

We use the attribute capital_gain introduced in Table 2.2 to define a new attribute gain_positive with categories - intervals $(0; 5000\rangle$ denoted as $(0; 5\rangle$, $(5000; 10000\rangle$ denoted as $(5; 10\rangle$, $(10000; 20000\rangle$ denoted as $(10; 20\rangle$, and > 20000 denoted as > 20. Frequencies of particular categories of the attribute capital_gain_positive are shown in Fig. 7.11. We can see that a frequency of the category $(0; 5\rangle$ is 1 584, which are 39 percent from all positive values of gain; similarly for the additional categories.

We assume three levels of extreme gain corresponding to three basic Boolean attributes gain_positive(> 20), gain_positive($(10; 20\rangle, > 20$) – shortly gain_positive(> 10), and gain_positive($(5; 10\rangle$, $(10; 20\rangle$, $> 20\rangle$) – shortly gain_positive(> 5).

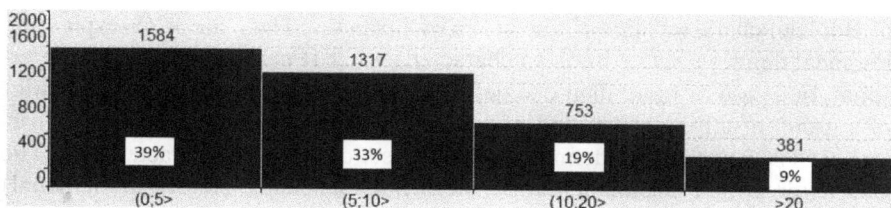

Figure 7.11: Frequencies of categories of attribute gain_positive.

These three levels can be seen as *Right cuts* gain_positive(α) of length 1–3 of the attribute gain_positive (the definition of right cuts is presented in Section 4.4.1).

We are interested in segments of persons with a remarkable chance to have an extreme gain. We assume that segments of such persons are defined by suitable Boolean attributes derived from attributes of the *Adult* dataset. A task to find all segments of persons with a remarkable chance to have an extreme gain can be formulated as a task to find all association rules

$$\mathscr{S}\mathscr{G} \approx \text{gain\_positive}(\alpha)$$

where $\mathscr{S}\mathscr{G}$ is a derived Boolean attribute, $\approx$ is a suitable 4ft-quantifier and gain_positive(α) is one of the above introduced right cuts.

We use a 4ft-quantifier $\approx_{3,500}$ of founded lift introduced in Section 7.1.2. This quantifier is defined by the condition

$$\frac{a(a+b+c+d)}{(a+b)(a+c)} \geq 3 \wedge a \geq 500$$

meaning that a relative frequency of persons satisfying gain_positive(α) among persons satisfying $\mathscr{S}\mathscr{G}$ is at least three-time higher than a relative frequency of persons satisfying gain_positive(α) among all persons described in the *Adult* dataset and that there are at least 500 persons satisfying both $\mathscr{S}\mathscr{G}$ and gain_positive(α).

It remains to define a set of relevant segments $\mathscr{S}\mathscr{G}$ and a set of right cuts gain_positive(α) in a form suitable for the 4ft-Miner procedure. We assume that segments $\mathscr{S}\mathscr{G}$ are defined by Boolean attributes in a form

$$\mathscr{B}(\text{Personal}) \wedge \mathscr{B}(\text{Family}) \wedge \mathscr{B}(\text{Society}) \wedge \mathscr{B}(\text{Employment}). \qquad (7.4)$$

Here $\mathscr{B}(\text{Personal})$, $\mathscr{B}(\text{Family})$, $\mathscr{B}(\text{Society})$, and $\mathscr{B}(\text{Employment})$ are Boolean characteristics of the groups of attributes Personal, Family, Society, and Employment introduced in Table 1.1 in Section 1.5.1.

Recall that sets of Boolean characteristics of groups of attributes are defined by the $\mathscr{R}\mathscr{P}$-expressions introduced in Section 4.4.4. We usually use the same name for a group of attributes and for an $\mathscr{R}\mathscr{P}$-expression defining a set

of Boolean characteristics of this group of attributes. Thus, the $\mathcal{RP}$-expression
Personal defines a set of Boolean characteristics $\mathcal{B}$(Personal) of the group Personal. In a case of more than one definitions of a set of Boolean characteristics
of a group of attributes we will of course use suitable different names for particular $\mathcal{RP}$-expressions defining different sets of Boolean characteristics. The
$\mathcal{RP}$-expression Personal defining a set of Boolean characteristics $\mathcal{B}$(Personal)
of the group of attributes Personal is presented in a tabular form in Fig. 7.12.

Name: Personal		Type: *Conjunction*		Length: $0-3$	
Attribute	Coefficients	Length	Gace	B/R	Class
sex	*Subsets*	$1-1$	*Pos*	B	–
age_exp	*Sequences*	$1-3$	*Pos*	B	–
education	*Sequences*	$1-4$	*Pos*	B	–

Figure 7.12: $\mathcal{RP}$-expression Personal.

Boolean characteristics $\mathcal{B}$(Personal) are defined as conjunctions of length
$0-3$ of basic Boolean attributes sex(α), age_exp(β), education(γ). The possible length 0 means that $\mathcal{B}$(Personal) can be missing in the conjunction (7.4).
A conjunction of length 1 is one of the basic Boolean attributes defined in
Fig. 7.12. Basic Boolean attributes sex(α) are defined such that coefficient α
is a subset of length 1, thus sex(*Female*) and sex(*Male*) are defined.

Basic Boolean attributes age_exp(β) are defined such that β is a set of 1–3
consecutive categories. There are four categories of the attribute age_exp (see
Section 2.2.2), thus there are 4+3+2 = 9 basic Boolean attributes age_exp(β).
Basic Boolean attributes education(γ) are defined such that γ is a set of 1–4 consecutive categories. There are 16 categories of the attribute education (see Table
2.1), thus there are 16+15+14+13 = 58 basic Boolean attributes education(γ).
All basic Boolean attributes are positive only and considered as Basic according
to columns Gace and B/R. For details see Section 4.4.4.

Sets of Boolean characteristics $\mathcal{B}$(Family), $\mathcal{B}$(Society), and $\mathcal{B}$(Employment)
of the groups of attributes Family, Society, and Employment are defined by the
$\mathcal{RP}$-expressions Family, Society, and Employment. They are presented in tabular forms in Fig. 7.13. A set of all relevant segments $\mathcal{SG}$ in the form (7.4) is
defined by an $\mathcal{RC}$-expression in a short tabular form according to Fig. 7.14.

A set of three right cuts gain_positive(α) is defined by a simple $\mathcal{RP}$-expression Extreme_gain is shown in Fig. 7.15.

We can summarize that we are interested in association rules

$$\mathcal{B}(\text{Personal}) \wedge \mathcal{B}(\text{Family}) \wedge \mathcal{B}(\text{Society}) \wedge \mathcal{B}(\text{Employment})$$

$$\approx_{3,500} \text{gain\_positive}(\alpha)$$

where the set of relevant antecedents is given by the $\mathcal{RC}$-expression $\mathcal{SG}$ introduced in Fig. 7.14 and the set of relevant succedents is given by the $\mathcal{PC}$-

Name: Family		Type: *Conjunction*		Length: $0-2$	
Attribute	Coefficients	Length	Gace	B/R	Class
marital_status	*Subsets*	$1-1$	*Pos*	B	—
relationship	*Subsets*	$1-1$	*Pos*	B	—

Name: Society		Type: *Conjunction*		Length: $0-3$	
Attribute	Coefficients	Length	Gace	B/R	Class
native_country	*Subsets*	$1-1$	*Pos*	B	—
race	*Subsets*	$1-1$	*Pos*	B	—
workclass	*Subsets*	$1-1$	*Pos*	B	—

Name: Employment		Type: *Conjunction*		Length: $0-2$	
Attribute	Coefficients	Length	Gace	B/R	Class
hours_per_week	*Subsets*	$1-1$	*Pos*	B	—
occupation	*Subsets*	$1-1$	*Pos*	B	—
income	*Subsets*	$1-1$	*Pos*	B	—

Figure 7.13: $\mathscr{R}\mathscr{P}$-expressions Family, Society, and Employment.

Name: $\mathscr{S}\mathscr{G}$	Length: $1-10$
Set of partial cedents	Length
Personal	$0-3$
Family	$0-2$
Society	$0-3$
Employment	$0-3$

Figure 7.14: $\mathscr{R}\mathscr{C}$-expression $\mathscr{S}\mathscr{G}$.

Name: Extreme_gain		Type: *Conjunction*		Length: 1	
Attribute	Coefficients	Length	Gace	B/R	Class
gain_positive	*Right cuts*	$1-3$	*Pos*	B	—

Figure 7.15: $\mathscr{R}\mathscr{P}$-expression Extreme_gain.

expression Extreme_gain shown in Fig. 7.15. We use the secured approach to missing information introduced in Section 7.1.5. The run of the 4ft-Miner procedure with these parameters resulted in 340 found true rules in 139 seconds. More than $1.68 * 10^6$ rules were tested. Five rules with the shortest antecedents are displayed in a short form in Fig. 7.16.

The rule $Segment_{M\_1} \approx_{3.812,521}$ gain_positive(> 10) where

$$Segment_{M\_1} = \text{age\_exp}(Middle\text{-}aged, Senior, Old) \wedge$$
$$\wedge \text{ education}(Bachelors, Masters, Prof\text{-}school, Doctorate) \wedge$$
$$\wedge \text{ marital\_status}(Married\text{-}civ\text{-}spouse) \wedge \text{ race}(White)$$

Lift	Hypothesis
3.082	education(*Masters,Prof-school*) >+< gain_positive(>*5*)
3.127	education(*Masters,Prof-school,Doctorate*) >+< gain_positive(>*5*)
3.152	sex(*Male*) & education(*Bachelors,Masters,Prof-school*) >+< gain_positive(> *10*)
3.493	sex(*Male*) & education(*Masters,Prof-school,Doctorate*) >+< gain_positive(>*5*)
3.302	sex(*Male*) & education(*Bachelors,Masters,Prof-school,Doctorate*) >+< gain_positive(> *10*)

Figure 7.16: Five rules concerning extreme gain with the shortest antecedents.

has the highest lift, i.e., 3.812. This rule does not deal with missing information and $\langle 521, 5366, 613, 42342 \rangle$ is its 4ft-table. The rule means that the relative frequency of persons satisfying gain_positive(> 10) among persons satisfying $Segment_{M\_1}$ is $\frac{521}{521+5366} / \frac{521+613}{521+5366+613+42342} = 3.812$-times higher than relative frequency of persons satisfying gain_positive(> 10) among all 48 842 persons. A confidence of this rule is $\frac{521}{521+5366} = 0.086$.

The sixteenth rule $Segment_{M\_16} \approx_{3.650,503}$ gain_positive(> 10) where

$$Segment_{M\_16} = \text{age\_exp}(\textit{Middle-aged, Senior, Old}) \wedge$$

$$\wedge \text{ education}(\textit{Bachelors, Masters, Prof-school, Doctorate}) \wedge$$

$$\wedge \text{ marital\_status}(\textit{Married-civ-spouse}) \wedge \text{ native\_country}(\textit{United-States})$$

deals with missing information in the secured way. A nine-fold table $9ft(Segment_{M\_16}, \text{gain\_positive}(> 10), Adult)$ of this rule is shown in the left part of Fig. 7.17. A secured completion of this nine-fold table is in the right part of Fig. 7.17.

Let us note that it holds 5 433 = 5 259 + 174 and 631 = 614 + 17, see Section 7.1.5. Note that in the case of the nine-fold table in the left part of Fig. 7.17 is the 4ft-table in the right part of Fig. 7.17 a secured completion both for the 4ft-

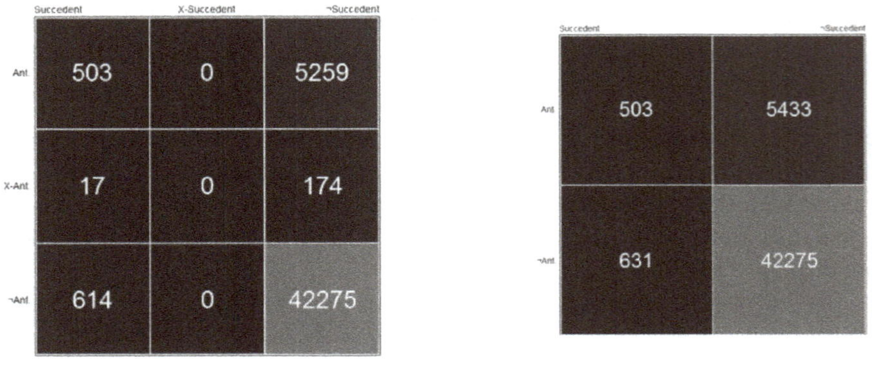

	Succedent	X-Succedent	¬Succedent
Ant	503	0	5259
X-Ant	17	0	174
¬Ant	614	0	42275

	Succedent	¬Succedent
Ant	503	5433
¬Ant	631	42275

Figure 7.17: $9ft(Segment_{M\_16}, \text{gain\_positive}(> 10), Adult)$ and its secured completion.

quantifier $\approx_{q,Base}$ and for the 4ft-quantifier $\Rightarrow_{p,Base}$. The confidence of the rule $Segment_{M\_16} \approx_{3.650,503}$ gain_positive(> 10) is $\frac{503}{503+5433} = 0.085$.

The set of 340 output rules has the following important features:

- Each rule contains a basic Boolean attribute education(α) in the antecedent where α is a coefficient with at least two categories.

- Each rule also contains one of basic Boolean attributes gain_positive(> 5), i.e., gain_positive($(5;10)$, $(10;20)$, > 20)) and gain_positive(> 10), i.e., gain_positive($(10;20)$, > 20). More information on the structure of the set of rules is available in Table 7.4.

- Let us note that the apriori algorithm deals with rules—couples of conjunctions of attribute-value pairs [20]. Thus, the two previous points mean that no output rule can be produced by the apriori algorithm without application of very complex transformations, see also [138].

- There is only one rule with the attribute income in antecedent. This is very probably caused by the fact that there are 16 281 missing values for this attribute, see Table 2.1.

- The length of antecedents varies from one 1 to 7, see Table 7.5.

- A confidence of rules varies from 0.070 to 0.190.

The rule

$$education(\textit{Masters, Prof-school, Doctorate}) \approx_{3,500} \text{gain\_positive}(> 5) \quad (7.5)$$

is true in the *Adult* data matrix. There are 18 rules with the basic Boolean attribute education(*Masters, Prof-school, Doctorate*) in antecedent, see Table 7.4.

Table 7.4: Structure of the resulting set of true rules $\mathscr{SG} \approx_{3,0,500}$ gain_positive(α).

Antecedent contains	Succedent contains	# of rules
education(*Masters, Prof-school*)	gain_positive(> 5)	4
education(*Masters, Prof-school, Doctorate*)	gain_positive(> 5)	18
education(*Bachelors, Masters, Prof-school*)	gain_positive(> 5)	105
	gain_positive(> 10)	14
education(*Bachelors, Masters) Prof-school, Doctorate*)	gain_positive(> 5)	127
	gain_positive(> 10)	40
education(*Assoc-acdm, Bachelors, Masters, Prof-school*)	gain_positive(> 5)	24
	gain_positive(> 10)	8
Total rules		340

Table 7.5: The numbers of rules for particular lengths of antecedent.

Antecedent length	# of rules	Antecedent length	# of rules
1	2	5	86
2	23	6	39
3	75	7	7
4	108	–	–

Thus, there is a question–if all the 17 rules in a form

$$\text{education}(\textit{Masters, Prof-school, Doctorate}) \wedge \varphi$$
$$\approx_{3,500} \text{gain\_positive}(> 5)$$

true in the *Adult* dataset cannot be considered as consequences of the rule (7.5). A similar question is discussed and partly answered in Chapter 17.

7.3.2 Conjunctions in succedent—very rich persons

A goal of this section is to show a usefulness of a possibility to use conjunctions of basic Boolean attributes in consequent. Note that this is not possible in arules [20]. Let us assume that we are interested in segments of very rich persons. We consider a person as very rich if she/he has both positive gain and large income. A task to find all segments of persons with a relatively high chance to be very rich can be formulated as a task to find all association rules

$$\mathscr{SG} \approx \text{gain\_positive}(\alpha) \wedge \text{income}(\textit{Large})$$

true in the *Adult* dataset where $\mathscr{SG}$ is a derived Boolean attribute, $\approx$ is a suitable 4ft-quantifier and gain_positive$(\alpha) \wedge$ income$(\textit{Large})$ is a Boolean attribute expressing that a person is very rich.

We know that a relative frequency of extreme gain in segments of persons used in the previous Section 7.3.1 is relatively low and that a confidence of rules resulting in the task solved in the previous section is about 0.1. This leads us to use a 4ft-quantifier $\Rightarrow_{0.1,500}$ of founded confidence introduced at the end of Section 7.1.2. This quantifier is defined by the condition

$$\frac{a}{a+b} \geq 0.1 \wedge a \geq 500$$

meaning that a relative frequency of very rich persons in a segment $\mathscr{SG}$ is at least 0.1 and that there is at least 500 very rich persons in the segment $\mathscr{SG}$.

We are interested in positive gain, not only in extreme gain as in the previous section. Thus, we use right cuts gain_positive(α) of length 1–4 instead of length 1–3. This means that we are interested in the basic Boolean attributes – gain_positive(> 20), gain_positive(> 10), and gain_positive(> 5) introduced in

Section 7.3.1 and also in a basic Boolean attribute gain_positive($\langle 0;5 \rangle$,$\langle 5;10 \rangle$, $\langle 10;20 \rangle$, $> 20 \rangle$) – shortly gain_positive(> 0).

We use a definition of a set of relevant segments $\mathscr{SG}$ similar to that used in the previous section. We assume that segments $\mathscr{SG}$ are defined by Boolean attributes in a form

$$\mathscr{B}(\text{Personal}) \wedge \mathscr{B}(\text{Family}) \wedge \mathscr{B}(\text{Society}) \wedge \mathscr{B}(\text{Employment}). \qquad (7.6)$$

Here $\mathscr{B}(\text{Personal})$, $\mathscr{B}(\text{Family})$, $\mathscr{B}(\text{Society})$, and $\mathscr{B}(\text{Employment})$ are Boolean characteristics of the groups of attributes Personal, Family, Society, and Employment. Sets of Boolean characteristics of groups of attributes Personal, Family, Society are defined in the same way as in the previous section. Thus, $\mathscr{RP}$-expression Personal introduced in Fig. 7.12 is used to define $\mathscr{B}(\text{Personal})$ and $\mathscr{RP}$-expressions Family and Society introduced in Fig. 7.13 are used to define $\mathscr{B}(\text{Family})$ and $\mathscr{B}(\text{Society})$ respectively. The set $\mathscr{B}(\text{Employment})$ is defined by the $\mathscr{RP}$-expression Employment introduced in Fig. 7.18. This differs from that in Fig. 7.13 by omitting attribute income since it is used in succedent. A set of all relevant segments $\mathscr{SG}$ in the form (7.6) is defined by an $\mathscr{RC}$-expression in a short tabular form according to Fig. 7.19.

Name: Employment		Type: *Conjunction*		Length: $0-2$	
Attribute	Coefficients	Length	Gace	B/R	Class
hours_per_week	*Subsets*	$1-1$	*Pos*	B	—
occupation	*Subsets*	$1-1$	*Pos*	B	—

Figure 7.18: $\mathscr{RP}$-expression Employment.

Name: $\mathscr{SG}$	Length: $1-10$
Set of partial cedents	Length
Personal	$0-3$
Family	$0-2$
Society	$0-3$
Employment	$0-2$

Figure 7.19: $\mathscr{RC}$-expression $\mathscr{SG}$.

A set of all relevant succedents gain_positive(α) $\wedge$ income(*Large*) is defined by an $\mathscr{RP}$-expression Very_rich shown in Fig. 7.20.

We used the secured approach to missing information introduced in Section 7.1.5. A run of the 4ft-Miner procedure with these parameters resulted in 406 found true rules in 117 seconds. Nearly $1.18 * 10^6$ rules were tested. The five rules with the shortest antecedents are displayed in a short form in Fig. 7.21.

Name: Very_rich			Type: *Conjunction*	Length: 2	
Attribute	Coefficients	Length	Gace	B/R	Class
gain_positive	*Right cuts*	$1-4$	*Pos*	B	–
income	*One category*	*Large*	*Pos*	B	–

Figure 7.20: $\mathscr{R}\mathscr{P}$-expression Very_rich.

PIM Hypothesis

0.101 education(*Bachelors,Masters*) & relationship(*Husband*) >+< gain_positive(*<=>20*) & income(*Large*)
0.100 education(*Bachelors,Masters,Prof-school*) & marital(*Married-civ-spouse*) >+< gain_positive(*>5*) & income(*Large*)
0.107 education(*Bachelors,Masters,Prof-school*) & marital(*Married-civ-spouse*) >+< gain_positive(*<=>20*) & income(*Large*)
0.102 education(*Bachelors,Masters,Prof-school*) & relationship(*Husband*) >+< gain_positive(*>5*) & income(*Large*)
0.109 education(*Bachelors,Masters,Prof-school*) & relationship(*Husband*) >+< gain_positive(*<=>20*) & income(*Large*)

Figure 7.21: Five rules concerning very rich people with the shortest antecedents.

The rule $Segment_{VR} \Rightarrow_{0.113,566}$ gain_positive$(> 0) \wedge$ income$(Large)$ denoted as $\mathscr{V}\mathscr{R}$ where

$$Segment_{VR} = \text{age\_exp}(Middle\text{-}aged, Senior, Old) \wedge$$
$$\wedge \text{ education}(Bachelors, Masters, Prof\text{-}school, Doctorate) \wedge$$
$$\wedge \text{ relationship}(Husband) \wedge \text{ native\_country}(United\text{-}States) \wedge$$
$$\wedge \text{ race}(White)$$

has the highest confidence, i.e., 0.113. The nine-fold table

$$9ft(Segment_{VR}, \text{gain\_positive}(> 0) \wedge \text{income}(Large), Adult)$$

of the rule $\mathscr{V}\mathscr{R}$, and its secured completion are shown in Fig. 7.22. Let us note that it holds $4465 = 4078 + 280 + 8 + 99$ and $2146 = 1099 + 1035 + 12$. Thus, the four-fold table $\langle 566, 4465, 2146, 41665 \rangle$ is a secured completion for the 4ft-quantifier $\Rightarrow_{0.1,500}$, see Section 7.1.5. The rule means that the relative frequency of persons satisfying gain_positive$(> 0) \wedge$ income$(Large)$ among persons satisfying $Segment_{VR}$ is $\frac{566}{566+4465} = 0.113$.

Lift of this rule is equal to $\frac{566}{566+4465} / \frac{566+2146}{566+4465+2146+41665} = 2.026$ if we use the above introduced secured completion for the 4ft-quantifier $\Rightarrow_{0.1,500}$. The secured completion for the $\approx_{q,Base}$ quantifier is $\langle 566, 4078 + 280 + 99, 1099 + 1035 + 12 + 8, 41665 \rangle = \langle 566, 4457, 2154, 41665 \rangle$ and thus a lift of the rule if equal to $\frac{566}{566+4457} / \frac{566+2154}{566+4457+2154+41665} = 2.023$ if we use the secured completion for the 4ft-quantifier $\approx_{q,Base}$.

Let us note that each of 406 output rules contains a basic Boolean attribute education(α) in the antecedent where α is a coefficient with at least two categories. In addition, each rule also contains one of basic Boolean attributes gain_positive(> 0), i.e., gain_positive$(\langle 0; 5\rangle, \langle 5; 10\rangle, \langle 10; 20\rangle, > 20\rangle)$ and gain_positive(> 5), i.e., gain_positive$(\langle 5; 10\rangle, \langle 10; 20\rangle, > 20)$. Thus, no output

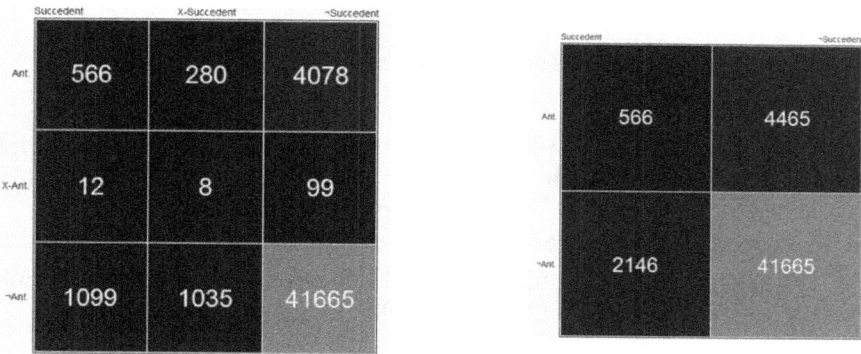

Figure 7.22: Nine-fold table of the rule $\mathscr{V}\mathscr{R}$ and its secured completion.

rule can be produced by the apriori algorithm without application of very complex transformations, see also [138].

The length of antecedents varies from 2 to 7, see Table 7.6. There is again the question if some rules with antecedent length 3 or 4 can be considered as consequences of some rule with shorter antecedent. A similar question is discussed and partly answered in Chapter 18.

Table 7.6: The numbers of rules for particular lengths of antecedent.

Antecedent length	# of rules	Antecedent length	# of rules
2	11	5	131
3	65	6	53
4	138	7	8

7.3.3 Disjunctions in succedent—rich persons

The task to find all segments of persons with a relatively high chance to be very rich was formulated as a task to find all association rules

$$\mathscr{S}\mathscr{G} \Rightarrow_{0.1,500} \text{gain\_positive}(\alpha) \wedge \text{income}(Large)$$

where $\mathscr{S}\mathscr{G}$ is a derived Boolean attribute. The highest confidence of the resulting rules is 0.113. This means that a chance to be very rich is not too much high.

Thus, we use a suitable definition of a rich person instead of the used definition of the very rich person. We consider a person as rich if she/he has positive gain or large income. We hope that there is a higher chance of being rich than a chance of being very rich. This led us to a task to find all association rules

$$\mathscr{S}\mathscr{G} \Rightarrow_{0.45,3000} \text{gain\_positive}(\alpha) \vee \text{income}(Large)$$

true in the *Adult* dataset where $\mathscr{S}\mathscr{G}$ is a derived Boolean attribute. The 4ft-quantifier $\Rightarrow_{0.45,3000}$ is defined by the condition

$$\frac{a}{a+b} \geq 0.45 \wedge a \geq 3000$$

meaning that a relative frequency of rich persons in a segment $\mathscr{S}\mathscr{G}$ is at least 0.45 and that there is at least 3000 rich persons in the segment $\mathscr{S}\mathscr{G}$.

We are again interested in positive gain. Thus, we use *Right cuts* gain_positive(α) of length 1–4. We use the same definition of a set of relevant segments $\mathscr{S}\mathscr{G}$ as in the the previous section. We assume that segments $\mathscr{S}\mathscr{G}$ are defined by Boolean attributes in a form

$$\mathscr{B}(\text{Personal}) \wedge \mathscr{B}(\text{Family}) \wedge \mathscr{B}(\text{Society}) \wedge \mathscr{B}(\text{Employment}). \qquad (7.7)$$

Here $\mathscr{B}(\text{Personal})$, $\mathscr{B}(\text{Family})$, $\mathscr{B}(\text{Society})$, and $\mathscr{B}(\text{Employment})$ are Boolean characteristics of the groups of attributes Personal, Family, Society, and Employment. $\mathscr{R}\mathscr{P}$-expression Personal introduced in Fig. 7.12 is used to define $\mathscr{B}(\text{Personal})$ and $\mathscr{R}\mathscr{P}$-expressions Family and Society introduced in Fig. 7.13 are used to define $\mathscr{B}(\text{Family})$ and $\mathscr{B}(\text{Society})$ respectively. The set $\mathscr{B}(\text{Employment})$ is defined by the $\mathscr{R}\mathscr{P}$-expression Employment introduced in Fig. 7.18. A set of all relevant segments $\mathscr{S}\mathscr{G}$ in the form (7.7) is defined by an $\mathscr{R}\mathscr{C}$-expression in a short tabular form according to Fig. 7.19. A set of all relevant succedents gain_positive(α) $\wedge$ income(*Large*) is defined by an $\mathscr{R}\mathscr{P}$-expression Rich shown in Fig. 7.23.

Name: Rich		Type: *Disjunction*		Length: $1-2$	
Attribute	Coefficients	Length	Gace	B/R	Class
gain_positive	*Right cuts*	$1-4$	*Pos*	B	–
income	*One category*	*Large*	*Pos*	B	–

Figure 7.23: $\mathscr{R}\mathscr{P}$-expression Rich.

We use again the secured approach to missing information introduced in Section 7.1.5. A run of the 4ft-Miner procedure with these parameters resulted in 387 true rules in 32 seconds. More than $384 * 10^3$ rules were tested. The five rules with the shortest antecedents are displayed in a short form in Fig. 7.24.

The rule $A_{Rich\_1} \Rightarrow_{0.567,3302}$ gain_positive(> 0) $\vee$ income(*Large*) where

$A_{Rich_1} = $ age_exp(*Middle-aged, Senior, Old*) $\wedge$

$\wedge$ education(*Bachelors, Masters, Prof-school, Doctorate*) $\wedge$

$\wedge$ marital_status(*Married-civ-spouse*) $\wedge$ hours(*Full-time, Over-time*)

has the highest confidence, i.e., 0.567. We denote this rule as *Rich*_1. The nine-fold table $9ft(A_{Rich_1}, $ gain_positive(> 0) $\vee$ income(*Large*), *Adult*) of the rule *Rich*_1 and its secured completion are shown in Fig. 7.25.

PIM Hypothesis

0.451 **sex**(*Male*) & **education**(*Bachelors,Masters,Prof-school,Doctorate*) >←< **gain_positive**(<=>20) | **income**(*Large*)
0.511 **education**(*Assoc-acdm,Bachelors,Masters*) & **marital**(*Married-civ-spouse*) >←< **gain_positive**(<=>20) | **income**(*Large*)
0.497 **education**(*Assoc-acdm,Bachelors,Masters*) & **marital**(*Married-civ-spouse*) >←< **gain_positive**(>5) | **income**(*Large*)
0.542 **education**(*Bachelors,Masters,Prof-school*) & **marital**(*Married-civ-spouse*) >←< **gain_positive**(<=>20) | **income**(*Large*)
0.529 **education**(*Bachelors,Masters,Prof-school*) & **marital**(*Married-civ-spouse*) >←< **gain_positive**(>5) | **income**(*Large*)

Figure 7.24: Five rules concerning rich people with the shortest antecedents.

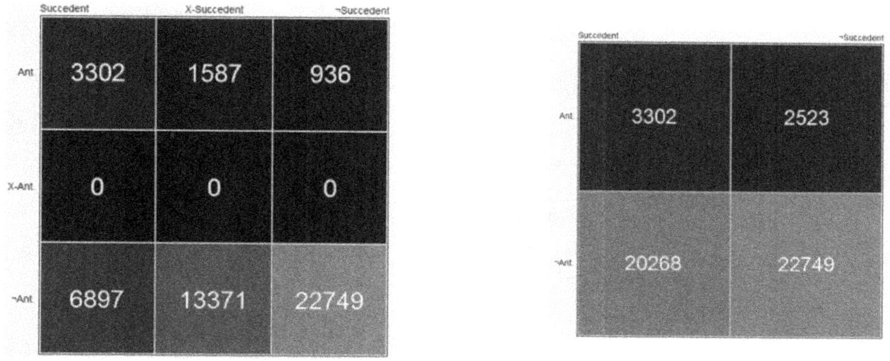

Figure 7.25: Nine-fold table of the rule *Rich_1* and its secured completion.

It holds $2523 = 936 + 1587$ and $20268 = 6897 + 13371$. Thus, the four-fold table $\langle 3302, 2523, 20268, 22749 \rangle$ is a secured completion for the 4ft-quantifier $\Rightarrow_{0.45,3000}$, see Section 7.1.5. The rule means that the relative frequency of persons satisfying gain_positive$(> 0) \vee$ income(*Large*) among persons satisfying A_{Rich_1} is $\frac{3302}{3302+2523} = 0.567$. Note that in the case of the nine-fold table in the left part of Fig. 7.25 is the 4ft-table in the right part of Fig. 7.17 a secured completion both for the 4ft-quantifier $\approx_{q,Base}$ and for the 4ft-quantifier $\Rightarrow_{p,Base}$. The lift of the rule $A_{Rich\_1} \Rightarrow_{0.567,3302}$ gain_positive$(> 0) \vee$ income(*Large*) is equal to $\frac{3302}{3302+2523} / \frac{3302+20268}{3302+2523+20268+22749} = 1.175$.

Let us emphasize, that each of 387 output rules contains a basic Boolean attribute education(α) in the antecedent where α is a coefficient with at least three categories. Thus, no output rule can be produced by the apriori algorithm without application of very complex transformations, see also [138].

There are five different succedents occurring in output rules. Their frequencies are introduced in Table 7.7. The length of antecedents varies from two to five, see Table 7.8.

There is again the question if some output rules can be considered as consequences of some other more simple output rules. It is shown in the next section that there are such rules that can be considered as the logical consequences of more simple output rules rule. An additional approach to this question is discussed in Section 17.1 and introduced in more details in Chapter 18.

Table 7.7: The numbers of rules for particular succedents.

Succedent	# of rules
income(*Large*)	8
gain_positive(> 20) $\lor$ income(*Large*)	10
gain_positive(> 10) $\lor$ income(*Large*)	44
gain_positive(> 5) $\lor$ income(*Large*)	132
gain_positive(> 0) $\lor$ income(*Large*)	193
Total	387

Table 7.8: The numbers of rules for particular lengths of antecedent.

Antecedent length	# of rules	Antecedent length	# of rules
2	27	4	172
3	140	5	48

7.3.4 Applying logical deduction—prime rules

Let us consider an association rule $\mathscr{A} \Rightarrow_{p,Base}$ income(*Large*) where

$$\mathscr{A} = \text{education}(Masters, Prof\text{-}school, Doctorate) \land$$
$$\land \text{ marital\_status}(Married\text{-}civ\text{-}spouse) \ .$$

Such rule with $\Rightarrow_{0.482,3214}$ instead of $\Rightarrow_{p,Base}$ results from the application of the 4ft-Miner procedure described above. Let us have association rules

$$\mathscr{A} \Rightarrow_{p,Base} \text{income}(Large) \tag{7.8}$$

and

$$\mathscr{A} \Rightarrow_{p,Base} \text{gain\_positive}(\alpha) \lor \text{income}(Large). \tag{7.9}$$

In addition, let us consider a general data matrix $\mathscr{M}$ and 4ft-tables

- $\mathscr{T}_1 = 4ft(\mathscr{A}, \text{income}(Large), \mathscr{M}) = \langle a_1, b_1, c_1, d_1 \rangle$

- $\mathscr{T}_2 = 4ft(\mathscr{A}, \text{gain\_positive}(\alpha) \lor \text{income}(Large), \mathscr{M}) = \langle a_2, b_2, c_2, d_2 \rangle$,

see Fig. 7.26 where we write incm(*Large*) instead of income(*Large*).
It holds:

- $a_2 \geq a_1$ since each row of M satisfying $\mathscr{A} \land$ income(*Large*) satisfies also $\mathscr{A} \land (\text{gain\_positive}(\alpha) \lor \text{income}(Large))$

- $a_1 + b_1 = a_2 + b_2 =$ the number of rows satisfying $\mathscr{A}$

- $a_2 \geq a_1$ and $a_1 + b_1 = a_2 + b_2$ implies $\frac{a_2}{a_2+b_2} \geq \frac{a_1}{a_1+b_1}$.

$\mathscr{M}$	incm(*Large*)	¬ incm(*Large*)
$\mathscr{A}$	a_1	b_1
¬$\mathscr{A}$	c_1	d_1

$\mathscr{M}$	gain_positive(α) ∨ incm(*Large*)	¬gain_positive(α) ∨ incm(*Large*)
$\mathscr{A}$	a_2	b_2
¬$\mathscr{A}$	c_2	d_2

Figure 7.26: 4ft-tables $\mathscr{T}_1 = \langle a_1, b_1, c_1, d_1 \rangle$ and $\mathscr{T}_2 = \langle a_2, b_2, c_2, d_2 \rangle$.

We can conclude that if $\frac{a_1}{a_1+b_1} \geq p \wedge a_1 \geq Base$, then also $\frac{a_2}{a_2+b_2} \geq p \wedge a_2 \geq Base$. In other words, if the rule (7.8) is true in a data matrix $\mathscr{M}$, then the rule (7.9) is sure also true in $\mathscr{M}$. Recall that we say in this case that the rule (7.9) logically follows from the rule (7.5). For details see Section 7.1.7 and Chapter 18.

Let us have also a rule

$$\mathscr{A} \Rightarrow_{p,Base} \text{gain\_positive}(\alpha') \vee \text{income}(Large). \tag{7.10}$$

If $\alpha \subsetneq \alpha'$, then truthfulness of the rule (7.9) in a data matrix $\mathscr{M}$ implies truthfulness also for the the rule (7.10). This is because each row of $\mathscr{M}$ satisfying gain_positive(α) satisfies also gain_positive(α'). We can conclude that the rule (7.10) logically follows from the rule (7.9).

Recall that we can put a value of the parameter *Prime* as *Yes*. In this case, only prime rules will be listed in the output. A rule $\mathscr{R}$ is prime if it is relevant and true and if it does not logically follow from another output rule. A default value of this parameter is *No*, see Section 7.1.8. We run the same task as in the previous section, however, with a value of the parameter *Prime* as *Yes*.

A run of the 4ft-Miner procedure with such parameters resulted in 256 prime rules in 31 seconds. There were 387 output rules in the previous section. This means that that 131 rules, i.e., 34 per cent of all rules, are not prime in our case. The strongest rule among the prime rules is the second strongest rule from the output rules in the previous section. The set of 256 prime rules has the following important features:

■ Each prime rule of course belongs to the set of 387 output rules introduced in the previous section.

■ Each rule from the set of 387 output rules introduced in the previous section is either prime or can be derived from one of the prime rules by simple deduction rules outlined at the beginning of this section. These deduction rules are shortly introduced also in Section 7.1.7 and described in Chapter 18.

■ The confidence of a not prime rule is the same or higher than the confidence of the prime rule from which the not prime rule can be derived. The same is true for the frequencies a from 4ft-tables of rules.

■ Using prime rules can reasonably decrease the number of output rules and to facilitate orientation in true rules. If we are interested in characteristics of a concrete not prime rule, we have to compute the characteristics separately by available tools.

■ There are five different succedents occurring in output prime rules. Their frequencies are introduced in Table 7.9.

■ The length of antecedents varies from 2 to 5, see Table 7.10.

Table 7.9: The numbers of rules for particular succedents.

Succedent	# of rules	
	not prime	prime
income(*Large*)	8	8
gain_positive(> 20) ∨ income(*Large*)	10	2
gain_positive(> 10) ∨ income(*Large*)	44	36
gain_positive(> 5) ∨ income(*Large*)	132	124
gain_positive(> 0) ∨ income(*Large*)	193	86
Total	387	256

Table 7.10: The numbers of rules for particular lengths of antecedent.

Antecedent length	# of rules	
	not prime	prime
2	27	13
3	140	83
4	172	119
5	48	41
Total	387	256

We have shown that the application of a logical deduction can remarkably decrease the number of output rules. However, a question remains if there are some other tools how to decrease the number of output rules without the loss of important rules. An approach to this problem is discussed in Chapter 17.

7.4   Applying 4ft-Miner in *Accidents* Dataset

The first goal of this section is to present possibilities of the 4ft-Miner procedure to mine rules, which can be seen as exception rules. The second goal is to present 4ft-quantifiers for histogram mode of the 4ft-Miner procedure introduced in Section 7.1.3. We use a histogram of the attribute Severity showing percentages of particular degrees of severity of accidents. Mining for association rules pointing to segments of accidents with a significantly higher percentage of accidents with a particular degree of severity is described in Section 7.4.1. Section 7.4.2 is devoted to association rules pointing to segments of accidents with a significantly lower percentage of accidents with a particular degree of severity. All resulting rules can be understood as exception rules according to Section 1.2.4.

Rules resulting from applications described in Section 7.4.2 are used to formulate a new analytical question to demonstrate possibilities of mining for "exceptions from exceptions". A possibility to get higher confidence by extending the given antecedent is shown in Section 7.4.3.

7.4.1   Exception rules—increasing columns of histogram

The histogram of the attribute Severity showing percentages of particular degrees of severity of accidents is available in Fig. 7.27. We can see that the percentages of fatal, serious and slight accidents are 1.9%, 19.6% and 78.5% respectively. There are two slim column in Fig. 7.27. The first one corresponds to the association rule

$$\text{Age}(\textit{56 - 65}) \ \wedge \ \text{Type}(\textit{Goods 7.5 tonnes+})$$
$$\Rightarrow_{0.109,108} \text{Severity}(\textit{Fatal}). \tag{7.11}$$

The confidence of this rule is 0.109 which means that among accidents satisfying Age(*56–65*) ∧ Type(*Goods 7.5 tonnes+*), there are 10.9% of fatal accidents which is 5.7 times higher than in the set of all accidents.

The second slim columns corresponds analogously to the association rule

$$\text{District}(\textit{Aberdeenshire}) \ \wedge \ \text{Type}(\textit{Motorcycle over 500cc})$$
$$\Rightarrow_{0.612,117} \text{Severity}(\textit{Serious}). \tag{7.12}$$

It is natural to consider the rules (7.11) and (7.12) as exceptions to the histogram of the attribute Severity. We show how the 4ft-Miner procedure with 4ft-quantifiers for histogram mode can be used to mine such exceptions. We solve the analytical question "*Are there any segments of accidents, which can be considered as exceptions to the histogram of the attribute Severity. We are interested in segments described by attributes Age, District, and Type.*"

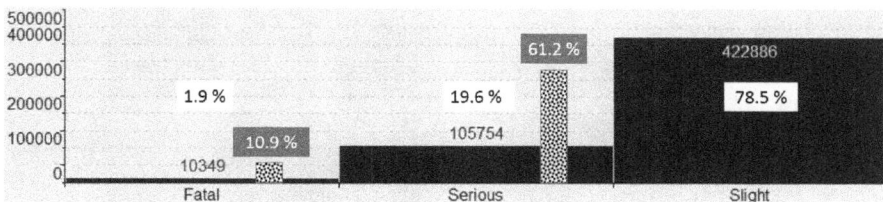

Figure 7.27: Percentages of degrees of severity with exceptions – increasing columns.

We use a run of the 4ft-Miner with a set of relevant antecedents given by the $\mathcal{RP}$-expression Exceptions to histogram and with a set relevant succedents given by the $\mathcal{RP}$-expression Histogram Severity available in Fig. 7.28.

Name: Exceptions to histogram		Type: *Conjunction*		Length: $1--3$	
Attribute	Coefficients	Length	Gace	B/R	Class
Age	*Subsets*	$1-1$	*Pos*	B	−
District	*Subsets*	$1-1$	*Pos*	B	−
Type	*Subsets*	$1-1$	*Pos*	B	−

Name: Histogram Severity		Type: *Conjunction*		Length: 1	
Attribute	Coefficients	Length	Gace	B/R	Class
Severity	*Subsets*	$1-1$	*Pos*	B	−

Figure 7.28: $\mathcal{RP}$-expressions Exceptions to histogram and Histogram Severity.

We use a 4ft-quantifier—a conjunction of 4ft-quantifiers $\sim_{\geq 2}$ (see equation (7.1) in Section 7.1.3) and $\oplus_{100}$ (see Table 7.1 in Section 7.1.2) which corresponds to the condition

$$\frac{a(a+b+c+d)}{(a+b)(a+c)} \geq 2 \wedge a \geq 100 \,. \tag{7.13}$$

Note that this conjunction is equivalent to the 4ft-quantifier $\approx_{2,100}$ of founded lift, see Section 7.1.2. We use the *Deleting* approach to missing information.

The run of the 4ft-Miner procedure with these parameters resulted in 21 true rules in 17 seconds. 19 554 rules were tested. The five rules with the highest lift are displayed in a short form in Fig. 7.29.

The first rule in Fig. 7.29 corresponds to rule (7.11) and the fifth rule in Fig. 7.29 corresponds to rule (7.12). 4ft-tables of rules (7.11) and (7.12) are available in Fig. 7.30. Here $Ant_{(7.11)}$ denotes the antecedent of the rule (7.11) and $Ant_{(7.12)}$ denotes the antecedent of the rule (7.12). We write "*Fatal*" instead of "Severity(*Fatal*)" and "*Serious*" instead of "Severity(*Serious*)".

Nr.	Id	Lift	Hypothesis
1	14	5.695	**Driver_Age_Band**(*56 - 65*) & **Type**(*21,Goods 7.5 tonnes +*) >+< **Severity**(*Fatal*)
2	9	5.385	**Driver_Age_Band**(*46 - 55*) & **Type**(*21,Goods 7.5 tonnes +*) >+< **Severity**(*Fatal*)
3	6	4.706	**Driver_Age_Band**(*36 - 45*) & **Type**(*21,Goods 7.5 tonnes +*) >+< **Severity**(*Fatal*)
4	21	4.557	**Type**(*21,Goods 7.5 tonnes +*) >+< **Severity**(*Fatal*)
5	18	3.122	**District**(*911,Aberdeenshire*) & **Type**(*5,Motorcycle over 500cc*) >+< **Severity**(*Serious*)

Figure 7.29: 4ft-Miner output - searching for segments increasing columns.

Accidents	*Fatal*	*¬ Fatal*
Ant$_{(7.11)}$	108	880
¬ Ant$_{(7.11)}$	10 227	527 214

Accidents	*Serious*	*¬ Serious*
Ant$_{(7.12)}$	117	74
¬ Ant$_{(7.12)}$	105 637	433 160

$4ft(Ant_{(7.11)}, Fatal, Accidents))$ $4ft(Ant_{(7.12)}, Serious, Accidents))$

Figure 7.30: 4ft-tables of rules (7.11) and (7.12).

We can conclude that the confidence of the rule (7.11) is equal to $\frac{108}{108+880} = 0,109$, which corresponds to the height of the first slim column in Fig. 7.27. The height of the first broad column corresponds to the percentage of the Fatal category which is 1,9%. We can conclude that the percentage of fatal accidents in the segment of accidents defined by the antecedent of the rule (7.11) is 5.7 times higher than the percentage of fatal accidents in the whole data matrix *Accidents*. This corresponds to the lift of the first rule in Fig. 7.29. An analogous conclusion is valid for the rule (7.12).

Note that a ratio of a height of a slim column (i.e., $\frac{a}{a+b}$) and a height of a broad column (i.e., $\frac{a+c}{a+b+c+d}$) corresponds to the lift $\frac{a(a+b+c+d)}{(a+b)(a+c)}$.

7.4.2   Exception rules—lowering columns of histogram

The rules (7.11) and (7.12) are exception rules pointing to circumstances causing an increase of columns of the histogram of the attribute Severity, see Fig. 7.27. We show that the 4ft-Miner procedure can be used also to find exception rules pointing to circumstances causing lowering of columns of the histogram of the attribute Severity. We use a run of the 4ft-Miner which differs from the run described in the previous section only such that we use 4ft-quantifier - a conjunction of 4ft-quantifiers $\sim_{\leq 0.6}$ (see equation (7.1) in Section 7.1.3) and $\oplus_{100}$ (see Table 7.1 in Section 7.1.2). This conjunction corresponds to the condition

$$\frac{a(a+b+c+d)}{(a+b)(a+c)} \leq 0.6 \land a \geq 100 . \tag{7.14}$$

This means that we are interested in rules pointing to segments of accidents lowering heights of columns of the histogram Severity such that the new height is maximally 0.6 of the height of the original column. Note that the lower lift $\frac{a(a+b+c+d)}{(a+b)(a+c)}$ of $\varphi \approx$ Severity(*category*), the more interesting rule $\varphi \approx$ Severity(*category*).

The run of the 4ft-Miner procedure with these parameters resulted in 17 true rules in 13 seconds. 19 554 rules were tested. The five rules with the lowest lift are displayed in a short form in Fig. 7.31.

A 4ft-table for the first rule

District(*Leeds*) $\land$ Type(*Bus_coach_17+*) $\Rightarrow_{0.09,126}$ Severity(*Serious*)   (7.15)

is available in Fig. 7.32. Here $Ant_{(7.15)}$ denotes the antecedent of the rule (7.15).

Nr.	Id	Lift	Hypothesis
1	3	0.436	**District**(*Leeds*) & **Type**(*Bus_coach_17+*) >+< **Severity**(*Serious*)
2	11	0.486	**District**(*Brighton and Hove*) & **Type**(*Bus_coach_17+*) >+< **Severity**(*Serious*)
3	1	0.514	**Driver_Age_Band**(*36 - 45*) & **Type**(*Bus_coach_17+*) >+< **Severity**(*Fatal*)
4	2	0.514	**Driver_Age_Band**(*46 - 55*) & **Type**(*Bus_coach_17+*) >+< **Severity**(*Fatal*)
5	13	0.533	**District**(*Exeter*) >+< **Severity**(*Serious*)

Figure 7.31: 4ft-Miner output - searching for segments lowering columns.

Accidents	Severity(*Serious*)	¬ Severity(*Serious*)
$Ant_{(7.15)}$	126	1 346
¬ $Ant_{(7.15)}$	105 628	431 889

Figure 7.32: 4ft-table of the rule (7.15).

We can conclude that the confidence of the rule (7.15) is equal to $\frac{126}{126+1346} = 0,086$. The height of the second broad column in Fig. 7.27 corresponds to the percentage of the *Serious* category which is 19,6%. We can conclude that the percentage of serious accidents in the segment of accidents defined by the antecedent of the rule (7.15) is 0.47 of the height of broad column corresponding the percentage of the Serious category in the whole data matrix *Accidents*. This corresponds to the lift of the first rule in Fig. 7.31.

7.4.3 Exception from exception—increasing confidence

The first four rules in Fig. 7.31 concern fatal or serious accidents of vehicles of type *Bus_coach_17+*. The confidences of these rules range from 0.010 to 0.095. All these rules are considered as exception rules pointing to circumstances causing lowering of columns of the histogram of the attribute Severity. Thus, it is natural to ask if there are some rules that can be considered as exceptions to these four rules. We will search for rules in a form

$$\text{Type}(\textit{Bus\_coach\_17+}) \wedge \mathscr{B}(\textit{Conditions})$$
$$\Rightarrow_{0.2,200} \text{Severity}(\alpha) \tag{7.16}$$

where $\mathscr{B}(\textit{Conditions})$ is a Boolean characteristic of the group Conditions of attributes introduced in Section 2.3.4. Severity(α) denotes Boolean attributes Severity(*Fatal*), Severity(*Fatal, Serious*), and Severity(*Serious*). Thus, we search for rules with more than twice higher confidence than our four rules with exceptionally low confidence.

We use two runs of the 4ft-Miner procedure. The set of relevant antecedents for both runs is defined by the $\mathscr{RC}$-expression Bus_Coach_17+_Condition available in a long tabular form in Table 7.11. Note that each relevant antecedent defined by the $\mathscr{RC}$-expression Bus_Coach_17+_Condition is a conjunction of

Table 7.11: $\mathscr{RC}$-expression Bus_Coach_17+_Condition.

Name: Bus_Coach_17+_Condition			Length: $1-7$		
Name: Bus_coach_17+		Type: *Conjunction*	Length: $1-1$		
Type	*One category*	*Bus_coach_17+*	*Pos*	*B*	−
Name: *Condition*		Type: *Conjunction*	Length: $1-6$		
Area	*Subsets*	$1-1$	*Pos*	*B*	−
Road_Type	*Subsets*	$1-1$	*Pos*	*B*	−
Speed_Limit	*Subsets*	$1-1$	*Pos*	*B*	−
Light	*Subsets*	$1-1$	*Pos*	*B*	−
Manoeuvre	*Subsets*	$1-1$	*Pos*	*B*	−
Vehicle_Location	*Subsets*	$1-1$	*Pos*	*B*	−

two partial cedents—a partial cedent Bus_coach_17+ which correspond to the Boolean attribute Type(*Bus_coach_17+*) and a partial cedent *Condition* which corresponds to $\mathscr{B}$(Conditions). In both runs, we use the 4ft-quantifier $\Rightarrow_{0.2,200}$ and the *Deleting* approach to missing information.

The set of relevant succedents in the first run is defined by the $\mathscr{RP}$-expression Severity_Lcut and by the $\mathscr{RP}$-expression Serious in the second run. Both $\mathscr{RP}$-expression are available in Fig. 7.33.

Name: Severity_Lcut		Type: *Conjunction*	Length: $1--1$		
Attribute	Coefficients	Length	Gace	B/R	Class
Severity	*Left cut*	$1-2$	*Pos*	*B*	−

Name: Serious		Type: *Conjunction*	Length: $1--1$		
Attribute	Coefficients	Length	Gace	B/R	Class
Severity	*One category*	*Serious*	*Pos*	*B*	−

Figure 7.33: $\mathscr{RP}$-expressions Severity_Lcut and Serious.

The first run of the 4ft-Miner procedure resulted in 25 true rules in less than one second. 1 854 rules were tested. A detailed output of the rule with the highest confidence is available in Fig. 7.34. Confidence of this rule is $\frac{582}{582+1829} = 0.241$.

Succedent of all output rules is in the form of Severity(*Fatal, Serious*), there is no succedent Severity(*Fatal*). All output rules contain the Boolean attribute Type(*Bus_coach_17+*) as required by the $\mathscr{RC}$-expression Bus_Coach_17+_Condition. Confidence of output rules ranges from 0.202 to 0.241. It holds:

■ 18 output rules contain Boolean attributes Manoeuvre(*Going ahead other*). Their confidences range from 0.215 to 0.240.

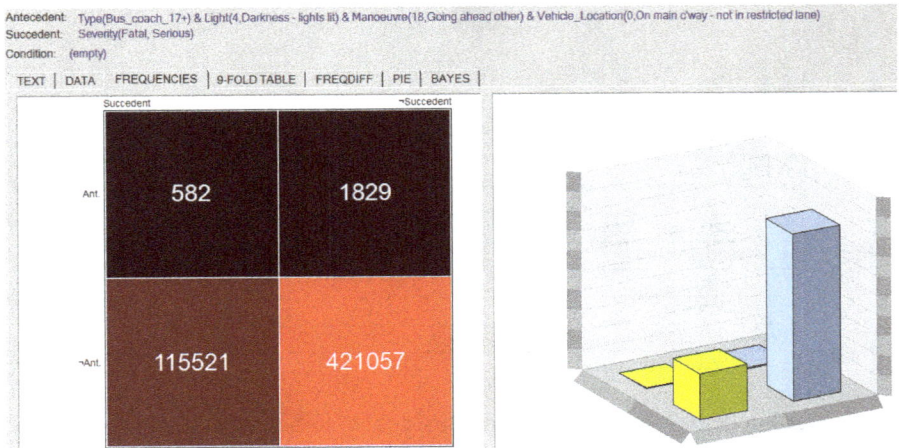

Figure 7.34: The strongest rule resulting from the first run of the 4ft-Miner.

■ Seven output rules contain Boolean attribute Manoeuvre(*Turning left*). Their confidences range from 0.202 to 0.211.

■ 16 output rules contain both Boolean attribute Manoeuvre(*Going ahead other*), and Light(*Darkness - lights lit*). Their confidences range from 0.236 to 0.240.

■ Eight output rules contain Boolean attributes Manoeuvre(*Going ahead other*), Light(*Darkness - lights lit*), and Area(*Urban*). Their confidences range from 0.236 to 0.240.

■ Four output rules contain all Boolean attributes Manoeuvre(*Going ahead other*), Light(*Darkness - lights lit*), Area(*Urban*), and Road_Type(*Single carriageway*). Their confidences range from 0.236 to 0.239.

The second run of the 4ft-Miner procedure resulted in 16 true rules in less than 3 seconds. 927 rules were tested. The confidence of the output rule ranges from 0.210 to 0.214. The structure of the output rules is similar to that of the first run.

Chapter 8

CF-Miner—Histograms

Most of the analytical software systems provide tools for producing histograms. However, to our best knowledge, there is no tool mining for conditional histograms informally introduced in Section 3.3. The goal of this chapter is to present the CF-Miner procedure which mines for various types of conditional histograms. The main features of the conditional histograms and of the CF-Miner procedure are introduced in Section 8.1.

Relatively simple examples of applications of the CF-Miner procedure are presented in Section 8.2. They concern the *Adult* dataset. The goal of these examples are to explain the main features and important properties of the CF-Miner procedure. Practically more interesting examples of applications of the CF-Miner procedure concern the *Accidents* dataset. They are available in Section 8.3. Additional examples are available in [137, 139].

8.1 CF-Miner and Related Notions

Important notions related to the CF-Miner are summarized in Section 8.1.1. An overview of the input and output of the CF-Miner procedure is given in Section 8.1.2. Generalised quantifiers used in the CF-Miner procedure are called *CF-quantifiers*. All the CF-quantifiers can be applied to a whole histogram or only to a part of a histogram called a *range of CF-quantifier*. A range of CF-quantifiers is introduced in Section 8.1.3. We use simple CF-quantifiers introduced in Section 8.1.4. We use also CF-quantifiers concerning steps in histograms described in Section 8.1.5. Let us note that there are also distribution CF-quantifiers based on distribution characteristics as arithmetic average, geometric average, nominal variation, variance, standard deviation and several additional. In addition, difference CF-quantifiers defined using conditions concerning differences of relevant histograms, and a given histogram are available.

8.1.1 Conditional histogram, CF-table and CF-pattern

A *conditional histogram* $A/[\chi, TypeHist]$ of an attribute A with categories $a_1, \ldots, a_t$ of a type $TypeHist$ at a data matrix $\mathcal{M}$ is a t-tuple $\langle h_1, \ldots, h_t \rangle$ of real numbers. Here χ is a Boolean attribute defining a sub-matrix of the data matrix $\mathcal{M}$. Particular items of t-tuple $\langle h_1, \ldots, h_t \rangle$ are given by a CF-table $CF(A, \mathcal{M}/\chi)$.

Recall that a CF-table $CF(A, \mathcal{M}/\chi)$ is a 2t-tuple of non-negative integers $\langle n_1, \ldots, n_t, m_1, \ldots, m_t \rangle$ introduced in Section 4.2. Here, n_i denotes the number of rows o of a data matrix $\mathcal{M}/\chi$ for which $A(o) = a_i$ and m_i denotes the number of rows o of a data matrix $\mathcal{M}$ for which $A(o) = a_i$. We assume $i = 1, \ldots, t$; see Section 4.2 and Fig. 8.1.

$\mathcal{M}$	a_1	$\ldots$	a_t
χ	n_1	$\ldots$	n_t
$True$	m_1	$\ldots$	m_t

Figure 8.1: CF-table $CF(A, \mathcal{M}/\chi) = \langle n_1, \ldots, n_t, m_1, \ldots, m_t \rangle$.

We understand a histogram $\langle h_1, \ldots, h_t \rangle$ of a type $TypeHist$ as a value of a function $Hist(CF(A, \mathcal{M}/\chi), TypeHist)$ introduced in Fig. 8.2. The function $Hist(CF(A, \mathcal{M}/\chi), TypeHist)$ has two arguments – a CF-table $CF(A, \mathcal{M}/\chi)$ according to Fig. 8.1 and a string $TypeHist$ with possible values *Abs*, *Proc*, and *Cat*. Note that it holds $n = n_1 + \cdots + n_t$.

A *CF-pattern* is an expression $\approx A/\chi$ where A is an attribute with t categories, χ is a Boolean attribute and a CF-quantifier $\approx$ is a condition concerning all possible CF-tables $CF(A, \mathcal{M}/\chi)$. Note that all such possible CF-tables are all possible 2t-tuple of non-negative integers $\langle n_1, \ldots, n_t, m_1, \ldots, m_t \rangle$ such that $n_i \geq m_i$ for $i = 1, \ldots t$.

A CF-pattern $\approx A/\chi$ is true in a data matrix $\mathcal{M}$ if a condition given by the CF-quantifier $\approx$ is satisfied for a CF-table $CF(A, \mathcal{M}/\chi)$.

$TypeHist$	$Hist(CF(A, \mathcal{M}/\chi), TypeHist)$ h_1	$\ldots$	h_t	Σ
Abs	n_1	$\ldots$	n_t	n
$Proc$	$100 * \frac{n_1}{n}$	$\ldots$	$100 * \frac{n_t}{n}$	100
Cat	$100 * \frac{n_1}{m_1}$	$\ldots$	$100 * \frac{n_t}{m_t}$	$-$

Figure 8.2: Heights $\langle h_1, \ldots, h_t \rangle$ of columns of a histogram.

Remark 8.1   We deal only with histograms $A/[\chi, Abs]$ of type *Abs*. Thus, we further usually write only A/χ instead of $A/[\chi, Abs]$. In addition, if there is not a danger of misunderstanding, we identify a CF-pattern $\approx A/\chi$ with a corresponding histogram A/χ.   ■

8.1.2   CF-Miner input and output

Input of the CF-Miner procedure consists of

- a categorical attribute A with t categories

- an $\mathscr{RC}$-expression *Cond* defining a set of relevant conditions χ

- a list *ListCFQ* of basic CF-quantifiers relevant to the attribute A

- a data matrix $\mathscr{M}$.

A set of relevant CF-patterns is defined as a set of all patterns $\approx A/\chi$ satisfying $\chi \in Cond$. CF-quantifier $\approx$ is defined as a conjunction of basic CF-quantifiers introduced in the list *ListCF*.

The CF-Miner procedure generates all relevant CF-patterns and verifies them in a given data matrix $\mathscr{M}$. Each generated CF-pattern $R \approx C/\chi$ is verified such that the CF-quantifier $\approx$ is applied to a CF-table $CF(A, \mathscr{M}/\chi)$. A condition related to a CF-quantifier $\approx$ is defined as a conjunction of all conditions corresponding to particular basic CF-quantifiers from the list *ListCFQ*. CF-patterns satisfying the condition given by the CF-quantifier $\approx$ are listed in an output.

8.1.3   Range of CF-quantifiers

A CF-quantifier $\approx$ defines a condition concerning a histogram $\langle h_1, \ldots, h_t \rangle$. This condition can be applied only to a part of a histogram concerning several consecutive categories $a_u, \ldots, a_v$. A range of a CF-quantifier $\approx$ for a histogram $\langle h_1, \ldots, h_t \rangle$ is given by a couple $\langle u, v \rangle$ of integers satisfying $1 \leq u < v \leq t$. This couple is called a *definition of a range of CF-quantifier*.

A *range $\langle u, v \rangle$ of a CF-quantifier $\approx$ for a histogram* $\langle h_1, \ldots, h_t \rangle$ is a vector $\langle h_u, \ldots, h_v \rangle$. Examples of applications of the range of CF-quantifiers are available, among others, in Section 8.2.3.

8.1.4   Simple frequencies CF-quantifiers

We use simple frequencies CF-quantifiers in a form

$$\langle \text{CFSF: } Sum, u, v, Abs, \circledast, Thr \rangle \tag{8.1}$$

where $\langle u, v \rangle$ is a definition of a range of CF-quantifier for a histogram $\langle h_1, \ldots, h_t \rangle$, $\circledast$ is one of the relations $<$, $\leq$, $=$, $\geq$, $>$ concerning real numbers, and Thr is a non-negative real number. A simple frequencies CF-quantifier

⟨CFSF: *Sum,u,v,Abs,⊛,Thr*⟩ defines a condition

$$n_u + \cdots + n_v \circledast Thr \tag{8.2}$$

concerning a CF-table $CF(A, \mathcal{M}/\chi) = \langle n_1, \ldots, n_t, m_1, \ldots, m_t \rangle$, see Fig. 8.1. We say that a *CF-quantifier* ⟨*CFSF: Sum,u,v,Abs,⊛,Thr*⟩ *is relevant for an attribute A with t categories* if $v \le t$.

A condition related to a simple frequencies CF-quantifier (8.1) is satisfied for a CF-table $CF(A, \mathcal{M}/\chi)$ if a relation (8.2) is valid. Let us note that there are additional simple frequencies CF-quantifiers concerning conditions related to minimum, maximum, and average values computed from histograms of all types *Abs*, *Proc* and *Cat*.

8.1.5 CF-quantifiers concerning steps in histogram

A *step in a histogram* $\langle h_1, \ldots, h_t \rangle$ is each couple $\langle h_i, h_{i+1} \rangle$ where $1 \le i \le t - 1$. A *height of a step* $\langle h_i, h_{i+1} \rangle$ *in a histogram* is defined as $|h_{i+1} - h_i|$. A *steps CF-quantifier* is an expression

$$\langle CFSt: \ Dir,u,v,MinH,Cons,TypeHist,\circledast,Thr\rangle \tag{8.3}$$

where *Dir* is one of the strings *Up, Down*; $\langle u, v \rangle$ is a definition of a range of CF-quantifier for a histogram $\langle h_1, \ldots, h_t \rangle$; *MinH* is a real non-negative number defining a minimal height of steps; *Cons* is one of the strings *Yes, No*; *TypeHist* is one of the types *Abs, Proc, Cat*; ⊛ is one of the relations $<, \le, =, \ge, >$ concerning integers; *Thr* is a non-negative integer. We say that a *CF-quantifier* ⟨*CFSt: Dir,u,v,MinH,Cons,TypeHist,⊛,Thr*⟩ *is relevant for an attribute A with t categories* if $v \le t$.

A steps CF-quantifier ⟨CFSt: *Dir,u,v,MinH,Cons,TypeHist,⊛,Thr*⟩ relevant for an attribute A defines a condition

$$StCF(Dir,u,v,MinH,Cons,\mathcal{H}) \circledast Thr \tag{8.4}$$

for a given CF-table $CF(A, \mathcal{M}/\chi)$. Here $\mathcal{H} = \langle h_1, \ldots, h_t \rangle$ is a histogram assigned to a CF-table $CF(A, \mathcal{M}/\chi)$ according to Fig. 8.2; *TypeHist* given by the steps CF-quantifier in question is used. $StCF(Dir, u, v, MinH, Cons, \mathcal{H})$ is a function defined in Fig. 8.3. It assigns the number of steps in $\mathcal{H} = \langle h_1, \ldots, h_t \rangle$ satisfying the conditions given by the parameters *Dir,u,v,MinH,Cons*. The function $StCF(Dir,u,v,MinH,Cons,\mathcal{H})$ concerns steps in a range $\langle u, v \rangle$ of a histogram $\mathcal{H} = \langle h_1, \ldots, h_t \rangle$. The steps are defined this way:

■ A couple $\langle h_i, h_{i+1} \rangle$ is a *step up of a minimal height MinH in a range* $\langle u, v \rangle$ *of a histogram* $\langle h_1, \ldots, h_t \rangle$ if it holds $h_{i+1} - h_i \ge MinH$ and $u \le i \le v - 1$.

Dir	Cons	Value of $StCF(Dir,u,v,MinH,Cons,\mathcal{H})$
Up	No	the number of steps up of a minimal height $MinH$ in a range $\langle u,v \rangle$ of a histogram $\mathcal{H} = \langle h_1,\ldots,h_t \rangle$
Up	Yes	the length of the longest sequence of steps up of a minimal height $MinH$ in a range $\langle u,v \rangle$ of a histogram $\langle h_1,\ldots,h_t \rangle$
Down	No	the number of steps down of a minimal height $MinH$ in a range $\langle u,v \rangle$ of a histogram $\mathcal{H} = \langle h_1,\ldots,h_t \rangle$
Down	Yes	the length of the longest sequence of steps down of a minimal height $MinH$ in a range $\langle u,v \rangle$ of a histogram $\mathcal{H} = \langle h_1,\ldots,h_t \rangle$

Figure 8.3: Values of $StCF(Dir,u,v,MinH,Cons,\mathcal{H})$.

■ A couple $\langle h_i, h_{i+1} \rangle$ is a *step down of a minimal height MinH in a range* $\langle u,v \rangle$ *of a histogram* $\langle h_1,\ldots,h_t \rangle$ if it holds $h_i - h_{i+1} \geq MinH$ and $u \leq i \leq v-1$.

■ If $Seq = \langle \ \langle h_k,h_{k+1} \rangle, \langle h_{k+1},h_{k+2} \rangle, \ldots, \langle h_{k+w},h_{k+w+1} \rangle \ \rangle$ is a sequence of steps up or steps down in a range $\langle u,v \rangle$ of a histogram $\langle h_1,\ldots,h_t \rangle$, then *w is a length of Seq.*

A condition related to a steps CF-quantifier (8.3) is satisfied for a CF-table $CF(A,\mathcal{M}/\chi)$ if a relation (8.4) is valid.

8.2 Applying CF-Miner in *Adult* Dataset

The goal of this section is to explain the main features of the CF-Miner procedure. We use relatively simple examples concerning the well known *Adult* dataset. We start with histograms of attributes age_5 and age_exp presented in Fig. 8.4. These histograms concern the whole data matrix *Adult*.

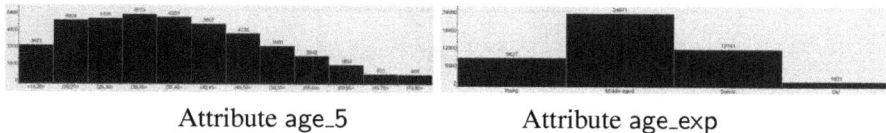

Attribute age_5   Attribute age_exp

Figure 8.4: Histograms of attributes age_5, and age_exp.

We see that both histograms are first increasing and then decreasing. Our goal is to find sub-matrices of the *Adult* data matrix at which histograms of attributes have a different shape. First, we show an example of searching sub-matrices at which histograms of the age_exp attribute increase, see Section 8.2.1. In Section 8.2.2, an example of searching for decreasing histograms is introduced. Mining for first decreasing and then increasing conditional histograms is presented in Section 8.2.3.

8.2.1 Increasing histograms

The conditional histogram $\mathcal{H}_{exmp} = \text{age\_exp}/Cond_{exmp}$ available in Fig. 8.5 is an example of increasing histogram. Here it holds

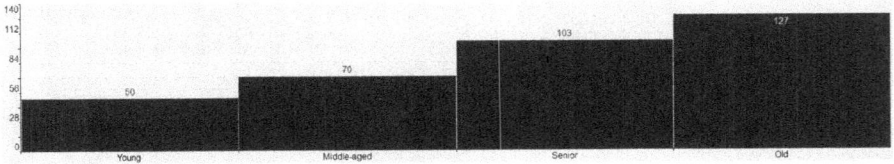

Figure 8.5: Example of increasing conditional histogram.

$$Cond_{exmp} = \text{age\_exp}/(\text{education}(HS\text{-}grad) \wedge \\ \wedge \text{hours-per-week}(Part\text{-}time) \wedge \text{relationship}(Not\text{-}in\text{-}family)) . \quad (8.5)$$

The histogram shows frequencies of particular categories of the attribute age_exp for a sub-matrix $Adult/Cond_{exmp}$. This histogram corresponds to a CF-table $CF(\text{age\_exp}, Adult/Cond_{exmp})$ introduced in Fig. 8.6. We see in the second column of the second row, that there are 50 young persons satisfying $Cond_{exmp}$ and 9 627 young persons in the whole data matrix $Adult$. This is the same for additional categories of the age_exp attribute.

We are interested in increasing histograms $\text{age\_exp}/\chi$ of the attribute age_exp in the $Adult$ dataset. Let us use a CF-table $CF(\text{age\_exp}, Adult/\chi)$ shown in Fig. 8.7.

Adult	Young	Middle-aged	Senior	Old
$Cond_{exmp}$	50	70	103	127
True	9 627	24 671	12 741	1 803

Figure 8.6: CF-table $CF(\text{age\_exp}, Adult/Cond_{exmp})$.

Adult	Young	Middle-aged	Senior	Old
χ	n_1	n_2	n_3	n_4
True	9 627	24 671	12 741	1 803

Figure 8.7: CF-table $CF(\text{age\_exp}, Adult/\chi)$.

We deal with histograms of type *Abs* which means that the heights of columns correspond to frequencies of particular categories in the data sub-matrix $Adult/\chi$. Thus, we are interested in histograms – quadruples $\langle n_1, n_2, n_3, n_4 \rangle$. We use the notion of a *step in histogram* introduced in Section 8.1.5 to formalize a fact that a histogram is increasing. Recall that a step-up of a minimal height 20 is each couple $\langle n_i, n_{i+1} \rangle$ such that $n_{i+1} - n_i \geq 20$ where $1 \leq i \leq 3$. Note that all steps in histogram at Fig. 8.5 are up and their minimal height is 20. This can be verified in the CF-table $CF(\text{age\_exp}, Adult/Cond_{exmp})$ shown in Fig. 8.6.

Let us further specify our task such that we are interested in increasing the conditional histograms age_exp/χ of the attribute age_exp in the *Adult* data matrix where a minimal difference of heights of columns is 20. This can be specified by a steps CF-quantifier $\approx_{Up3}$ defined as

$$\approx_{Up3} = \langle \text{CFSt: } Up, 1, 4, 20, Yes, Abs, =, 3 \rangle, \tag{8.6}$$

see Section 8.1.5. The CF-quantifier (8.6) defines a condition

$$StCF(Up, 1, 4, 20, Yes, \mathcal{H}_{Task}) = 3 \tag{8.7}$$

according to equation (8.4). Here $\mathcal{H}_{Task} = \langle n_1, n_2, n_3, n_4 \rangle$ is a histogram assigned to a CF-table $CF(\text{age\_exp}, Adult/\chi)$. Note that the type of histogram $\mathcal{H}_{Task}$ is *Abs*. A value of the function $StCF(Up, 1, 4, 20, Yes, \mathcal{H}_{Task})$ is defined by the second row of the table in Fig. 8.3 as the length of the longest sequence of steps up of a minimal size 20 in a range $\langle 1, 4 \rangle$ of the histogram $\mathcal{H}_{Task} = \langle n_1, n_2, n_3, n_4 \rangle$. Let us note that there are several equivalent forms of the CF-quantifier (8.6).

An additional requirement in the histograms age_exp/χ we are interested in, is that there are at least 300 of rows of data matrix *Adult* satisfying χ. This can be expressed by a simple frequencies CF-quantifier $\approx_{Sum300}$ defined as

$$\approx_{Sum300} = \langle \text{CFSF: } Sum, 1, 4, Abs, \geq, 300 \rangle, \tag{8.8}$$

see Section 8.1.4. This CF-quantifier defines a condition

$$n_1 + n_2 + n_3 + n_4 \geq 300 . \tag{8.9}$$

We can summarize that we use a CF-quantifier $\approx$ given by a list $ListCFQ = [\approx_{Up3}, \approx_{Sum300}]$ of two basic CF-quantifiers. This list defines a CF-quantifier

$$\approx_{Up3Sum300} = \approx_{Up3} \wedge \approx_{Sum300} \tag{8.10}$$

which corresponds to a condition

$$StCF(Up, 1, 4, 20, Yes, \mathcal{H}_{Task}) = 3 \wedge n_1 + n_2 + n_3 + n_4 \geq 300 . \tag{8.11}$$

We use conditions χ in a usual form

$$\mathcal{B}(\text{Personal}) \wedge \mathcal{B}(\text{Family}) \wedge \mathcal{B}(\text{Society}) \wedge \mathcal{B}(\text{Employment}) \wedge \mathcal{B}(\text{Capital})$$

where $\mathcal{B}(\text{Personal})$, $\mathcal{B}(\text{Family})$, $\mathcal{B}(\text{Society})$, $\mathcal{B}(\text{Employment})$, and $\mathcal{B}(\text{Capital})$, are Boolean characteristics of the groups of attributes Personal, Family, Society, Employment, and Capital. We use an $\mathcal{RP}$-expression Personal_2 defining a set of Boolean characteristics $\mathcal{B}(\text{Personal})$ of the group of attributes Personal presented in a tabular form in Fig. 8.8. Sets of Boolean characteristics

Name: Personal_2		Type: *Conjunction*		Length: $0 - 3$	
Attribute	Coefficients	Length	Gace	B/R	Class
sex	*Subsets*	$1 - 1$	*Pos*	B	—
education	*Sequences*	$1 - 4$	*Pos*	B	—

Name: Capital		Type: *Conjunction*		Length: $0 - 3$	
Attribute	Coefficients	Length	Gace	B/R	Class
capital_gain	*Subsets*	$1 - 1$	*Pos*	B	—
capital_loss	*Subsets*	$1 - 1$	*Pos*	B	—

Figure 8.8: $\mathscr{R}\mathscr{P}$-expressions Personal_2 and Capital.

Name: $\mathscr{H}\mathscr{C}\mathscr{O}\mathscr{N}\mathscr{D}$	Length: $1 - 13$
Set of partial cedents	Length
Personal_2	$0 - 3$
Family	$0 - 2$
Society	$0 - 3$
Employment	$0 - 3$
Capital	$0 - 2$

Figure 8.9: $\mathscr{R}\mathscr{C}$-expression $\mathscr{H}\mathscr{C}\mathscr{O}\mathscr{N}\mathscr{D}$.

$\mathscr{B}$(Family), $\mathscr{B}$(Society), and $\mathscr{B}$(Employment) of the groups of attributes Family, Society, and Employment are defined by the $\mathscr{R}\mathscr{P}$-expressions Family, Society, and Employment presented in tabular forms in Fig. 7.13. The set $\mathscr{B}$(Capital) of Boolean characteristics of the group Capital of attributes is defined by the $\mathscr{R}\mathscr{P}$-expression Capital available in Fig. 8.8. A set of relevant conditions is given by an $\mathscr{R}\mathscr{C}$-expression $\mathscr{H}\mathscr{C}\mathscr{O}\mathscr{N}\mathscr{D}$ introduced in Fig. 8.9.

A run of the CF-Miner procedure with the above introduced parameters resulted in 31 true CF-patterns in 117 seconds. More than $1.7 * 10^6$ CF-patterns were tested. The CF-expression

$$\approx_{Up3Sum300} \text{age\_exp}/Cond_{exmp}$$

is an example of an output pattern. The CF-table $CF(\text{age\_exp}, Adult/Cond_{exmp})$ is available in Fig. 8.6 and the conditional histogram age_exp/$Cond_{exmp}$ is available in Fig. 8.5. The set of 31 output CF-patterns can be divided into two subsets. The first subset consists of two CF-patterns:

■ the above introduced $\approx_{Up3Sum300} \text{age\_exp}/Cond_{exmp}$

■ $\approx_{Up3Sum300} \text{age\_exp}/Cond_{exmp} \wedge \text{capital\_loss}(None)$, a histogram age_exp/$Cond_{exmp} \wedge \text{capital\_loss}(None)$ has a shape similar to that of $\approx_{Up3Sum300} \text{age\_exp}/Cond_{exmp}$ in Fig. 8.5.

The second subset consists of 29 CF-patterns (we write *"Part-time"* instead of "hours-per-week(Part-time)"; similarly for *"Widowed"*):

■ a CF-pattern $\approx_{Up3Sum300}$ age_exp/*Part-time* $\wedge$ *Widowed*

■ 28 CF-patterns $\approx_{Up3Sum300}$ age_exp/*Part-time* $\wedge$ *Widowed* $\wedge$ ω, where ω is a conjunction of 1–4 basic Boolean attributes.

A conditional histogram $\mathcal{H}_{PartWid}$ = age_exp/*Part-time* $\wedge$ *Widowed* is available in the left part of Fig. 8.10. Note that its shape is different from a shape of a histogram $\mathcal{H}_{Widowed}$ = age_exp/marital_status(*Widowed*) available in the right part of Fig. 8.10. A shape of each of 28 histograms age_exp/*Part-time* $\wedge$ *Widowed* $\wedge$ ω is similar to the shape of the histogram $\mathcal{H}_{PartWid}$.

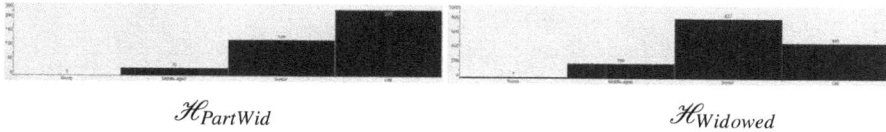

$\mathcal{H}_{PartWid}$ $\qquad\qquad\qquad\qquad$ $\mathcal{H}_{Widowed}$

Figure 8.10: Histograms $\mathcal{H}_{PartWid}$ and $\mathcal{H}_{Widowed}$.

Let us emphasize that the number of output histograms (given by output CF-patterns) strongly depends on the minimal height of steps and on the minimal number of rows satisfying the condition χ. Minimal height of steps is given by parameter *MinH* of the steps CF-quantifier $\langle$CFSt: *Dir, u, v, MinH, Cons, TypeHist*, $\circledast$, *Thr*$\rangle$ The minimal number of rows satisfying the condition χ is given by a value of the parameter *Thr* of the simple frequencies quantifier $\langle$CFSF: *Sum, u, v, Abs*, $\circledast$, *Thr*$\rangle$.

The numbers of output increasing histograms for several combinations of minimal height *MinH* of steps and minimal number *Sum* of rows satisfying χ are available in Table 8.1. Note that decreasing minimal *MinH* means that the number of verifications increases.

Table 8.1: The numbers of output histograms for minimal heights and *Sum*.

Minimal *Sum*	Minimal *MinH*	Verifications	Time of solution	# of histograms
500	20	1 079 694	73 s	0
500	10	1 079 694	73 s	16
400	20	1 327 666	89 s	6
400	10	1 327 666	89 s	22
300	20	1 719 055	117 s	31
300	10	1 719 055	117 s	67

8.2.2 Decreasing histograms

We mine with decreasing histograms of the attribute age_exp in this section. We use a simple CF-quantifier $\approx_{Down3}$ defined as

$$\approx_{Down3} \;=\; \langle \text{CFSt: } Down, 1, 4, 50, Yes, Abs, =, 3 \rangle. \tag{8.12}$$

We use $MinH = 50$ in $\approx_{Down3}$ instead of $MinH = 20$ in $\approx_{Up3}$ introduced in (8.6). We use a simple CF-quantifier $\approx_{Sum5000}$ defined as

$$\approx_{Sum5000} \;=\; \langle \text{CFSF: } Sum, 1, 4, Abs, \geq, 5000 \rangle, \tag{8.13}$$

instead of a simple frequencies CF-quantifier (8.8). We can summarize that we use a CF-quantifier $\approx$ given by a list $ListCFQ = [\approx_{Down3}, \approx_{Sum5000}]$ of two basic CF-quantifiers. This list defines a CF-quantifier

$$\approx_D \;=\; \approx_{Down3} \;\wedge\; \approx_{Sum5000} \tag{8.14}$$

which corresponds to a condition

$$StCF(Down, 1, 4, 50, Yes, \mathcal{H}_{Task}) = 3 \;\wedge\; n_1 + n_2 + n_3 + n_4 \geq 5000 \,. \tag{8.15}$$

A set of relevant conditions χ is defined in the same way as for the increasing histograms. This means that relevant conditions are given by the $\mathscr{RC}$-expression $\mathscr{HCOND}$ introduced in Fig. 8.9.

A run of the CF-Miner procedure with the above introduced parameters resulted in 621 true CF-patterns in 12 seconds. More than $46 * 10^3$ CF-patterns were tested. The set of 621 output patterns have the following important features. (We write *"Never-married"* instead of "marital(*Never-married*)", etc.)

- There are CF-patterns $\approx_D$ age_exp/*Never-married*, $\approx_D$ age_exp/*Own-child*, and $\approx_D$ age_exp/*Part-time* with CF-tables according to Fig. 8.11.

Adult	Young	Middle-aged	Senior	Old	Σ
marital(*Never-married*)	8 229	6 851	924	113	16 117
relationship(*Own-child*)	5 230	2 122	217	12	7 581
hours-per-week(*Part-time*)	2 819	1 272	978	844	5 913
True	9 627	24 671	12 741	1 803	48 842

Figure 8.11: CF-tables of three above introduced CF-patterns.

- Each output CF-pattern contains exactly one basic Boolean attribute from the attributes marital(*Never-married*), relationship(*Own-child*), and hours-per-week(*Part-time*).

Histogram age_exp/*Never-married*

Histogram age_exp/*Own-child*

Histogram age_exp/*Part-time*

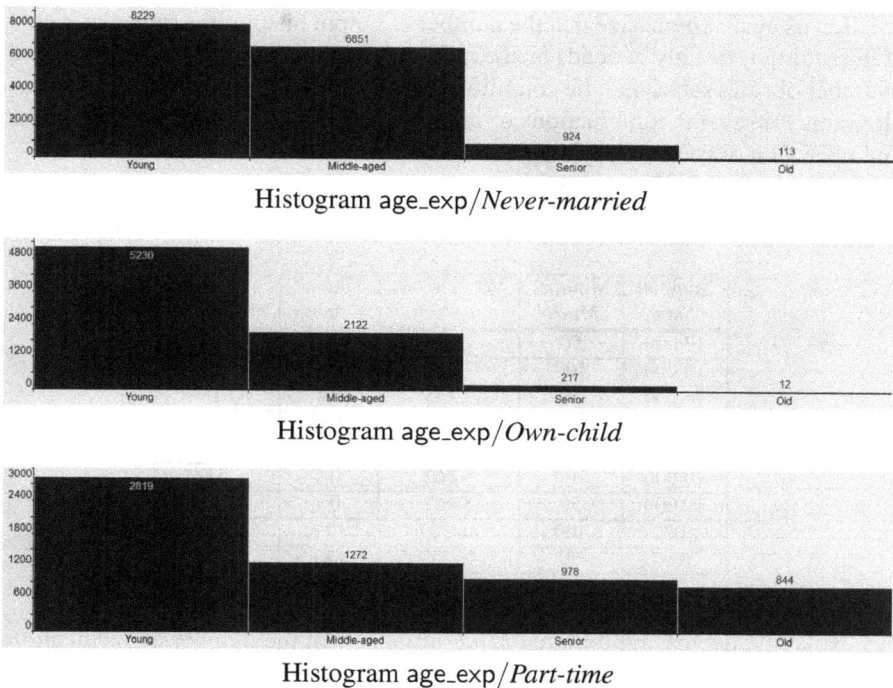

Figure 8.12: Three above introduced histograms.

- There are 570 CF-patterns $\approx_D$ age_exp/*Never-married* $\land$ α. A shape of histograms age_exp/*Never-married* $\land$ α is similar to that of age_exp/*Never-married* available in Fig. 8.12.

- There are 41 CF-patterns $\approx_D$ age_exp/*Own-child* $\land$ β. A shape of histograms age_exp/*Own-child* $\land$ β is similar to that of age_exp/*Own-child* available in Fig. 8.12.

- There are 7 CF-patterns of the form $\approx_D$ age_exp/*Part-time* $\land$ γ. A shape of histograms age_exp/*Part-time* $\land$ γ is similar to that of age_exp/*Part-time* available in Fig. 8.12.

A question is if, under some conditions, a validity of CF-patterns

$$\approx_D \text{ age\_exp}/\textit{Never-married} \land \alpha$$

cannot be deduced from a validity of CF-pattern

$$\approx_D \text{ age\_exp}/\textit{Never-married} .$$

Similar questions arise for CF-patterns $\approx_D$ age_exp/*Own-child* $\land$ β and $\approx_D$ age_exp/*Part-time* $\land$ γ, see Section 17.3.

Let us again emphasize that the number of output histograms (given by output CF-patterns) strongly depends on the minimal height of steps and on the minimal number of rows satisfying the condition χ. The numbers of output increasing histograms for several combinations of minimal height *MinH* of steps and minimal number *Sum* of rows satisfying χ are available in Table 8.2.

Table 8.2: The numbers of decreasing histograms for minimal heights and *Sum*.

Minimal *Sum*	Minimal *MinH*	Verifications	Time of solution	# of histograms
17 000	50	1 426	0 s	0
16 000	100	2 135	0 s	1
16 000	50	2 135	0 s	1
15 000	100	2 591	0 s	3
15 000	50	2 591	0 s	3
10 000	100	8 162	0 s	31
10 000	50	8 162	0 s	31
5000	100	46 377	12 s	599
5000	50	46 377	12 s	621

Note that decreasing minimal *MinH* implies that the number of verifications increases. The value 0 in the column "Time of solution" means that the time of solution was less than 0.5 seconds.

8.2.3 First decreasing and then increasing histograms

The goal of this section is to demonstrate possibilities of application of ranges of CF-quantifiers. Let us emphasize that real practical interestingness of results is not the main goal. However, particular results can be interesting as a "part of a mosaic". We start with a histogram age_5/*All* showing the numbers of persons for particular categories of attribute age_5, see Fig. 8.13.

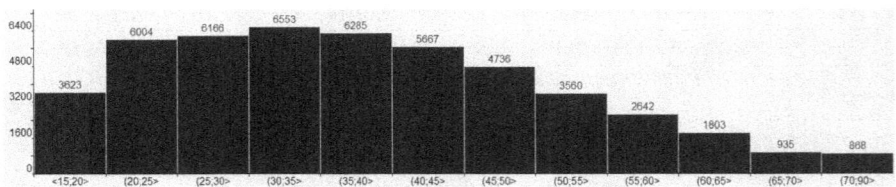

Figure 8.13: Histogram age_5/*All*.

This histogram first increases (first three steps are up) and then decreases (last eight steps are down). The question is whether a sub-matrix *Adult*/χ of the *Adult* data matrix exists such that histogram age_5/χ first decreases and then increases. Attribute age_5 has 12 categories. A conditional histogram age_5/χ is thus a 12-tuple $\mathscr{H} = \langle h_1, \ldots, h_{12} \rangle$ of integer numbers, see Section 8.1.1. There are 11 steps in such a histogram. This histogram is considered as first decreasing and then increasing if the first r steps are down, and the remaining steps are up.

First, we assume that $1 \leq r \leq 10$ – there is at least one step down at the beginning of a histogram and at least one step up at the end of a histogram. We will deal with steps of minimal height 1. The fact that first r steps in an age_5/χ histogram are down can be expressed by a steps CF-quantifier

$$\approx_{D,r} = \langle \text{CFSt: } Down, 1, r+1, 1, Cons, Abs, =, r \rangle . \tag{8.16}$$

A steps CF-quantifier (8.16) defines a condition

$$StCF(Down, 1, r+1, 1, Cons, \mathscr{H}) = r \tag{8.17}$$

saying that $h_1 - h_2 \geq 1, \ldots, h_r - h_{r+1} \geq 1$. This really means that first r steps in histogram $\mathscr{H} = \langle h_1, \ldots, h_{12} \rangle$ are down with minimal height 1. The fact that the remaining steps are up can be expressed by a steps CF-quantifier

$$\approx_{U,r} = \langle \text{CFSt: } Up, r+1, 12, 1, Cons, Abs, =, 11-r \rangle , \tag{8.18}$$

which defines a condition

$$StCF(Up, r+1, 12, 1, Cons, \mathscr{H}) = 11 - r \tag{8.19}$$

saying that $h_{r+2} - h_{r+1} \geq 1, \ldots, h_{12} - h_{11} \geq 1$. This means that last $11 - r$ steps are up with minimal height 1. A steps CF-quantifier $\approx_{D,U,r}$ defined as

$$\approx_{D,U,r} = \approx_{D,r} \wedge \approx_{U,r} , \tag{8.20}$$

means that first r steps in a histogram $\mathscr{H} = \langle h_1, \ldots, h_{12} \rangle$ are down and that the remaining steps are up for $1 \leq r \leq 10$.

CF-quantifiers $\approx_{D,U,0} = \langle \text{CFSt: } Up, 1, 12, 1, Cons, Abs, =, 11 \rangle$ and $\approx_{D,U,11} = \langle \text{CFSt: } Down, 1, 12, 1, Cons, Abs, =, 11 \rangle$ meaning that first 0 steps are down (i.e., all steps are up) and first 11 steps are down, i.e., all steps are down respectively will also be used. We can conclude that the quantifier $\approx_{D,U,r}$ expresses that first r steps are down and that the remaining steps are up for $0 \leq r \leq 11$.

Finally, we use a simple CF-quantifier $\approx_{Sum200}$ defined as

$$\approx_{Sum200} = \langle \text{CFSF: } Sum, 1, 12, Abs, \geq, 200 \rangle. \tag{8.21}$$

We can summarize that we use a CF-quantifier $\approx_r$ defined as

$$\approx_r = \approx_{D,U,r} \wedge \approx_{Sum200} . \tag{8.22}$$

A set of relevant conditions χ is defined in the same way as for the increasing and decreasing histograms in Sections 8.2.1 and 8.2.2. This means that relevant conditions are given by the $\mathcal{RC}$-expression $\mathcal{HCOND}$ introduced in Fig. 8.9.

We have run all 12 CF-Miner applications with CF-quantifiers $\approx_r$ for $r = 1, \ldots, 12$. However, for $r = 0, \ldots, 5$, there are no histograms satisfying specified conditions. Thus, we redefined CF-quantifiers $\approx_r$ for $r = 0, \ldots, 5$ such that a minimal height of a step was set to 0. This means that a steps CF-quantifier $\approx_{D,r}$ is for $r = 0, \ldots, 5$ defined not by the CF-quantifier (8.16) but by the CF-quantifier

$$\approx_{D,r} = \langle \text{CFSt: } Down, 1, r+1, 0, Cons, Abs, =, r \rangle, \tag{8.23}$$

saying that $h_1 \geq h_2, \ldots, h_r \geq h_{r+1}$ (i.e., *not increasing steps*). A steps CF-quantifier $\approx_{U,r}$ is for $r = 0, \ldots, 5$ defined not by the CF-quantifier (8.18) but by the CF-quantifier

$$\approx_{U,r} = \langle \text{CFSt: } Up, r, 12, 0, Cons, Abs, =, 11-r \rangle, \tag{8.24}$$

saying that $h_{r+1} \geq h_r, \ldots, h_{12} \geq h_{11}$.

The time of solutions of particular runs ranges from 119 to 161 seconds depending on the number of found CF-pattern. This is because a communication with metabase when storing found CF-patterns. 662 995 CF-patterns were tested in each particular run. Results of runs for $r = 0, \ldots, 5$ are available in Table 8.3. The numbers of decreasing and increasing steps are introduced together with the number of found CF-patterns. In addition, an example of resulting CF-pattern $\approx_r$ age_5/χ is presented together with the number of rows *Sum* satisfying the condition χ. Histograms age_5/χ corresponding to examples of resulting CF-patterns are available in Fig. 8.14.

Note that we often write "*Male*" instead of "sex(*Male*)", "*United-States*" instead of "native-country(*United-States*)", "*Widowed*" instead of "marital-status(*Widowed*)", "*Not-in-family*" instead of "relationship(*Not-in-family*)", "*Own-child*" instead of "relationship(*Own-child*)", "*Part-time*" instead of "hours-per-week(*Part-time*)", "*White*" instead of race(*White*)", etc.

Several comments to the presented results follow.

Down-0,Up-11: All nine histograms with 11 steps up contain basic Boolean attribute marital-status(*Widowed*). Four of these histograms $\langle h_1, \ldots, h_{12} \rangle$ satisfy $h_1 = h_2 = h_3 = h_4 = 0$.

Down-1,Up-10: All six CF-patterns with 10 steps up contain basic Boolean attribute marital-status(*Widowed*). Histogram age_5/(*Male* $\wedge$ *United-States* $\wedge$ *Widowed*) from the previous point also satisfies conditions for this point. Again, four of these six histograms $\langle h_1, \ldots, h_{12} \rangle$ satisfy $h_1 = h_2 = h_3 = h_4 = 0$.

Down-2,Up-9: All four CF-patterns with nine steps up are among six CF-patterns with 10 steps up with minimal height 0. Thus, the same example

Table 8.3: Mining for histograms starting with 0–5 steps down.

# of steps		# of CF-	Example of resulting $\approx_r$ age_5/χ	
Down	Up	patterns	Condition χ	*Sum*
0	11	9	*Male $\wedge$ United-States $\wedge$ Widowed*	262
1	10	6	*United-States $\wedge$ Widowed $\wedge$* *$\wedge$ Not-in-family $\wedge$ Part-time*	282
2	9	4	*United-States $\wedge$ Widowed $\wedge$* *$\wedge$ Not-in-family $\wedge$ Part-time*	282
3	8	8	*Male $\wedge$ education(9th,10th, 11th)$\wedge$ White $\wedge$* *$\wedge$ capital_gain(None) $\wedge$ Part-time*	390
4	7	22	occupation(*Handlers-cleaners*) $\wedge$ *Part-time*	300
5	6	83	education(*9th,10th,11th,12th*) $\wedge$ *$\wedge$ United-States $\wedge$ Newer-married $\wedge$* *$\wedge$ relationship(Own-child) $\wedge$ Part-time*	695

of a CF-pattern is introduced. However, the corresponding histogram is presented only once in Fig. 8.14. All four histograms $\langle h_1, \ldots, h_{12} \rangle$ satisfy $h_1 = h_2 = h_3 = h_4 = 0$; see also second histogram in Fig. 8.14.

Down-3,Up-8: All eight CF-patterns with 8 steps up contain basic Boolean attribute hours-per-week(*Part-time*). The histogram with the highest *Sum* = 390 is presented. Four histograms from these eight histograms satisfy $h_1 = h_2 = h_3 = h_4 = 0$. Three remaining histograms satisfy $h_4 = h_5 = \cdots = h_{12} = 0$.

Down-4,Up-7: All 22 CF-patterns with 7 steps up contain basic Boolean attribute hours-per-week(*Part-time*). Nine histograms from these 22 histograms satisfy $h_4 = h_5 = \cdots = h_{12} = 0$ or $h_5 = \cdots = h_{12} = 0$. The histogram with the highest *Sum* = 300 from remaining 13 histograms is presented.

Down-5,Up-6: All 83 CF-patterns with 6 steps up contain basic Boolean attribute hours-per-week(*Part-time*) and 74 of them contain attribute education. The histogram with the highest *Sum* = 695 is presented. All histograms have a shape similar to that of histogram "Down-5,Up-6" presented in Fig. 8.14. Most of them even satisfy $h_4 = h_5 = \cdots = h_{12} = 0$ or $h_5 = \cdots = h_{12} = 0$.

We can conclude that most of the found histograms have a "light tail" and are not too much interesting. However, the first two may be considered as deserving further consideration.

Results of runs of the CF-Miner procedure for $r = 6, \ldots, 11$ are available in Table 8.4. Histograms age_5/χ corresponding to examples of resulting CF-patterns are available in Figs. 8.15 and 8.16.

age_5/(*Male* ∧ *United-States* ∧ *Widowed*)

age_5/(*United-States* ∧ *Widowed* ∧ *Not-in-family* ∧ *Part-time*)

age_5/(*Male* ∧ education(*9th,10th,11th*) ∧ *White* ∧ capital_gain(*None*) ∧
∧ *Part-time*)

age_5/(occupation(*Handlers-cleaners*) ∧ *Part-time*)

age_5/(education(*9th,10th,11th,12th*) ∧ *United-States* ∧ *Newer-married* ∧
∧ relationship(*Own-child*) ∧ *Part-time*)

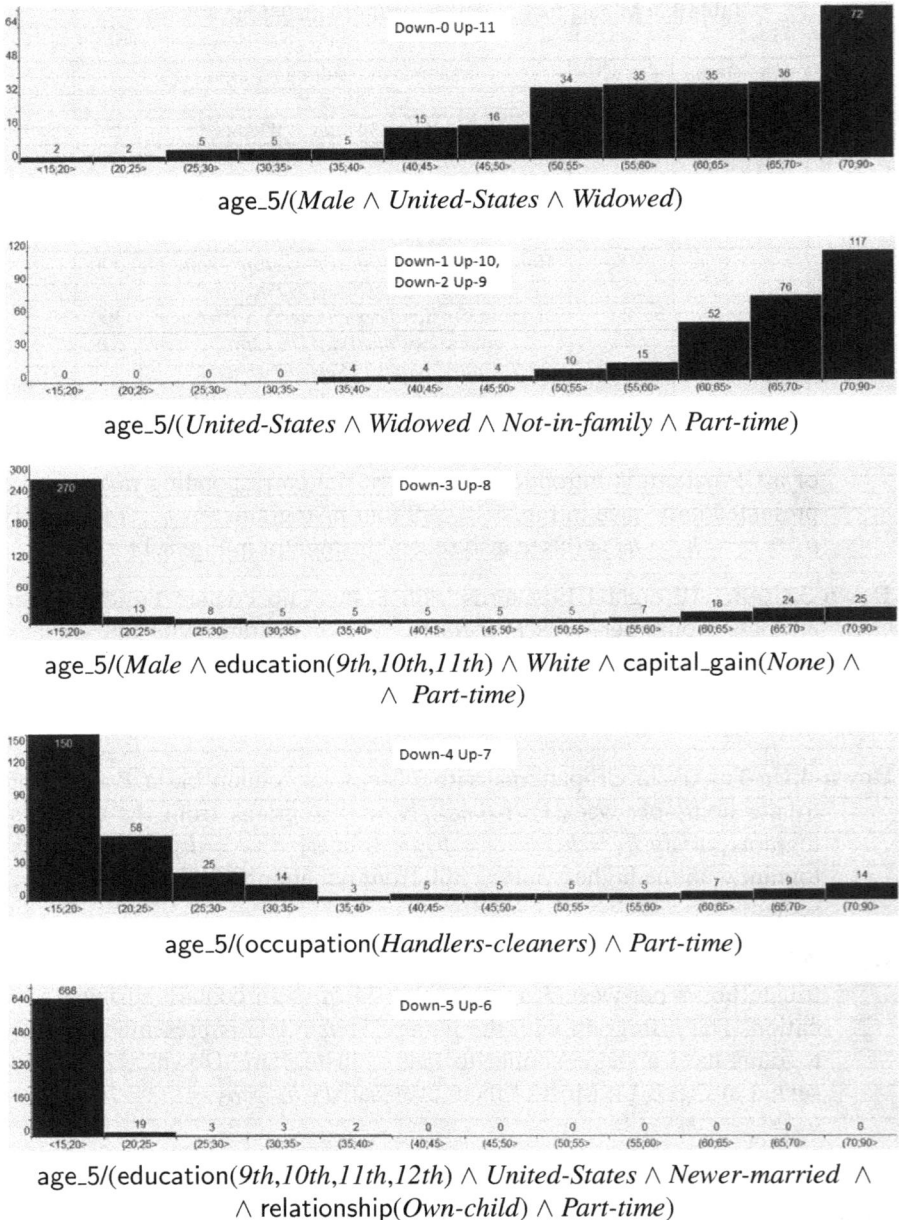

Figure 8.14: Examples of histograms starting with 0–5 not increasing steps.

Histograms corresponding to CF-patterns with the highest *Sum* are presented for each combination of the numbers of down and up steps. Several additional comments to the presented results follow.

Table 8.4: Mining for histograms starting with 6–11 steps down.

# of steps		# of CF-	Example of resulting $\approx_r$ age_5/χ	
Down	Up	patterns	Condition χ	*Sum*
6	5	8	*Male* ∧ education(*9th,10th,11th*) ∧ ∧ *United-States* ∧ *Part-time*	1 066
7	4	16	*Part-time*	5 913
8	3	45	*Newer-married* ∧ *Part-time*	3 254
9	2	22	*Male* ∧ education(*9th,10th,11th*) ∧ ∧ *Own-child*	2 127
10	1	1 208	*Own-child*	7 581
11	0	319	*Female* ∧ capital_gain(*None*) ∧ education(*11th,12th,HS-grad, Some-college*) ∧ *United-States*	8 931

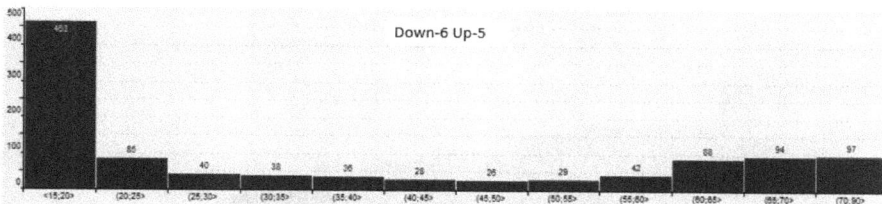

age_5/(*Male* ∧ education(*9th,10th,11th*) ∧ *United-States* ∧ *Part-time*)

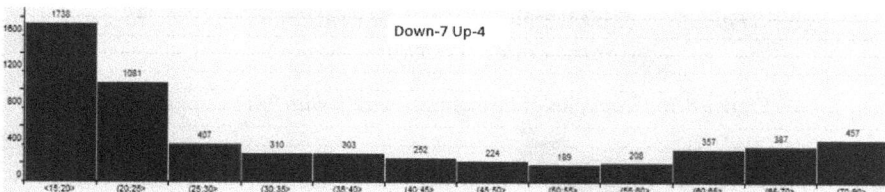

age_5/*Part-time*

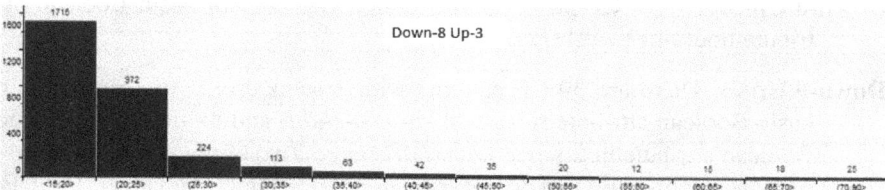

age_5/(*Newer-married* ∧ *Part-time*)

Figure 8.15: Examples of histograms starting with 6–8 steps down.

Down-6,Up-5: All eight CF-patterns with 5 steps up contain basic Boolean attribute hours-per-week(*Part-time*).

age_5/(*Male* ∧ education(*9th,10th, 11th*) ∧ *Own-child*)

age_5/*Own-child*

age_5/(*Female* ∧ capital_gain(*None*) ∧ education(*11th,12th,HS-grad, Some-college*) ∧ *United-States*)

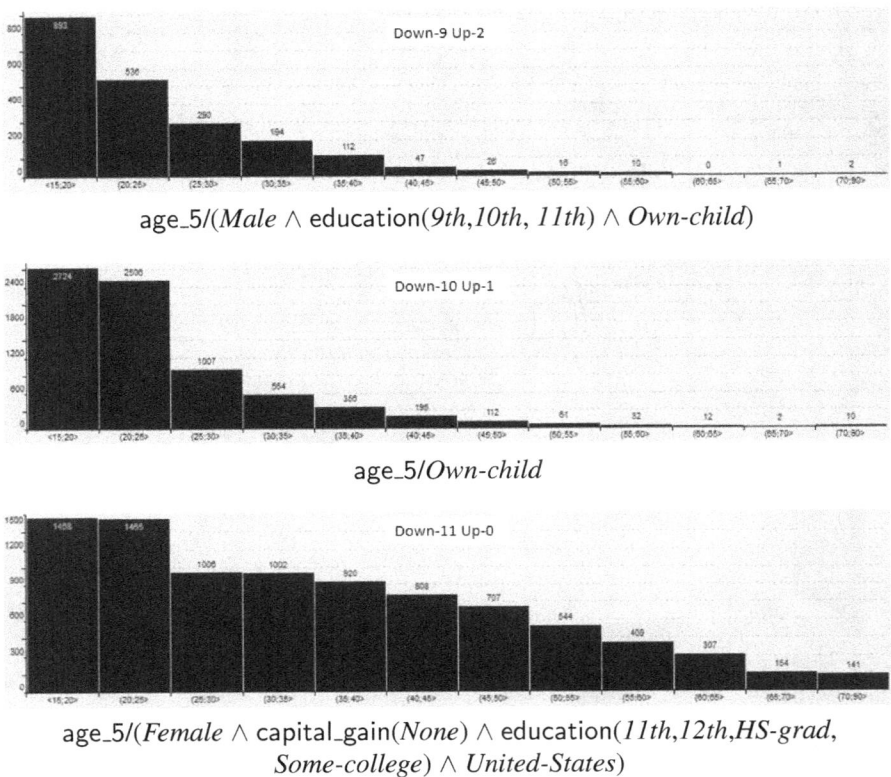

Figure 8.16: Examples of histograms starting with 9–11 steps down.

Down-7,Up-4: All 16 CF-patterns with 4 steps up contain basic Boolean attribute hours-per-week(*Part-time*).

Down-8,Up-3: All 45 CF-patterns with 3 steps up contain basic Boolean attribute hours-per-week(*Part-time*).

Down-9,Up-2: There are 39 CF-patterns with 2 steps up. 21 of them contain basic Boolean attribute relationship(*Own-child*) and do not contain basic Boolean attribute hours-per-week(*Part-time*). Two of them contain hours-per-week(*Part-time*) and do not contain relationship(*Own-child*). 16 CF-patterns contain neither relationship(*Own-child*) nor hours-per-week(*Part-time*).

Down-10,Up-1: There are 1 028 CF-patterns with one step up. 877 of them contain basic Boolean attribute relationship(*Own-child*). 301 of them do not contain relationship(*Own-child*) and contain basic Boolean attribute marital-status(*Never-married*). 30 of them contain neither relationship(*Own-child*) nor marital-status(*Never-married*).

Down-11,Up-0: There are 319 CF-patterns with 1 step up. 133 of them contain basic Boolean attribute marital-status(*Never-married*). 278 of them contain basic Boolean attribute education(α) where α contains at least two from categories *10th, 11th, 12th, HS-grad, Some-college*.

We can conclude that most of the found CF-patterns contain at least one of the basic Boolean attributes hours-per-week(*Part-time*), relationship(*Own-child*) and marital-status(*Never-married*). Note that histograms age_5/*Part-time* and age_5/*Own-child* are shown in Figs. 8.15 and 8.16, combinations Down_7_Up_4 and Down_10_Up_1 respectively. Let us emphasize that these basic Boolean attributes are also important when interpreting results of mining for decreasing histograms of attribute age_exp, see Section 8.2.2, Figs. 8.11 and 8.12.

Again, a question is if, under some conditions, a validity of CF-patterns

$$\approx_r \text{age\_5/hours-per-week}(\textit{Part-time}) \wedge \kappa$$

cannot be deduced from a validity of a CF-pattern

$$\approx_r \text{age\_5/hours-per-week}(\textit{Part-time}) \, .$$

The same holds for CF-patterns $\approx_r$ age_5/relationship(*Own-child*) $\wedge$ κ and $\approx_r$ age_5/marital-status(*Never-married*) $\wedge$ κ.

8.3 Applying CF-Miner in *Accidents* Dataset

The general trend of the numbers of accidents in the period 2005–2015 is decreasing, see Fig. 8.17. Thus, it is natural to ask questions concerning segments of accidents with various properties concerning trends of accidents in the whole period 2005–2015. Four such analytical questions are formulated and solved in Sections 8.3.1–8.3.4.

8.3.1 Large segments of accidents with decreasing trend

The total number of accidents has a decreasing trend in the period 2005–2015 as shown in Fig. 8.17. Thus, there is a natural question if there are large segments

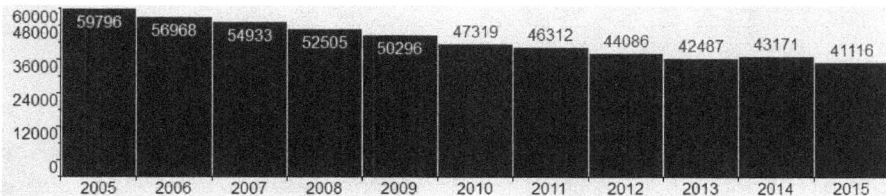

Figure 8.17: Decresing trend of all accidents in period 2005–2015.

Name: Type_Sex		Type: *Conjunction*		Length: $0-3$	
Attribute	Coefficients	Length	Gace	B/R	Class
Type	*Subsets*	$1-1$	*Pos*	B	−
Sex	*Subsets*	$1-1$	*Pos*	B	−

Figure 8.18: $\mathcal{RP}$-expression Type_Sex.

of accidents defined by a type of vehicle and a sex of a driver with a decreasing trend similar to the trend of the number of all accidents. We use the CF-Miner procedure to answer this question. Relevant large segments are segments defined by *RP*-expression Type_Sex available in Fig. 8.18.

A segment is considered as large if it consists of at least 5000 accidents. Thus we use the CF-quantifier *Sum* $\geq$ 5000. A decreasing trend similar to the trend of the number of all accidents is defined such that the histogram Year/χ has at least nine steps down from the ten possible steps, see Fig. 8.17. Here χ is a relevant condition defined by the *RC*-expression Type_Sex. Thus, we use a CF-quantifier $\langle$CFSt: *Down*, 1, 11, 1, *No, Abs*, $\geq$, 9$\rangle$ meaning that we are interested in segments χ such a conditional histogram of the attribute Year for this segment has at least nine steps down with minimal height 1 and we do not require consecutive steps. The condition defined by this CF-quantifier is satisfied also by the histogram in Fig. 8.17.

The run of the CF-Miner procedure with these parameters resulted in 11 conditional histograms in less than 1 second. 22 conditional histograms were tested. The results in a concise form are available in Fig. 8.19.

Nr.	Id	Sum	Hypothesis
1	1	386346	**Year / Type**(*Car*)
2	2	365266	**Year / Sex**(*Male*)
3	3	233149	**Year / Type**(*Car*) **& Sex**(*Male*)
4	4	137270	**Year / Sex**(*Female*)
5	5	126102	**Year / Type**(*Car*) **& Sex**(*Female*)
6	6	40935	**Year / Type**(*Bus_coach_17+*)
7	7	36104	**Year / Type**(*Bus_coach_17+*) **& Sex**(*Male*)
8	8	20639	**Year / Type**(*Motorcycle over 500cc*)
9	9	19534	**Year / Type**(*Motorcycle over 500cc*) **& Sex**(*Male*)
10	10	7304	**Year / Type**(*Motorcycle 50cc and under*)
11	11	5604	**Year / Type**(*Motorcycle 50cc and under*) **& Sex**(*Male*)

Figure 8.19: Segments with similar histograms of Year attribute as in Fig. 8.17.

The largest segment is given by the Boolean attribute Type(car), it consists of 386 346 accidents. The corresponding histogram Year/Type(*car*) is shown in Fig. 8.20.

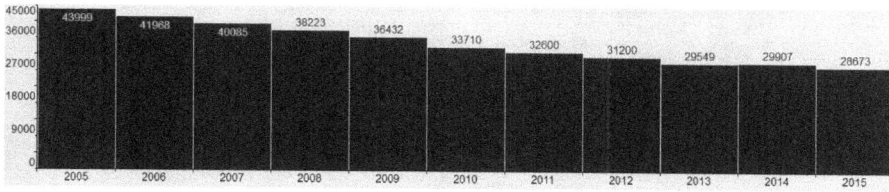

Figure 8.20: Histogram Year/Type(*car*).

The additional segments are analogous. We get the same results if we use the CF-quantifier *Sum* ≥ 10. In this case, 58 conditional histograms were tested. Note that the run of the CF-Miner requiring ten decreasing steps gives no result. We can conclude that the only conditional histograms with at least nine steps down and a condition given by the attributes Type and Sex are the segment introduced in Fig. 8.19. They concern cars, buses and coaches with 17 and more seats, motorcycles over 500cc, and motorcycles 50cc and under.

8.3.2   Exceptions to generally decreasing trend

We have just seen that the general trend of accidents is decreasing and that the general trend is valid also for various large segments of accidents. Thus, a question arise as to whether there are segments of accidents with an increasing trend in the whole period 2005–2015. Such segments can be naturally considered as exceptions to the general decreasing trend.

Thus, we are interested in increasing conditional histograms Year/χ with ten steps up. This leads to a CF-quantifier $\langle$CFSt: $Up, 1, 11, 1, Yes, Abs, \geq, 10\rangle$. We will investigate segments of accidents consisting of at least 3 000 accidents, and this means that we use the CF-quantifier *Sum* ≥ 3000.

We are interested in segments defined by attributes Type and Vehicle_Age from the group Vehicle, see Section 2.3.5, Sex, Age, Driver_IMD, and Journey from the group Driver, see Section 2.3.3 and Area, Road_Type, Speed_Limit, Light, Vehicle_Location, and Manoeuvre from the group Conditions, see Section 2.3.4. We use the $\mathcal{RP}$-expressions Vehicle, Driver and Conditions introduced in Fig. 8.21.

Note that coefficients – sequences of various lengths are used for attributes Vehicle_Age, Age, and Driver_IMD. Also, we allow to use maximally one attribute in $\mathcal{RP}$-expressions Driver and Conditions. Finally, the set of all relevant conditions χ is defined by the $\mathcal{RC}$-expression Vehicle_Driver_Conditions in Table 8.5.

The run of the CF-Miner procedure with these parameters resulted in 20 decreasing conditional histograms in 63 second; 61 059 conditional histograms were tested. The CF-patterns for the five largest segments are listed in Fig. 8.22.

Name: Vehicle		Type: *Conjunction*		Length: 0 − 2	
Attribute	Coefficients	Length	Gace	B/R	Class
Type	*Subsets*	1 − 1	*Pos*	B	−
Vehicle_Age	*Sequences*	1 − 5	*Pos*	B	−

Name: Driver		Type: *Conjunction*		Length: 0 − 1	
Attribute	Coefficients	Length	Gace	B/R	Class
Sex	*Subsets*	1 − 1	*Pos*	B	−
Age	*Sequences*	1 − 3	*Pos*	B	−
Driver_IMD	*Sequences*	1 − 3	*Pos*	B	−
Journey	*Subsets*	1 − 1	*Pos*	B	−

Name: Conditions		Type: *Conjunction*		Length: 0 − 1	
Attribute	Coefficients	Length	Gace	B/R	Class
Area	*Subsets*	1 − 1	*Pos*	B	−
Road_Type	*Subsets*	1 − 1	*Pos*	B	−
Speed_Limit	*Subsets*	1 − 1	*Pos*	B	−
Light	*Subsets*	1 − 1	*Pos*	B	−
Vehicle_Location	*Subsets*	1 − 1	*Pos*	B	−
Manoeuvre	*Subsets*	1 − 1	*Pos*	B	−

Figure 8.21: $\mathscr{RP}$-expressions Vehicle, Driver and Conditions.

Table 8.5: $\mathscr{RC}$-expression Vehicle_Driver_Conditions in a short tabular form.

Name: Vehicle_Driver_Conditions	Length: 0 − 4
Set of partial cedents	Length
Vehicle	0 − 2
Driver	0 − 1
Conditions	0 − 1

Nr.	Id	Sum	Hypothesis
1	18	6351	**Year** / **Type**(*Car*) & **Speed_limit**(*20*)
2	5	6208	**Year** / **Sex**(*Male*) & **Speed_limit**(*20*)
3	20	5097	**Year** / **Type**(*Car*) & **Journey**(*5 Other/Not known*) & **Speed_limit**(*20*)
4	3	5006	**Year** / **Driver_Age_Band**(*26 - 35..46 - 55*) & **Speed_limit**(*20*)
5	16	4959	**Year** / **Vehicle_Age**(>= *13*) & **Driver_Age_Band**(*46 - 55*) & **Light**(*1,Daylight*)

Figure 8.22: The five largest segments for $\mathscr{RC}$-expression Vehicle_Driver_Conditions.

Note that the expression Age(*26-35..46-55*) in the fourth row in Fig. 8.22 is an abbreviation of the basic Boolean attribute Age(*26-35,36-45,46-55*). Similarly, the expression Vehicle_Age(*>=13*) in the fifth row is an abbreviation of the basic Boolean attribute Vehicle_Age(*13,14,15,16-20,>20*).

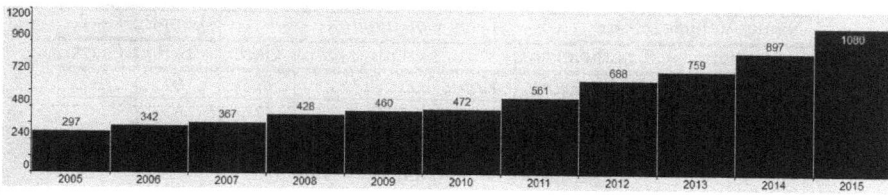

Figure 8.23: Histogram Year/Type(*Car*) ∧ Speed_Limit(*20*).

The histogram Year/(Type(*Car*) ∧ Speed_Limit(*20*)) is shown in Fig. 8.23. The corresponding segment consists of 6 351 accidents. We can conclude that this histogram is a strong exception to the increasing histogram Year/Type(*car*) shown in Fig. 8.20.

8.3.3 Exceptions to a concrete decreasing trend

One of the decreasing histograms found in Section 8.3.1 is the histogram Year/Type(*Motor cycle over 500 cc*), see the eight row in Fig. 8.19. This histogram is shown in Fig. 8.24.

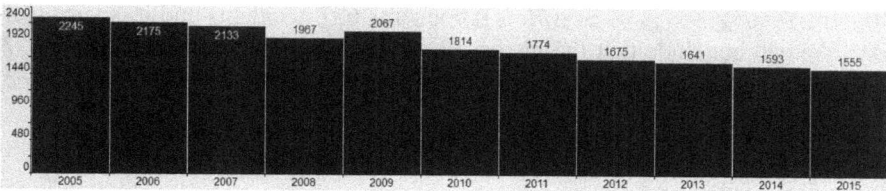

Figure 8.24: Histogram Year/Type(*Motor cycle over 500 cc*).

However, no exception to this histogram was found in the previous section. Thus, we are interested whether there are sub-segments of the segment of accidents of the type Motorcycle 500cc+ with an increasing or at least non-decreasing trend. We show, how this question can be solved.

We are interested in non-decreasing conditional histograms Year/χ with ten steps up. This leads to a CF-quantifier $\langle$CFSt: $Up, 1, 11, 0, Yes, Abs, \geq, 10\rangle$ meaning that we search histograms with ten steps up; however, minimal height of the step is zero. We will investigate segments of accidents consisting of at least 500 accidents, and thus we use the CF-quantifier $Sum \geq 500$.

We use the $\mathscr{RP}$-expression Vehicle_500cc introduced in Fig. 8.25, $\mathscr{RP}$-expression Driver already defined in Fig. 8.21 and $\mathscr{RC}$-expression Vehicle_500cc_Driver introduced in Table 8.6.

The run of the CF-Miner procedure with these parameters resulted into two non-decreasing conditional histograms in eight seconds. The first resulting his-

Name: Vehicle_500cc		Type: *Conjunction*		Length: $1-2$	
Attribute	Coefficients	Length	Gace	B/R	Class
Type	*One category*	*Motor cycle 500cc+*	*Pos*	B	–
Vehicle_Age	*Sequences*	$1-5$	*Pos*	R	–

Figure 8.25: $\mathscr{R}\mathscr{P}$-expression Vehicle.

Table 8.6: $\mathscr{R}\mathscr{C}$-expression Vehicle_500cc_Driver in a short tabular form.

Name: Vehicle_Driver_Conditions	Length: $0-4$
Set of partial cedents	Length
Vehicle_500cc	$1-2$
Driver	$0-1$

togram Year/χ_{500cc_Exc} where

$$\chi_{500cc\_Exc} = \text{Vehicle\_Age}(\textit{12,13,14,15,16-20}) \wedge$$
$$\wedge \text{ Type}(\textit{Motor cycle over 500 cc}) \wedge \text{Age}(\textit{46-55})$$

concerns 711 accidents. It is shown in Fig. 8.26. The second resulting histogram Year/($\chi_{500cc\_Exc} \wedge$ Sex(*Male*)) concerns 687 accidents and it is very similar. We can conclude that the histograms Year/χ_{500cc_Exc} and Year/($\chi_{500cc\_Exc} \wedge$ Sex(*Male*)) are exceptions to the histogram Year/Type(*Motor cycle over 500 cc*) presented in Fig. 8.24.

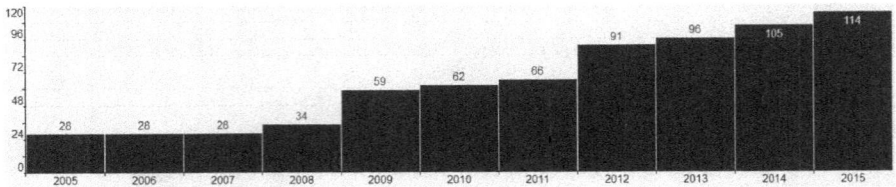

Figure 8.26: Histogram Year/χ_{500cc_Exc}

8.3.4 Exceptions to exception to generally decreasing trend

The segment defined by a Boolean attribute

$$\text{Car\_speed\_20} = \text{Type}(\textit{car}) \wedge \text{Speed\_limit}(\textit{20})$$

is the largest found segment with increasing trend – an exception to generally decreasing trend, see Section 8.3.2. Thus a question arises, whether there is a local authority defined by one of the attributes District, Highway, Police which can be considered as an exception to this exception.

We do not assume that there is decreasing trend for the whole period. We consider an authority $AUTH$ as an exception from the exception given by the increasing trend for a segment given by Boolean attribute Type(*car*) $\land$ Speed_limit(*20*) if the histogram for a segment given by a Boolean attribute Car_speed_20 $\land$ $AUTH$ has at least three consecutive steps down in the last five years. This leads to a CF-quantifier $\langle$CFSt: *Down*, 7, 11, 1, *Yes*, *Abs*, $\geq$, 3$\rangle$ meaning that we search histograms with at least three consecutive steps down in the range $\langle 7, 11 \rangle$. We will investigate segments of accidents consisting of at least 500 accidents, and thus we use the CF-quantifier *Sum* $\geq$ 500.

We use the $\mathscr{RP}$-expression Car_limit_20 introduced in Fig. 8.27, $\mathscr{RP}$-expression Authorities defined in Fig. 8.28 and $\mathscr{RC}$-expression Car_limit_20_Authorities introduced in Table 8.7.

Name: Car_limit_20		Type: *Conjunction*		Length: $2-2$	
Attribute	Coefficients	Length	Gace	B/R	Class
Type	*One category*	car	Pos	B	—
Speed_Limit	*One category*	20	Pos	B	—

Figure 8.27: $\mathscr{RP}$-expression Car_limit_20.

Name: Authorities		Type: *Conjunction*		Length: $1-1$	
Attribute	Coefficients	Length	Gace	B/R	Class
District	*Subsets*	$1--1$	Pos	B	—
Highway	*Subsets*	$1--1$	Pos	B	—
Police	*Subsets*	$1--1$	Pos	B	—

Figure 8.28: $\mathscr{RP}$-expression Authorities.

Table 8.7: $\mathscr{RC}$-expression Car_limit_20_Authorities in a short tabular form.

Name: Car_limit_20_Authorities	Length: $0-4$
Set of partial cedents	Length
Car_limit_20	$2-2$
Authorities	$1-1$

The run of the CF-Miner procedure with these parameters resulted in four conditional histograms satisfying the given conditions, see Fig. 8.29. The largest segment contains 511 accidents, and it is defined by a Boolean attribute

$$\text{Car\_speed\_20\_Great\_Manchester} = \text{Type}(car) \land$$
$$\land \text{ Speed\_limit}(20) \land \text{Police}(Great\ Manchester) \ .$$

The histogram Year/Car_speed_20_Great_Manchester is presented in Fig. 8.30. It has three consecutive steps down in the last five years. Thus, we consider Po-

Nr.	Id	Sum	Hypothesis
1	3	511	**Year / Type**(*Car*) & **Speed_limit**(*20*) & **Police**(*Greater Manchester*)
2	4	311	**Year / Type**(*Car*) & **Speed_limit**(*20*) & **Police**(*Thames Valley*)
3	1	116	**Year / Type**(*Car*) & **Speed_limit**(*20*) & **District**(*Newcastle upon Tyne*)
4	2	116	**Year / Type**(*Car*) & **Speed_limit**(*20*) & **Highway**(*E08000021,Newcastle upon Tyne*)

Figure 8.29: Four authorities – exceptions to exceptions.

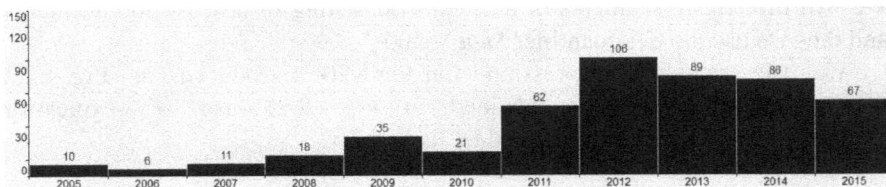

Figure 8.30: Histogram Year/Car_speed_20_Great_Manchester.

lice(Great Manchester) as an exception from the increasing trend for the segment given by Boolean attribute Type(*car*) ∧ Speed_limit(*20*).

Chapter 9

KL-Miner—Pairs of Categorical Attributes

There are many methods and analytical software systems providing tools for the analysis of relations of two categorical attributes. The goal of this chapter is to present the KL-Miner procedure which approaches the analysis of relations of two categorical attributes in the GUHA-style. The main features of the Kl-Miner procedure and related notions are introduced in Section 9.1. Examples of applications of the KL-Miner procedure in the *STULONG* dataset are presented in Section 9.2. Examples of applications of the KL-Miner procedure in the *Hotel* dataset are described in Section 9.3. Additional examples are available in [85, 140].

9.1 KL-Miner and Related Notions

Note that the KL-Miner procedure is informally introduced in Section 3.4. An overview of its input and output is given in Section 9.1.1. Four types of frequencies can be used in contingency tables the procedure uses to evaluate a relation of two categorical attributes. They are defined in Section 9.1.2. Generalized quantifiers the KL-Miner procedure deals with are called *KL-quantifiers*. Two types of KL-quantifiers are used in examples of applications of the KL-Miner procedure. One of them can be applied to a whole contingency table or only to its part called the *range of KL-quantifier*. A range of KL-quantifier is defined in Section 9.1.3. Simple frequencies KL-quantifiers are introduced in Section 9.1.4. Advanced KL-quantifiers are described in Section 9.1.5.

9.1.1 KL-Miner input and output

The KL-Miner procedure deals with *KL-patterns* $R \approx C/\chi$. Here $\approx$ is a KL-quantifier, R and C are categorical attributes, and χ is a Boolean attribute defined in Section 1.4.3. Recall that a KL-pattern is an assertion on a relation of attributes R and C at a sub-matrix $\mathcal{M}/\chi$ of an analysed data matrix $\mathcal{M}$. There are the following parameters defining a set of relevant KL-patterns to be generated and verified:

- a set *Row* of relevant row attributes R

- a set *Col* of relevant column attributes C

- an $\mathcal{RC}$-expressions defining a set *Cond* of relevant conditions χ

- a list *ListKL* of basic KL-quantifiers, each basic KL-quantifier is relevant to each pair $\langle R,C \rangle$ of attributes $R \in Row$ and $C \in Col$.

A set of relevant KL-patterns is defined as a set of all patterns $R \approx C/\chi$ satisfying $R \in Row$, $C \in Col$, $\chi \in Cond$. KL-quantifier $\approx$ is defined as a conjunction of basic KL-quantifiers introduced in the list *ListKL*.

The KL-Miner procedure generates all relevant KL-patterns and verifies them in a given data matrix $\mathcal{M}$. Each generated KL-pattern $R \approx C/\chi$ is verified such that the KL-quantifier $\approx$ is applied to a KL-table $KL(R,C,\mathcal{M}/\chi)$. KL-patterns satisfying the condition given by the KL-quantifier $\approx$ are listed in an output.

9.1.2 Four types of frequencies

There are four types of frequencies *TypeFreq* for contingency table: *Abs*, *ProcRow*, *ProcCol*, and *ProcCond*. All frequencies are given by a KL-table $KL(R,C,\mathcal{M}/\chi)$ introduced in Section 4.2. It is understood as a data matrix $\{n_{i,j}\}_{i=1,\ldots,K}^{j=1,\ldots,L}$ of non-negative integers. Here, $n_{i,j}$ denotes the number of rows o of a data matrix $\mathcal{M}/\chi$ satisfying $R(o) = r_i$ and $C(o) = c_j$ where $i = 1,\ldots,K$ and $j = 1,\ldots,L$. $\mathcal{M}$ is an analysed data matrix. This KL-table is together with marginal frequencies introduced also in Fig. 9.1.

M/χ	c_1	$\ldots$	c_L	
r_1	$n_{1,1}$	$\ldots$	$n_{1,L}$	$n_{1,*}$
$\vdots$	$\vdots$	$\ddots$	$\vdots$	$\vdots$
r_K	$n_{K,1}$	$\ldots$	$n_{K,L}$	$n_{K,*}$
	$n_{*,1}$	$\ldots$	$n_{*,L}$	n

Figure 9.1: KL-table $KL(R,C,\mathcal{M}/\chi)$ together with marginal frequenies.

It holds

$$n_{i,*} = \Sigma_{k=1}^{L} n_{i,k} \qquad n_{*,j} = \Sigma_{k=1}^{K} n_{k,j} \qquad n = \Sigma_{i=1,j=1}^{K,L} n_{i,j} \quad . \tag{9.1}$$

A contingency table of type *Abs*, *ProcRow*, *ProcCol*, and *ProcCond* is understood as a result of application of a function $Freq(\{n_{i,j}\}, TypeFreq)$ to a data matrix $\{n_{i,j}\}$ of frequencies, we assume $Freq(\{n_{i,j}\}, TypeFreq) = \{t_{i,j}\}_{i=1,\ldots,K}^{j=1,\ldots,L}$. A value of the function depends on a parameter *TypeFreq*, see Fig. 9.2.

TypeFreq	$t_{i,j}$
Abs	$t_{i,j} = n_{i,j}$
ProcRow	$t_{i,j} = 100 * \frac{n_{i,j}}{n_{i,*}}$
ProcCol	$t_{i,j} = 100 * \frac{n_{i,j}}{n_{*,j}}$
ProcCond	$t_{i,j} = 100 * \frac{n_{i,j}}{n}$

Figure 9.2: Contingency tables of type *Abs*, *ProcRow*, *ProcCol*, and *ProcCond*.

9.1.3 Range of KL-quantifiers

A range of a KL-quantifier $\approx$ for a contingency table $\{t_{i,j}\}_{i=1,\ldots,K}^{j=1,\ldots,L}$ is given by a quadruple $\langle p,q,r,s \rangle$ of positive integers satisfying $1 \leq p < q \leq K$ and $1 \leq r < s \leq L$. This quadruple is called a *definition of a range of KL-quantifier*.

A range $\langle p,q,r,s \rangle$ *of a KL-quantifier* $\approx$ *for a contingency table* $\{t_{i,j}\}_{i=1,\ldots,K}^{j=1,\ldots,L}$ is a data matrix $\{t_{i,j}\}_{i=p,\ldots,q}^{j=r,\ldots,s}$ consisting of frequencies from the contingency table.

9.1.4 Simple frequencies KL-quantifiers

We use simple frequencies KL-quantifiers in a form

$$\langle \text{KLSF: } Sum, p, q, r, s, Abs \rangle \circledast Thr \tag{9.2}$$

or

$$\langle \text{KLSF: } Sum, p, q, r, s, ProcCond \rangle \circledast Thr \ . \tag{9.3}$$

Here p, q, r, s are positive integers such that $\langle p,q,r,s \rangle$ is a definition of a range, $\circledast$ is one of the relations $<, \leq, =, \geq, >$ concerning real numbers, and Thr is a non-negative real number.

We say that a KL-quantifier $\langle \text{KLSF: } Sum, p, q, r, s, Abs \rangle \circledast Thr$ is *relevant to a pair* $\langle R, C \rangle$ *of attributes R and C where R has K categories and C has L categories* if it holds both $q \leq K$ and $s \leq L$. There is an analogous definition for a KL-quantifier $\langle \text{KLSF: } Sum, p, q, r, s, ProcCond \rangle \circledast Thr$.

The simple frequencies KL-quantifier $\langle \text{KLSF: } Sum, p, q, r, s, Abs \rangle \circledast Thr$ defines a condition

$$n_{p,r} + \cdots + n_{p,s} + \cdots + n_{q,r} + \cdots + n_{q,s} \circledast Thr \ . \tag{9.4}$$

We write $Sum_{\langle p,q,r,s \rangle} \geq Thr$ instead of $\langle KLSF:\ Sum,p,q,r,s,Abs \rangle \geq Thr$.

The KL-quantifier $\langle KLSF:\ Sum,p,q,r,s,ProcCond \rangle \circledast Thr$ defines a condition

$$100 * \frac{n_{p,r} + \cdots + n_{p,s} + \cdots + n_{q,r} + \cdots + n_{q,s}}{n} \circledast Thr \ . \qquad (9.5)$$

Here $n_{p,r},\ldots,n_{p,s},\ldots,n_{q,r},\ldots,n_{q,s}$ and n are frequencies from KL-table $KL(R,C,\mathcal{M}/\chi)$, see Fig. 9.1.

Let us note that there are additional simple frequencies KL-quantifiers concerning conditions related to minimum, maximum and average values computed from histograms of all types of frequencies Abs, $ProcRow$, $ProcCol$, and $ProcCond$ types of frequencies introduced in Section 9.1.2.

9.1.5 Advanced KL-quantifiers

We will use advanced KL-quantifiers in a form of a condition

$$AdvKL \circledast Thr \qquad (9.6)$$

where $AdvKL$ is τb or $|\tau b|$. Here τb denotes a Kendall's coefficient defined for the KL-table introduced in Fig. 9.1 such that $\tau_b = \frac{2(P-Q)}{\sqrt{(n^2 - \sum_{k=1}^{K} n_{k,*}^2)(n^2 - \sum_{l=1}^{L} n_{*,l}^2)}}$, where it holds that $P = \sum_{k=1}^{K} \sum_{l=1}^{L} n_{k,l} \sum_{i=k+1}^{K} \sum_{j=l+1}^{L} n_{i,j}$ and $Q = \sum_{k=1}^{K} \sum_{l=1}^{L} n_{k,l} \sum_{i=k+1}^{K} \sum_{j=1}^{l-1} n_{i,j}$. The expression $|\tau_b|$ denotes an absolute value of τ_b.

Finally, $\circledast$ is one of relations $<, \leq, =, \geq, >$ concerning real numbers and Thr is a real number satisfying $Thr \in \langle 0;1 \rangle$ for $|\tau_b|$ and $Thr \in \langle -1;1 \rangle$ for τ_b. A condition related to an advanced KL-quantifier 9.6 is satisfied for a contingency table $\{n_{i,j}\}_{i=1,\ldots,K}^{j=1,\ldots,L}$ if a relation $AdvKL \circledast Thr$ is valid. A range of KL-quantifiers is not applicable to the $AdvKL \circledast Thr$ KL-quantifiers. Each $AdvKL \circledast Thr$ KL-quantifier is relevant to each pair $\langle R,C \rangle$ of attributes.

Let us note that there are additional advanced KL-quantifiers based on Cramer's V coefficient, Chi-square test, conditional entropy, mutual information and additional theoretical measures of dependence.

9.2 Applying KL-Miner in *STULONG* Dataset

Let us emphasize that the goal of the presented example is neither to get new medical knowledge nor to present a qualified application of statistical analysis. The only goal is to briefly demonstrate the possibilities of the KL-Miner procedure. We use Kendall's quantifier based on Kendall's coefficient τ_b, introduced in Section 9.1.5. Let us note that τ_b gives values from $\langle -1,1 \rangle$ with the following interpretation: $\tau_b > 0$ indicates positive ordinal dependence, $\tau_b < 0$ indicates negative ordinal dependence, $\tau_b = 0$ indicates ordinal independence, $|\tau_b = 1|$ indicates that a column attribute C in question is a function of a row attribute R, see Section 9.1.1.

First, we show an example of searching conditions under which there is a high ordinal dependence between systolic and diastolic blood pressures. This is available in Section 9.2.1. The second example concerns searching ordinal attributes *Atr* and conditions χ such that τ_b indicates almost ordinal independence between *Atr* and systolic or diastolic blood pressures. This example is presented in Section 9.2.2.

9.2.1 Conditions indicating high ordinal dependence

Recall that we deal with the *Entry* data matrix which is a part of the *STULONG* dataset, see Section 2.4. The attributes Systolic and Diastolic belong to the group Blood pressure, see Section 2.4.7. A contingency table KL(Systolic, Diastolic, *Entry*) is shown in Fig. 9.3.

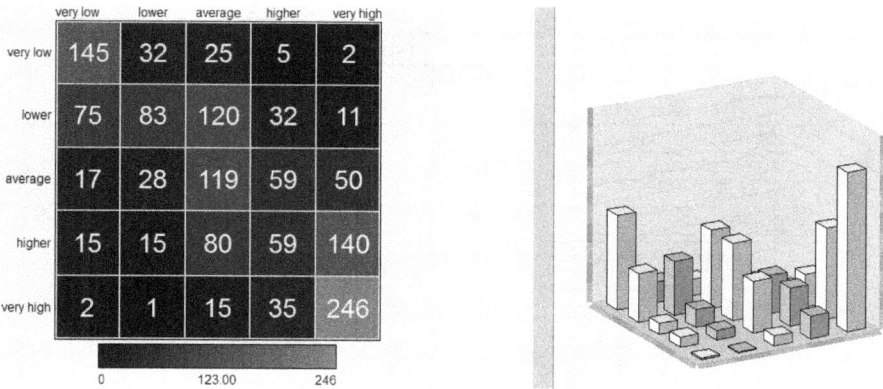

Figure 9.3: Contingency table KL(Systolic, Diastolic, *Entry*), $\tau_b = 0.64$.

It holds $\tau_b = 0.64$ for the contingency table KL(Systolic, Diastolic, *Entry*). Thus, there is a natural question if there is a data matrix *Entry*/χ such that a value of τ_b for a contingency table KL(Systolic, Diastolic, *Entry*/χ) is remarkably higher. We require $\tau_b \geq 0.8$ and at least 100 rows in a data matrix *Entry*/χ. This question can be answered by an application of the KL-Miner procedure with the following input parameters (see Section 9.1.1):

The Systolic attribute is the only relevant row attribute and the Diastolic attribute is the only relevant column attribute. A KL-quantifier $\approx^\tau_{0.8,100}$ defined as $\tau_B \geq 0.8 \wedge Sum_{\langle 1,K,1,L \rangle} \geq 100$ is used.

The set *Cond* of relevant conditions is given as a set of conjunctions of Boolean characteristics of all groups of attributes of the *Entry* data matrix except the Blood pressure group. Characteristics of particular groups are conjunctions of basic Boolean attributes defined by the $\mathscr{RP}$-expressions Personal, Anamnesis, Risks, Measurement, Alcohol consumption, and Biochemical examination introduced in Figs. 9.4 and 9.5.

Name: Personal		Type: *Conjunction*		Length: $0-4$	
Attribute	Coefficients	Length	Gace	B/R	Class
Education	*Subsets*	$1-1$	*Pos*	B	−
MaritalStatus	*Subsets*	$1-1$	*Pos*	B	−
Responsibility	*Subsets*	$1-1$	*Pos*	B	−
Age	*Subsets*	$1-1$	*Pos*	B	−

Name: Anamnesis		Type: *Conjunction*		Length: $0-5$	
Attribute	Coefficients	Length	Gace	B/R	Class
Diabetes	*Subsets*	$1-1$	*Pos*	B	−
Hyperlipidemia	*Subsets*	$1-1$	*Pos*	B	−
Hypertension	*Subsets*	$1-1$	*Pos*	B	−
Ictus	*Subsets*	$1-1$	*Pos*	B	−
Myocardial infarction	*Subsets*	$1-1$	*Pos*	B	−

Name: Risks		Type: *Conjunction*		Length: $0-5$	
Attribute	Coefficients	Length	Gace	B/R	Class
FamilyAnamnesisRisk	*Subsets*	$1-1$	*Pos*	B	−
HypertensionRisk	*Subsets*	$1-1$	*Pos*	B	−
CholesterolRisk	*Subsets*	$1-1$	*Pos*	B	−
ObesityRisk	*Subsets*	$1-1$	*Pos*	B	−
SmokingRisk	*Subsets*	$1-1$	*Pos*	B	−

Name: Measurement		Type: *Conjunction*		Length: $0-3$	
Attribute	Coefficients	Length	Gace	B/R	Class
BMI	*Sequences*	$1-2$	*Pos*	B	−
Subscapularis	*Sequences*	$1-2$	*Pos*	B	−
Triceps	*Sequences*	$1-2$	*Pos*	B	−

Figure 9.4: $\mathscr{RP}$-expressions Personal, Anamnesis, Risks, and Measurement.

We require that the set *Cond* is a conjunction of 1–5 basic Boolean attributes. This means that the set *Cond* is defined by the $\mathscr{RC}$-expression Cond introduced in Table 9.1.

There are four resulting true KL-patterns; 1 565 503 KL-patterns were tested in eight minutes and four seconds of a run of the KL-Miner procedure with the above described parameters. The contingency table $KL(\text{Systolic}, \text{Diastolic}, Entry/\chi_1)$ of the strongest resulting KL-pattern Systolic $\approx_{0.8,100}^{\tau}$ Diastolic$/\chi_1$ with $\tau_b = 0.82$ is shown in Fig. 9.6. There are 103

Name: Alcohol consumption	Type: *Conjunction*	Length: $0-4$		
Attribute	Coefficients	Length	Gace	B/R	Class
Alcohol	*Sequences*	$1-2$	*Pos*	B	–
Beer	*Sequences*	$1-2$	*Pos*	B	–
Vine	*Sequences*	$1-2$	*Pos*	B	–
Liquors	*Sequences*	$1-2$	*Pos*	B	–

Name: Biochemical examination	Type: *Conjunction*	Length: $0-3$		
Attribute	Coefficients	Length	Gace	B/R	Class
Urine	*Subsets*	$1-2$	*Pos*	B	–
Cholesterol	*Sequences*	$1-2$	*Pos*	B	–
Triglycerides	*Sequences*	$1-2$	*Pos*	B	–

Figure 9.5: $\mathscr{RP}$-expressions Alcohol consumption and Biochemical examination.

Table 9.1: $\mathscr{RC}$-expression Cond.

Name: Cond	Length: $1-5$
Set of partial cedents	Length
Personal	$0-4$
Anamnesis	$0-5$
Risks	$0-5$
Measurement	$0-3$
Alcohol consumption	$0-4$
Biochemical examination	$0-3$

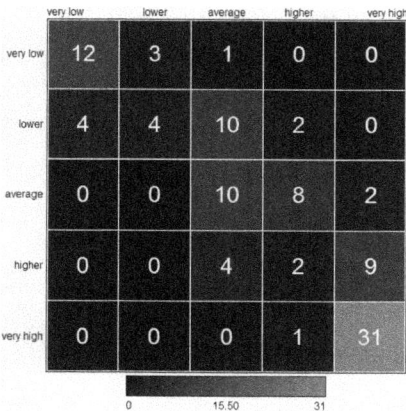

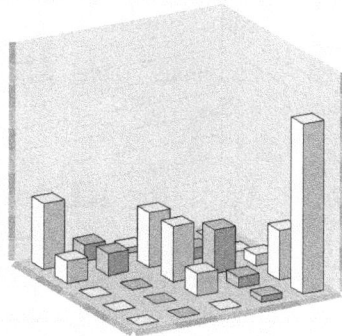

Figure 9.6: Contingency table $KL(\text{Systolic}, \text{Diastolic}, Entry/\chi_1)$, $\tau_b = 0.82$.

rows in the data matrix $Entry/\chi_1$. The Boolean attribute χ_1 denotes the condition

Hyperlipidemia(*no*) ∧ CholesterolRisk(*no*) ∧ Alcohol(*regularly*) ∧
∧ Beer(*up to 1 litre / day, more than 1 litre / day*) .

9.2.2 Conditions indicating almost ordinal independence

In the previous task, we were interested in a data matrix $Entry/\chi$ such that a value of τ_b for a contingency table KL(Systolic, Diastolic, $Entry/\chi$) is higher than 0.8. Now we are interested in a data matrix $Entry/\chi$ and an attribute Atr such that a value of τ_b for a contingency table KL(Atr, Diastolic, $Entry/\chi$) is around zero. Note that this means that Atr and Diastolic are almost ordinal independent.

We use attributes Subscapularis, Triceps, Cholesterol, and Triglycerides. We require $|\tau_b| \leq 0.00001$ and at least 500 rows in a data matrix $Entry/\chi$. This question can be answered by an application of the KL-Miner procedure with the following input parameters.

The set of relevant row attributes consists of attributes Subscapularis, Triceps, Cholesterol, and Triglycerides. The Diastolic attribute is the only relevant column attribute. The set *Cond* of relevant conditions is the same as in the previous task. This means that it is defined by the $\mathscr{RC}$-expression Cond introduced in Table 9.1.

A KL-quantifier $\approx^{|\tau|}_{\leq 0.00001, 500}$ defined as $|\tau_B| \leq 0.00001 \land Sum_{\langle 1, K, 1, L \rangle} \geq 500$ is used. 150 768 KL-patterns were tested in four seconds of a run of the KL-Miner with such parameters. There are six resulting true KL-patterns. Five of them are in a form Cholesterol $\approx^{\tau}_{\leq 0.00001, 500}$ Diastolic/χ and one in a form Triceps $\approx^{\tau}_{\leq 0.00001, 500}$ Diastolic/χ. There is exactly one KL-pattern with $\tau_b = 0$. It is the KL-pattern Cholesterol $\approx^{\tau}_{\leq 0.00001, 500}$ Diastolic, $Entry/\chi_2$ where the Boolean attribute χ_2 denotes the condition

MaritalStatus(*married*) $\land$ Ictus(*no*) $\land$ BMI(*normal, overweight*) $\land$

$\land$ Subscapular(*average, high*) $\land$ Urine(*normal, sugar positive*) .

The contingency table KL(Cholesterol, Diastolic, $Entry/\chi_2$) with $\tau_b = 0$ is shown in Fig. 9.7.

9.3 Applying KL-Miner in *Hotel* Dataset

The *Hotel* dataset is a fictive dataset described in Section 2.5. It is generated by the ReverseMiner procedure introduced in Chapter 16. We use this dataset to present an application of a range of KL-quantifier.

9.3.1 Applying range of KL-quantifier

The following application of the KL-Miner procedure is inspired by the contingency table KL(QSatisfaction, DayOfWeek, *HotelPlusExternal*) shown in Fig. 9.8. There are 80 rows satisfying both QSatisfaction(*average*) and DayOfWeek(*Mon*) etc. We can conclude that there are only 330 rows of the data matrix *HotelPlusExternal* satisfying both QSatisfaction(*average, high*)

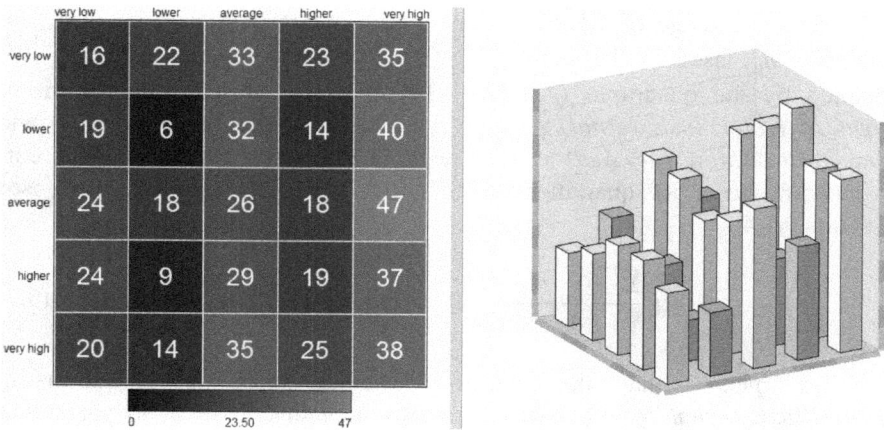

Figure 9.7: Contingency table $KL($Cholesterol, Diastolic, $Entry/\chi_2)$, $\tau_b = 0$.

Figure 9.8: $KL($QSatisfaction, DayOfWeek, $HotelPlusExternal)$.

and DayOfWeek(*Mon, Tue, Wed*). In other words, there are 330, i.e., 16.5 percent of all stays starting on Monday, Tuesday, or Wednesday which are evaluated as average or high what concerns satisfaction.

Thus, a question arises whether there is a sub-matrix *HotelPlusExternal*/χ such that at least 50 percent of its rows satisfy both QSatisfaction(*average, high*) and DayOfWeek(*Mon, Tue, Wed*). We are interested in data matrices with at least 250 rows.

The number of rows of the sub-matrix *HotelPlusExternal*/χ satisfying both QSatisfaction(*average, high*) and DayOfWeek(*Mon, Tue, Wed*) is equal to the sum $\sum_{i=2,3}^{j=1,2,3} n_{i,j}$ where

$$\{n_{i,j}\}_{i=1,\ldots,3}^{j=1,\ldots,7} = KL(\text{QSatisfaction}, \text{DayOfWeek}, HotelPlusExternal/\chi),$$

see Section 9.1.2. The categories of the attributes QSatisfaction and DayOfWeek are numbered according to Fig. 9.8.

Recall the range of KL-quantifiers introduced in Section 9.1.3. The frequencies $n_{i,j}$ used in the sum $\sum_{i=2,3}^{j=1,2,3} n_{i,j}$ constitute the range $\{n_{i,j}\}_{i=2,3}^{j=1,2,3}$ defined by the quadruple $\langle 2,3,1,3 \rangle$. The fact that at least 50 percent of rows of *HotelPlusExternal*/χ satisfy both QSatisfaction(*average, high*) and DayOfWeek(*Mon, Tue, Wed*) can be expressed by the condition (9.7) corresponding to a KL-quantifier $\langle KLSF: Sum, 2, 3, 1, 3, ProcCond \rangle \geq 50$, see Section 9.1.4.

$$100 * \frac{\sum_{i=2,3}^{j=1,2,3} n_{i,j}}{n} \geq 50 \text{ where } n = \sum_{i=1,...,3}^{j=1,...,7} n_{i,j} \qquad (9.7)$$

This means, that the question whether there is a sub-matrix *HotelPlusExternal*/χ with at least 250 rows such that at least 50 percent of rows satisfy both QSatisfaction(*average, high*) and DayOfWeek(*Mon, Tue, Wed*) can be solved by a run of the KL-Miner procedure with the following parameters: The QSatisfaction attribute is the only relevant row attribute. The DayOfWeek is the only relevant column attribute. A KL-quantifier $\approx_{range}$ defined as

$$\langle KLSF: Sum, 2, 3, 1, 3, ProcCond \rangle \geq 50 \wedge Sum_{\langle 2,3,1,3 \rangle} \geq 250 \qquad (9.8)$$

is used together with $\mathscr{RP}$-expressions Guest, Domicile, Meteo, Stay, Price, and Check-in introduced in Fig. 9.9 and the $\mathscr{RC}$-expression Cond_range shown in Table 9.2.

Table 9.2: $\mathscr{RC}$-expression Cond_range.

Name: Cond_range	Length: $1 - 16$
Set of partial cedents	Length
Guest	$0 - 2$
Domicile	$0 - 1$
Meteo	$0 - 2$
Stay	$0 - 5$
Price	$0 - 3$
Check-in	$0 - 3$

13 405 KL-patterns were tested in one second of a run of the KL-Miner with the above described parameters. All of the 16 resulting true KL-patterns concern data sub-matrices *HotelPlusExternal*/χ where χ is a basic Boolean attribute TypeOfVisits(*business*) or a conjunction containing TypeOfVisits(*business*). The KL-Table KL(QSatisfaction, DayOfWeek, *HotelPlusExternal*/*business*) where "*business*" abbreviates "TypeOfVisits(*business*)" is available in Fig. 9.10. Note that if we denote the KL-Table KL(QSatisfaction, DayOfWeek,

Name: Guest		Type: *Conjunction*		Length: $0-2$	
Attribute	Coefficients	Length	Gace	B/R	Class
Age	*Sequences*	$1-2$	*Pos*	B	—
Sex	*Subsets*	$1-1$	*Pos*	B	—

Name: Domicile		Type: *Conjunction*		Length: $0-1$	
Attribute	Coefficients	Length	Gace	B/R	Class
State	*Subsets*	$1-1$	*Pos*	B	—
City	*Subsets*	$1-1$	*Pos*	B	—
Foreigner	*Subsets*	$1-1$	*Pos*	B	—

Name: Meteo		Type: *Conjunction*		Length: $0-2$	
Attribute	Coefficients	Length	Gace	B/R	Class
Sky	*Subsets*	$1-1$	*Pos*	B	—
Temperature	*Sequences*	$1-2$	*Pos*	B	—

Name: Stay		Type: *Conjunction*		Length: $0-5$	
Attribute	Coefficients	Length	Gace	B/R	Class
Nights	*Subsets*	$1-1$	*Pos*	B	—
SaturdayNight	*Subsets*	$1-1$	*Pos*	B	—
Persons	*Subsets*	$1-1$	*Pos*	B	—
PersonNights	*Sequences*	$1-2$	*Pos*	B	—
TypeOfVisits	*Subsets*	$1-1$	*Pos*	B	—

Name: Price		Type: *Conjunction*		Length: $0-3$	
Attribute	Coefficients	Length	Gace	B/R	Class
PriceRoom	*Sequences*	$1-2$	*Pos*	B	—
PriceFood	*Sequences*	$1-2$	*Pos*	B	—
PriceTotal	*Sequences*	$1-2$	*Pos*	B	—

Name: Check-in		Type: *Conjunction*		Length: $0-3$	
Attribute	Coefficients	Length	Gace	B/R	Class
Month	*Cyclical*	$1-3$	*Pos*	B	—
Season	*Subsets*	$1-1$	*Pos*	B	—
Year	*Subsets*	$1-1$	*Pos*	B	—

Figure 9.9: $\mathscr{RP}$-expressions Guest, Domicile, Meteo, Stay, Price, and Check-in.

$HotelPlusExternal/business)$ as $\{n_{i,j}\}_{i=1,\ldots,3}^{j=1,\ldots,7}$, then it holds

$$\frac{\sum_{i=2,3}^{j=1,2,3} n_{i,j}}{\sum_{i=1,\ldots,3}^{j=1,\ldots,7} n_{i,j}} = \frac{210}{410} = 0.51 \ . \qquad (9.9)$$

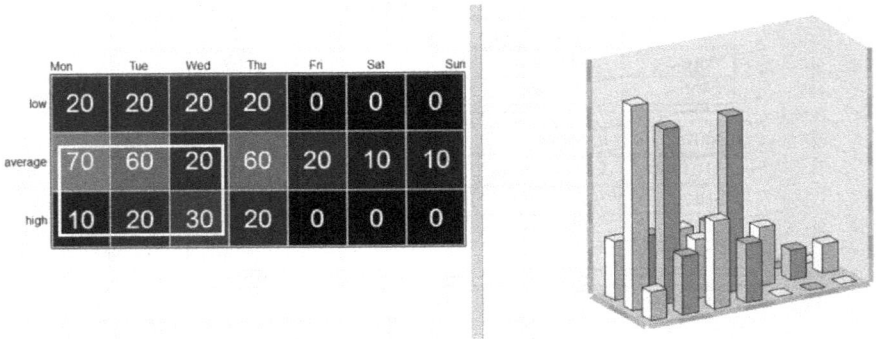

Figure 9.10: KL(QSatisfaction, DayOfWeek, *HotelPlusExternal*/Type(*business*)).

Chapter 10

SD4ft-Miner—Couples of GUHA Association Rules

Main features of the SD4ft-Miner procedure are introduced in Section 10.1 together with related notions. Examples of application of the SD4ft-Miner procedure in *Accidents* dataset are described in Section 10.2. Additional examples are available in [121, 128, 132] and in [76] where the name SDS-Miner is used.

10.1   SD4ft-Miner and Related Notions

The SD4ft-Miner procedure and couples of association rules it deals with are informally introduced in Section 3.5. The procedure deals with SD4ft-patterns in one of the forms $\alpha \bowtie \beta : \varphi \approx \psi / \chi$, $\alpha \bowtie \alpha \wedge \beta : \varphi \approx \psi / \chi$, and $\alpha \bowtie \neg \alpha : \varphi \approx \psi / \chi$ mentioned in Section 3.5. A used form is determined by a parameter *Mode*. The examples presented in this chapter concern only the form $\alpha \bowtie \beta : \varphi \approx \psi / \chi$ which corresponds to the value $\alpha \bowtie \beta$ of the parameter *Mode*.

The remaining used parameters and output of the procedure are introduced in Section 10.1.1. Generalised quantifiers the SD4ft-Miner procedure deals with are called *SD4ft-quantifiers*. The SD4ft-quantifiers used in the presented examples are introduced in Section 10.1.2.

10.1.1   SD4ft-Miner input and output

There are the following parameters defining a set of relevant SD4ft-patterns $\alpha \bowtie \beta : \varphi \approx \psi / \chi$ to be generated and verified:

■ an $\mathscr{RC}$-expression defining a set *Ant* of relevant antecedents φ

■ an $\mathscr{RC}$-expression defining a set *Succ* of relevant succedents ψ

■ an $\mathscr{RC}$-expression defining a set *Cond* of relevant conditions χ

■ an $\mathscr{RC}$-expression defining a set $\mathscr{S}_\alpha$ of relevant Boolean attributes α

■ an $\mathscr{RC}$-expression defining a set $\mathscr{S}_\beta$ of relevant Boolean attributes β

■ a list *ListSD4ft* of basic SD4ft-quantifiers

■ a parameter *Mode* with possible values $\alpha \bowtie \beta$, $\alpha \bowtie \alpha \wedge \beta$ and $\alpha \bowtie \neg \alpha$; only the value $\alpha \bowtie \beta$ is used in the examples below.

A set of relevant SD4ft-patterns is for parameter $Mode = \alpha \bowtie \beta$, defined as a set of all patterns $\alpha \bowtie \beta : \varphi \approx \psi / \chi$ satisfying $\varphi \in Ant$, $\psi \in Succ$, $\chi \in Cond$, $\alpha \in \mathscr{S}_\alpha$, $\beta \in \mathscr{S}_\beta$. SD4ft-quantifier $\approx$ is defined as a conjunction of basic SD4ft-quantifiers introduced in the list *ListSD4ft*. In addition, it must hold that φ, ψ, χ have no common attributes and there is no attribute used both in one of φ, ψ, χ and in α or β. However, α and β can have common attributes. The $\mathscr{RC}$-expression defining a set *Cond* can be missing.

The SD4ft-Miner procedure generates all relevant SD4ft-patterns and verifies them in a given data matrix $\mathscr{M}$. Each generated SD4ft-pattern is verified such that the SD4ft-quantifier $\approx$ is applied to a couple of contingency tables given by the parameter *Mode* (i.e., $\alpha \bowtie \beta$ in examples below). SD4ft-patterns satisfying the condition given by the SD4ft-quantifier $\approx$ are listed in an output.

10.1.2   SD4ft-quantifiers

An SD4ft-pattern $\alpha \bowtie \beta : \varphi \approx \psi / \chi$ is true in a data matrix $\mathscr{M}$ if a condition related to a SD4ft-quantifier $\approx$ is satisfied for a couple of 4ft-tables $T1 = \langle a_1, b_1, c_1, d_1 \rangle$ and $T2 = \langle a_2, b_2, c_2, d_2 \rangle$ introduced in Fig. 10.1. This means that conditional association rules $\varphi \approx \psi / \chi \wedge \alpha$ and $\varphi \approx \psi / \chi \wedge \beta$ are compared.

SD4ft-quantifier is usually a conjunction of several basic SD4ft-quantifiers. Examples of important basic SD4ft-quantifiers are available in Table 10.1.

It holds: $B_1 > 0, B_2 > 0$ are positive integer numbers, $0 \leq p_1 \leq 1, 0 \leq p_2 \leq 1$, $0 \leq s_1 \leq 1, 0 \leq s_2 \leq 1$ are real numbers, $q \geq 0$ is a real number $\circledast$ is one of the relations $<, \leq, =, \geq, >$ concerning real numbers.

$\mathscr{M}/\chi \wedge \alpha$	ψ	$\neg\psi$
φ	a_1	b_1
$\neg\varphi$	c_1	d_1

$\mathscr{M}/\chi \wedge \beta$	ψ	$\neg\psi$
φ	a_2	b_2
$\neg\varphi$	c_2	d_2

Figure 10.1: $Mode = \alpha \times \beta$: $T1 = \langle a_1, b_1, c_1, d_1 \rangle$ and $T2 = \langle a_2, b_2, c_2, d_2 \rangle$.

Table 10.1: Simple SD4ft-quantifiers.

#	Name	Expression	$\approx (a,b,c,d) = 1$ if and only if
1	$Base_1$	–	$a_1 \circledast B_1$
2	$Base_2$	–	$a_2 \circledast B_2$
3	$Conf_1$	$Conf_1 \circledast p_1$	$\frac{a_1}{a_1+b_1} \circledast p_1 \wedge a_1 + b_1 > 0$
4	$Conf_2$	$Conf_2 \circledast p_2$	$\frac{a_2}{a_2+b_2} \circledast p_2 \wedge a_2 + b_2 > 0$
5	$RConf$	$RConf \circledast q$	$\frac{a_1}{a_1+b_1} / \frac{a_2}{a_2+b_2} \circledast q \wedge a_1 + b_1 > 0 \wedge a_2 + b_2 > 0$
6	$DiffConf$	$DiffConf \circledast q$	$\frac{a_1}{a_1+b_1} - \frac{a_2}{a_2+b_2} \circledast q \wedge a_1 + b_1 > 0 \wedge a_2 + b_2 > 0$

Conditions in the last column with the symbol $\circledast$ are satisfied if relations given by $\circledast$ are satisfied. A conjunction

$$RConf \geq 2 \wedge a_1 \geq 100 \wedge a_2 \geq 100$$

is a typical example of SD4ft-quantifier.

10.2 Applying SD4ft-Miner in *Accidents* Dataset

Six analytical questions concerning the *Accidents* dataset are formulated and solved in this section to demonstrate possibilities of the SD4ft-Miner procedure. First, tasks concerning differences among districts and differences between particular districts and the whole dataset are introduced in Sections 10.2.1–10.2.3. Then, three tasks concerning differences and similarities between male and female drivers are described in Sections 10.2.4–10.2.6.

10.2.1 Differences among districts

Recall that there are 416 districts, see Section 2.3.6. We are interested in differences among particular districts what concerns relative frequencies of fatal or serious accidents. An example of such a difference is presented in Fig. 10.2 in a form of a detailed SD4ft-Miner output.

There are two conditional association rules are in Fig. 10.2:

$$\mathscr{R}_1 : \text{Sex}(Male) \Rightarrow_{0.043;90} \text{Severity}(Fatal)/\text{District}(Highland) \tag{10.1}$$

$$\mathscr{R}_2 : \text{Sex}(Male) \Rightarrow_{0.011;62} \text{Severity}(Fatal)/\text{District}(Westminster) . \tag{10.2}$$

A 4ft-table of rule $\mathscr{R}_1$ is $\langle 90, 2022, 16, 806 \rangle$ and a 4ft-table of rule $\mathscr{R}_2$ is $\langle 62, 5382, 8, 1184 \rangle$. Thus, a confidence $Conf(\mathscr{R}_1)$ of the rule $\mathscr{R}_1$ is 0.043 and a confidence $Conf(\mathscr{R}_2)$ of the rule $\mathscr{R}_2$ is 0.011. It holds $\frac{Conf(\mathscr{R}_1)}{Conf(\mathscr{R}_2)} = 3.74$ and this means that a relative frequency of fatal accidents of men in district Highland is 3.74 times higher than in district Westminster.

This leads us to the analytical question "*Which segments S of accidents and which couples of different districts $\langle D_1;D_2 \rangle$ satisfy that relative frequency of fatal*

Antecedent: Sex(Male)
Succedent: Severity(Fatal)
First set: District(Highland)
Second set: District(Westminster)
Condition: (empty)

TEXT | DATA | FIRST SET | SECOND SET | F+S SET | DIFF ABS | DIFF REL |

	Succedent	¬Succedent	Succedent	¬Succedent
Antecedent	90	2022	62	5382
¬Antecedent	16	806	8	1184

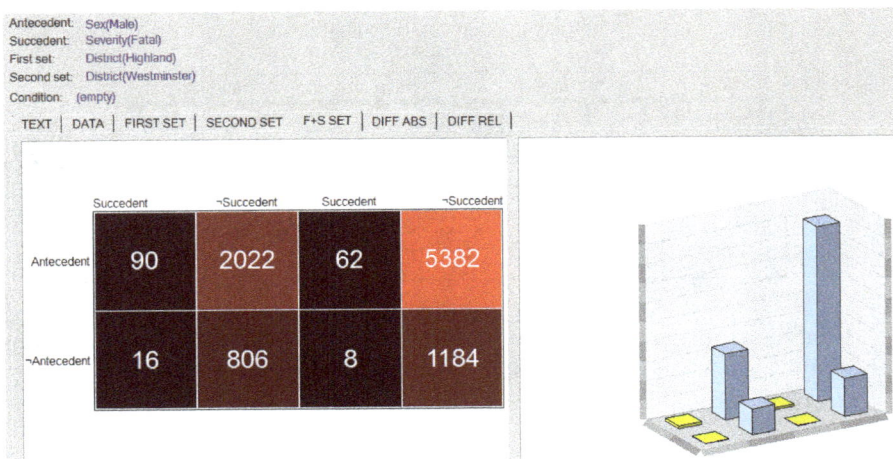

Figure 10.2: Example of SD4ft-Miner detailed output.

and serious accidents in segment S for district D_1 is at least two times higher than for D_2?" However, we are interested in districts with at least 2 000 accidents corresponding to attribute District2+, see Section 2.3.6. This can be formalized such that we are interested in true SD4ft-patterns

$$\text{District2+}(D_1) \bowtie \text{District2+}(D_2) : S \approx \text{Severity}(\omega) . \qquad (10.3)$$

We are interested in segments S of accidents specified by the type of vehicle and by the sex and the age of a driver. Let us emphasize that the attribute District2+ has 63 categories and thus $63^2 - 63 = 3\ 906$ couples of conditional association rules

$$S \approx \text{Severity}(\omega)/\text{District2+}(D_1) \quad \text{and} \quad S \approx \text{Severity}(\omega)/\text{District2+}(D_2)$$

concerning different districts must be generated and verified. Here both S and Severity(ω) are not concrete Boolean attributes but general forms of Boolean attributes. Let us note that in the case of using all 416 districts, there are $416^2 - 416 = 172\ 640$ couples of general forms of conditional association rules for verification.

This leads us to a definition of a rather narrow set of relevant segments. A set of relevant antecedents corresponding to a set of relevant segments S is specified by the $\mathcal{RP}$-expression Type_Driver available in Fig. 10.3. A set of relevant succedents is defined by the $\mathcal{RP}$-expression Fatal_Serious presented also in Fig. 10.3. Both sets $\mathcal{S}_\alpha$ of relevant Boolean attributes α and $\mathcal{S}_\beta$ of relevant Boolean attributes β are specified by the $\mathcal{RP}$-expression Districts_Couples available also in Fig. 10.3. Finally, SD4ft-quantifier defined by a conjunction $RConf \geq 2 \wedge a_1 \geq 100 \wedge a_2 \geq 100$ is applied in mode $\alpha \times \beta$.

Name: Type_Driver		Type: *Conjunction*		Length: $1-2$	
Attribute	Coefficients	Length	Gace	B/R	Class
Type	*Subsets*	$1-1$	*Pos*	B	—
Sex	*Subsets*	$1-1$	*Pos*	B	—
Age	*Sequences*	$4-4$	*Pos*	B	—

Name: Fatal_Serious		Type: *Conjunction*		Length: $1-1$	
Attribute	Coefficients	Length	Gace	B/R	Class
Severity	*Left cuts*	$1-2$	*Pos*	B	—

Name: Districts_Couples		Type: *Conjunction*		Length: $1-1$	
Attribute	Coefficients	Length	Gace	B/R	Class
District2+	*Subsets*	$1-1$	*Pos*	B	—

Figure 10.3: $\mathscr{R}\mathscr{P}$-expressions Type_Driver, Fatal_Serious, and Districts_Couples.

A run of the SD4ft-Miner with these parameters resulted in 196 true SD4ft-patterns. 1 062 432 SD4ft-patterns were tested in nine minutes and 14 seconds. A detailed output of the SD4ft-pattern with the highest value of *RConf* is available in Fig. 10.4. There are two conditional association rules in Fig. 10.4:

$$\mathscr{R}\mathscr{A}_1 : \text{Type}(Car) \wedge Male \Rightarrow_{0.318;612} \text{Severity}(Fatal, Serious)/Liverpool \quad (10.4)$$

$$\mathscr{R}\mathscr{A}_2 : \text{Type}(Car) \wedge Male \Rightarrow_{0.125;127} \text{Severity}(Fatal, Serious)/Plymouth.$$
$$(10.5)$$

Here we write "*Liverpool*" instead of "District 2+(*Liverpool*)", etc. $\langle 612, 1320, 637, 1850 \rangle$ and $\langle 127, 887, 171, 1151 \rangle$ are 4ft-tables of rules $\mathscr{R}\mathscr{A}_1$

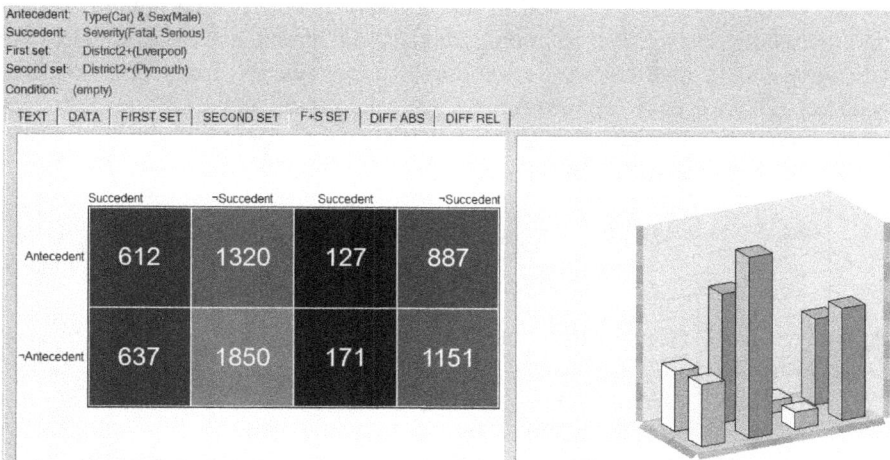

Figure 10.4: Detailed output of the SD4ft-pattern with the highest value of *RConf*.

and $\mathscr{R}\mathscr{A}_2$ respectively. Thus, confidence $Conf(\mathscr{R}\mathscr{A}_1)$ of the rule $\mathscr{R}\mathscr{A}_1$ is 0.318 and confidence $Conf(\mathscr{R}\mathscr{A}_2)$ of the rule $\mathscr{R}\mathscr{A}_2$ is 0.125. It holds $\frac{Conf(\mathscr{R}\mathscr{A}_1)}{Conf(\mathscr{R}\mathscr{A}_2)} = 2.53$ and this means that a relative frequency of fatal and serious accidents in Liverpool is 2.53 times higher than in Plymouth.

Recall that there are 196 true output SD4ft-patterns. A detailed interpretation of all results is out of range of this section. We only present selected additional interesting results in Table 10.2. Here it holds:

■ Severity for all output patterns, not only patterns presented in Table 10.2 is given by Boolean attribute Severity(*Fatal, Serious*).

■ We write *"Car"* instead of *"Type(Car)"*, *"Age(26–65)"* instead of *"Age(26–35,36–45,46–55,56–65)"*, etc. in Table 10.2.

Table 10.2: Selected interesting output SD4ft-patterns.

Segment	District_1	District_2	Conf_1	Conf_2	RConf
Car ∧ Age(26–65)	Wirral	Plymouth	0.295	0.118	2.529
Male ∧ ∧ Age(36–75)	Aberdeedshire	Plymouth	0.329	0.134	2.462
Bus_coach_17+ ∧ ∧ Male	Glasgow City	Leeds	0.224	0.094	2.385
Male ∧ ∧ Age(36–75)	Aberdeenshire	Croydon	0.329	0.151	2.178
Female	Wirral	Cornwall	0.255	0.127	2.012

10.2.2 Confidence in districts higher than in the whole dataset

In this section, we are interested in the differences between particular districts and the whole dataset. We will compare all 416 districts, see Section 2.3.6. We are interested in districts with remarkably higher relative frequencies of fatal accidents under specific circumstances. An example of such a difference is presented in Fig. 10.5 in a form of a detailed SD4ft-Miner output.

There are two association rules in Fig. 10.5:

$$\mathscr{R}\mathscr{B}_1 : Car \wedge \text{Age}(36\text{-}75) \Rightarrow_{0.080;16} \text{Severity}(Fatal)/Harborough \quad (10.6)$$

$$\mathscr{R}\mathscr{B}_2 : Car \wedge \text{Age}(36\text{-}75) \Rightarrow_{0.017;2430} \text{Severity}(Fatal) . \quad (10.7)$$

Again, we write *"Car"* instead of *"Type(Car)"* and *"Age(36–75)"* instead of *"Age(36–45,46–55,56–65,66–75)"*.

$\langle 16, 185, 25, 530 \rangle$ and $\langle\, 2\,430, 142\,609, 7\,919, 386\,031 \,\rangle$ are 4ft-tables of rules $\mathscr{R}\mathscr{B}_1$ and $\mathscr{R}\mathscr{B}_2$ respectively. Thus, confidence $Conf(\mathscr{R}\mathscr{B}_1)$ of the rule $\mathscr{R}\mathscr{B}_1$ is 0.080 and confidence $Conf(\mathscr{R}\mathscr{B}_2)$ of the rule $\mathscr{R}_2$ is 0.017. It holds $\frac{Conf(\mathscr{R}\mathscr{B}_1)}{Conf(\mathscr{R}\mathscr{B}_2)} = 4.75$ and this means that a relative frequency of fatal accidents of

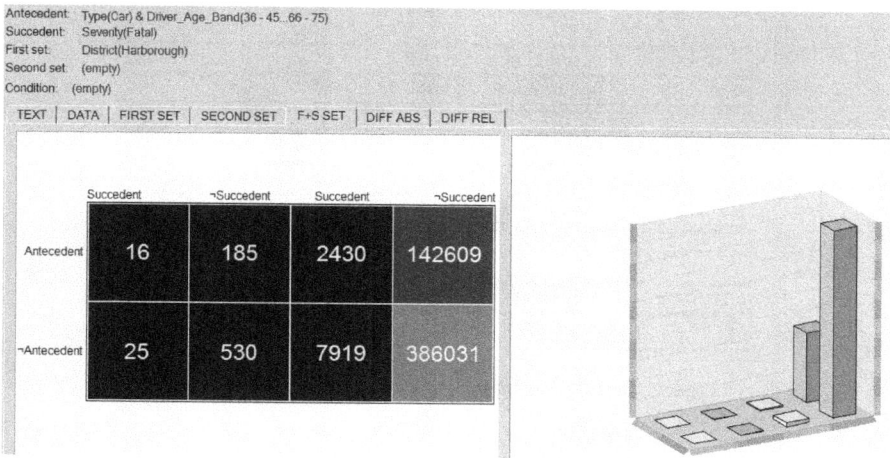

Figure 10.5: Example of remarkable higher frequency of fatal accidents.

cars with driver's age 36–75 in district Harborough is 4.75 times higher than in the whole dataset.

This leads us to the analytical question "*Which segments S of accidents and which districts D satisfy that relative frequency of fatal accidents in segment S for district D is at least 3 times higher than for the whole dataset?*" This can be formalized such that we are interested in true SD4ft-patterns

$$\text{District}(D) \bowtie True : S \approx \text{Severity}(Fatal) \qquad (10.8)$$

We use similar parameters for the SD4ft-Miner as in the previous section. A set of relevant antecedents corresponding to a set of relevant segments S is specified by the $\mathscr{RP}$-expression Type_Driver available in Fig. 10.3. A set of relevant succedents is defined by the $\mathscr{RP}$-expressions Fatal presented in Fig. 10.6. A set $\mathscr{S}_\alpha$ of relevant Boolean attributes α is defined by the $\mathscr{RP}$-expression Districts available in Fig. 10.6. Finally, a set $\mathscr{S}_\beta$ of relevant Boolean attributes β is specified by the $\mathscr{RP}$-expression True available also in Fig. 10.6.

Finally, mode $\alpha \times \beta$ and SD4ft-quantifier defined by a conjunction

$$RConf \geq 3 \wedge a_1 \geq 10 \wedge a_2 \geq 10$$

will be used.

A run of the SD4ft-Miner with these parameters resulted in 41 true SD4ft-patterns. 69 888 SD4ft-patterns were tested in 36 seconds. A detailed output of the SD4ft-pattern with the highest value of $RConf$ is available already in Fig. 10.5. A detailed interpretation of all results is out of range of this section. We again present only selected additional interesting results in Table 10.3. Here it holds:

Name: Fatal		Type: *Conjunction*		Length: $1-1$	
Attribute	Coefficients	Length	Gace	B/R	Class
Severity	*One category*	*Fatal*	*Pos*	*B*	−

Name: Districts		Type: *Conjunction*		Length: $1-1$	
Attribute	Coefficients	Length	Gace	B/R	Class
District	*Subsets*	$1-1$	*Pos*	*B*	−

Name: True		Type: *Conjunction*		Length: $1-1$	
Attribute	Coefficients	Length	Gace	B/R	Class
True	*One category*	1	*Pos*	*B*	−

Figure 10.6: $\mathscr{RP}$-expressions Fatal, Districts, and True.

■ *Conf_2* denotes a confidence of the rule for the whole dataset.

■ We write *"Male"* instead of *"Sex(Male)"*, *"Age($\geq$ 46)"* instead of *"Age(46–55,56–65,66–75,Over 75)"*, etc. in Table 10.3.

Table 10.3: Selected SD4ft-patterns with extremely high value of *RConf*.

Segment	District	$Conf\_1$	$Conf\_2$	*RConf*
Male ∧ Age($\geq$ 46)	*Monmouthshire*	0.103	0.024	4.257
Male ∧ Age($\geq$ 46)	*Horsham*	0.098	0.024	4.035
Male ∧ Age(*36–75*)	*Vale Royal*	0.085	0.023	3.607
Age($\geq$ 46)	*Broxbourne*	0.069	0.021	3.212

10.2.3 Confidence in districts lower than in the whole dataset

Here we are interested in districts with remarkable lower relative frequencies of fatal accidents under specific circumstances. An example of such a difference is presented in Fig. 10.7 in a form of a detailed SD4ft-Miner output. There are two association rules in Fig. 10.7:

$$\mathscr{RC}_1 : Car \wedge Male \Rightarrow_{0.007;10} \text{Severity}(Fatal)/Westminster \qquad (10.9)$$

$$\mathscr{RC}_2 : \mathscr{RC}_1 : Car \wedge Male \Rightarrow_{0.023;5321} \text{Severity}(Fatal) . \qquad (10.10)$$

Again, we write *"Car"* instead of *"Type(Car)"*, etc.

$\langle 10, 1\ 403, 60, 5\ 163 \rangle$ and $\langle 5\ 321, 227\ 828, 5\ 028, 300\ 812 \rangle$ are 4ft-tables of rules $\mathscr{RC}_1$ and $\mathscr{RC}_2$ respectively. Thus, confidence $Conf(\mathscr{RC}_1)$ of the rule $\mathscr{RC}_1$ is 0.007 and confidence $Conf(\mathscr{RB}_2)$ of the rule $\mathscr{RB}_2$ is 0.023. It holds $\frac{Conf(\mathscr{RC}_2)}{Conf(\mathscr{RC}_1)} = \frac{\frac{5321}{5321+227828}}{\frac{10}{10+1043}} = 3.22$ and this means that a relative frequency of fatal accidents of cars with male drivers in district Westminster is 3.22 times lower than in the whole dataset.

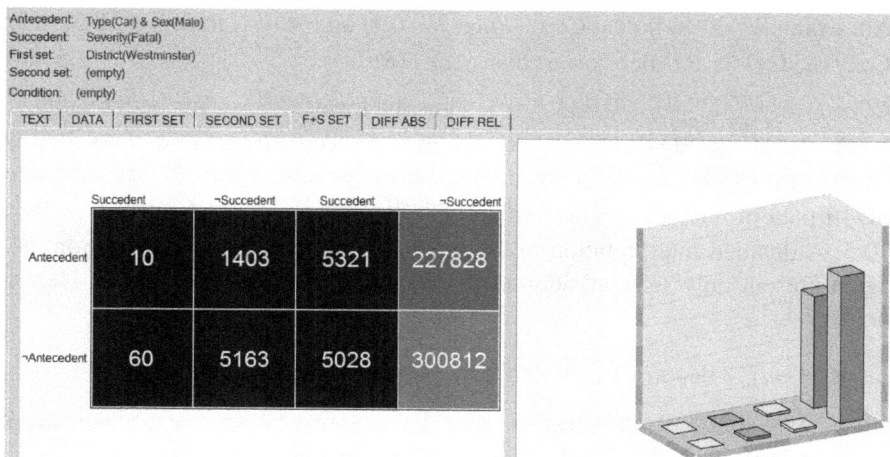

Antecedent:	Type(Car) & Sex(Male)
Succedent:	Severity(Fatal)
First set:	District(Westminster)
Second set:	(empty)
Condition:	(empty)

TEXT | DATA | FIRST SET | SECOND SET | F+S SET | DIFF ABS | DIFF REL

	Succedent	¬Succedent	Succedent	¬Succedent
Antecedent	10	1403	5321	227828
¬Antecedent	60	5163	5028	300812

Figure 10.7: Example of remarkable lower frequency of fatal accidents.

This leads us to the analytical question *"Which segments S of accidents and which districts D satisfy that relative frequency of fatal accidents in segment S for district D is at least 2.5 times lower than for the whole dataset?"* This can be formalized such that we are interested in true SD4ft-patterns

$$\text{District}(D) \bowtie True : S \approx \text{Severity}(Fatal) . \tag{10.11}$$

We use similar parameters for the SD4ft-Miner as in the previous section. A set of relevant antecedents corresponding to a set of relevant segments S is specified by the $\mathscr{R}\mathscr{P}$-expression Type_Driver available in Fig. 10.3. A set of relevant succedents is defined by the $\mathscr{R}\mathscr{P}$-expressions Fatal presented in Fig. 10.6. A set $\mathscr{S}_\alpha$ of relevant Boolean attributes α is defined by the $\mathscr{R}\mathscr{P}$-expression Districts available in Fig. 10.6. Finally, a set $\mathscr{S}_\beta$ of relevant Boolean attributes β is specified by the $\mathscr{R}\mathscr{P}$-expression True available also in Fig. 10.6. Finally, mode $\alpha \times \beta$ and SD4ft-quantifier defined by conjunction

$$RConf \leq 0.4 \wedge a_1 \geq 10 \wedge a_2 \geq 10 ,$$

will be used.

Let us note that we deal with couples of rules

$$Rule_1 : S \approx \text{Severity}(Fatal)/\text{District}(D) \tag{10.12}$$

$$Rule_2 : S \approx \text{Severity}(Fatal) . \tag{10.13}$$

We are interested in couples $\langle Rule_1, Rule_2 \rangle$ such that confidence $Conf(Rule_1)$ of $Rule_1$ is at least 2.5 times lower than confidence $Conf(Rule_2)$ of $Rule_2$. This means that $\frac{Conf(Rule_2)}{Conf(Rule_1)} \geq 2.5$ which is the same as $RConf = \frac{Conf(Rule_0 1)}{Conf(Rule_2)} \leq 0.4.$

An example: if it holds $Conf(Rule_1) = 0.1$ and $Conf(Rule_2) = 0.3$, then $Conf(Rule_1)$ is 3 times lower than $Conf(Rule_2)$, $\frac{Conf(Rule_2)}{Conf(Rule_1)} = 3 \geq 2.5$ and $RConf = \frac{Conf(Rule_0 1)}{Conf(Rule_2)} = 0.333 \leq 0.4$. Thus the lower $RConf$, the better value.

A run of the SD4ft-Miner with the above specified parameters resulted in 23 true SD4ft-patterns. 69 888 SD4ft-patterns were tested in 37 seconds. The SD4ft-pattern with the highest value of $\frac{Conf(Rule_2)}{Conf(Rule_1)}$ is available already in Fig. 10.5. A detailed interpretation of all results is out of range of this section. We again present only selected additional interesting results in Table 10.4. Here it holds:

- $Conf\_2$ denotes a confidence of the rule for the whole dataset.

- We write *"Male"* instead of *"Sex(Male)"*, *"Age($\leq$ 45)"* instead of *"Age(16–20,51–25,26–35,36–45)"*, etc. in Table 10.4.

Table 10.4: Selected SD4ft-patterns with extremely low values of $RConf$.

Segment	District	$Conf\_1$	$Conf\_2$	$\frac{Conf\_2}{Conf\_1}$
Male $\wedge$ Age($\leq$ 45)	*Edinburgh, City of*	0.009	0.025	2.726
Male $\wedge$ Age(21–55)	*Islington*	0.009	0.025	2.701
Male $\wedge$ Age(21–55)	*Brighton and Hove*	0.009	0.025	2.698
Male	*Portsmouth*	0.009	0.024	2.637

10.2.4 Relative frequency of accidents higher for male drivers

Our goal is to show that the SD4ft-Miner procedure can be used to search not only differences among subsets of objects defined by particular categories of an attribute but also similarities. We show three examples of analysis of differences and similarities between male and female drivers. We solve three mutually related analytical questions:

QM_F1 *"For which segments S of accidents is a relative frequency of fatal or serious accidents remarkably higher for male than for female drivers?"*

QM_F2 *"For which segments S of accidents is a relative frequency of fatal or serious accidents remarkably higher for female than for male drivers?"*

QM_F3 *"For which segments S of accidents is a relative frequency of fatal or serious accidents for female and for male drivers approximately the same?"*

All three questions can be formalized such that we are interested in true SD4ft-patterns

$$Sex(Male) \bowtie Sex(Female) : S \approx Severity(\omega) \qquad (10.14)$$

where S belongs to a suitably defined set of relevant segments, Severity(ω) belongs to a suitably defined set of Boolean attributes derived from the Severity attribute and $\approx$ is a suitable SD4ft-quantifier.

The set of relevant segments S as well as the set of relevant Boolean attributes Severity(ω) derived from the Severity attribute will be defined in the same way for all three relevant questions. $\mathcal{RP}$-expressions Driver, Vehicle, Date_Time, and Conditions available in Fig. 10.8 and $\mathcal{RC}$-expression AccidentsQM_F available in Table 10.5 are used to define the set of relevant segments. Details on the involved attributes are available in Sections 2.3.3, 2.3.2 and 2.3.4.

Name: Driver		Type: *Conjunction*		Length: $0-1$	
Attribute	Coefficients	Length	Gace	B/R	Class
Age	*Sequences*	$1-2$	*Pos*	B	—
Journey	*Subsets*	$1-1$	*Pos*	B	—

Name: Vehicle		Type: *Conjunction*		Length: $0-1$	
Attribute	Coefficients	Length	Gace	B/R	Class
Type	*Subsets*	$1-1$	*Pos*	B	—

Name: Date_Time		Type: *Conjunction*		Length: $0-1$	
Attribute	Coefficients	Length	Gace	B/R	Class
DayOfWeek	*Cyclical*	$1-3$	*Pos*	B	—
Hour	*Cyclical*	$1-3$	*Pos*	B	—
Month	*Cyclical*	$1-3$	*Pos*	B	—

Name: Conditions		Type: *Conjunction*		Length: $0-1$	
Attribute	Coefficients	Length	Gace	B/R	Class
Area	*Subsets*	$1-1$	*Pos*	B	—
Light	*Subsets*	$1-1$	*Pos*	B	—
Manoeuvre	*Subsets*	$1-1$	*Pos*	B	—
Road_Type	*Subsets*	$1-1$	*Pos*	B	—
Speed_Limit	*Subsets*	$1-1$	*Pos*	B	—
Vehicle_Location	*Subsets*	$1-1$	*Pos*	B	—

Figure 10.8: $\mathcal{RP}$-expressions Driver, Vehicle, Date_Time, and Conditions.

Table 10.5: $\mathcal{RC}$-expression AccidentsQM_F.

Name: AccidentsQM_F	Length: $1-2$
Set of partial cedents	Length
Driver	$0-1$
Vehicle	$0-1$
Date_Time	$0-1$
Conditions	$0-1$

A set of relevant Boolean attributes Severity(ω) is defined by the $\mathscr{RP}$-expression Fatal_Serious, see Fig. 10.3. $\mathscr{RP}$-expressions Male and Female available in Fig. 10.9 are used to define sets $\mathscr{S}_\alpha$ and $\mathscr{S}_\beta$.

Name: Male		Type: *Conjunction*		Length: $0-1$	
Attribute	Coefficients	Length	Gace	B/R	Class
Sex	*One category*	*Male*	*Pos*	B	—

Name: Female		Type: *Conjunction*		Length: $0-1$	
Attribute	Coefficients	Length	Gace	B/R	Class
Sex	*One category*	*Female*	*Pos*	B	—

Figure 10.9: $\mathscr{RP}$-expressions Male and Female.

The question QM_F1 is answered in this section, answers to the questions QM_F2 and QM_F3 are available in Sections 10.2.5 and 10.2.6. Let us emphasize that our goal is only to demonstrate the possibilities of the SD4ft-Miner procedure. Our goal is not to present a deeper analysis of answers to these questions. There are many related partial analytical questions to be formulated and solved. This encompasses modifications of definitions of sets of relevant SD4ft-patterns including definitions of SD4ft-quantifiers. We use the above introduced $\mathscr{RP}$-expressions and $\mathscr{RC}$-expressions and SD4ft-quantifiers defined with the goal to get a reasonable number of output patterns.

The question QM_F1 – "*For which segments S of accidents is a relative frequency of fatal or serious accidents remarkably higher for male than for female drivers?*" is solved by a run of the SD4ft-Miner with the following parameters. A set of relevant antecedents corresponding to a set of relevant segments S is specified by the $\mathscr{RC}$-expression AccidentsQM_F available in Table 10.5. A set of relevant succedents is defined by the $\mathscr{RP}$-expressions Fatal_Serious presented in Fig. 10.3. A set $\mathscr{S}_\alpha$ of relevant Boolean attributes α is specified by the $\mathscr{RP}$-expression Male and a set $\mathscr{S}_\beta$ of relevant Boolean attributes β is specified by the $\mathscr{RP}$-expression Female available in Fig. 10.9. We are interested in SD4ft-patterns

$$\text{Sex}(Male) \bowtie \text{Sex}(Female) : S \approx \text{Severity}(\omega)$$

where the SD4ft-quantifier $\approx$ is defined as a conjunction

$$RConf \geq 3.5 \wedge a_1 \geq 50 \wedge a_2 \geq 20,$$

and mode $\alpha \times \beta$ is used. Note that the ratio between minimal values for a_1 and a_2 corresponds to the ratio of the numbers of accidents with male and female drivers.

A run of the SD4ft-Miner with the above specified parameters resulted in 19 true SD4ft-patterns. 22 660 SD4ft-patterns were tested in 74 seconds.

Table 10.6: Selected SD4ft-patterns – frequency of accidents higher for male drivers.

Segment	Severity	$Conf_{Male}$	$Conf_{Female}$	$\frac{Conf_{Male}}{Conf_{Female}}$
Journey(*Part of work*) ∧ ∧ Area(*Rural*)	*Fatal*	0.030	0.006	4.602
Month(*Aug, Sep, Oct*) ∧ ∧ Maneouvre(*Going ahead left-hand bend*)	*Fatal*	0.041	0.011	3.692
Journey(*Part of work*) ∧ ∧ Hour(*8,9,10*)	*Fatal*	0.019	0.005	3.648

The SD4ft-pattern with the highest value of $RConf = \frac{Conf_{Male}}{Conf_{Female}}$ is available in Table 10.6 together with additional chosen interesting SD4ft-patterns. Here $Conf_{Male}$ denotes a confidence of rules $S \approx \text{Severity}(\omega)/\text{Sex}(Male)$ and $Conf_{Female}$ denotes a confidence of rules $S \approx \text{Severity}(\omega)/\text{Sex}(Female)$.

10.2.5 Relative frequency of accidents higher for female drivers

The analytical question QM_F2 – "*For which segments S of accidents is a relative frequency of fatal or serious accidents remarkably higher for female than for male drivers?*" is solved in this section by a run of the SD4ft-Miner procedure with the following parameters.

A set of relevant antecedents corresponding to a set of relevant segments S is specified by the $\mathscr{RC}$-expression AccidentsQM_F available in Table 10.5. A set of relevant succedents is defined by the $\mathscr{RP}$-expressions Fatal_Serious presented in Fig. 10.3. A set $\mathscr{S}_\alpha$ of relevant Boolean attributes α is specified by the $\mathscr{RP}$-expression Female and a set $\mathscr{S}_\beta$ of relevant Boolean attributes β is specified by the $\mathscr{RP}$-expression Male available in Fig. 10.9. We are interested in SD4ft-patterns

$$\text{Sex}(Female) \bowtie \text{Sex}(Male) : S \approx \text{Severity}(\omega)$$

where the SD4ft-quantifier $\approx$ is defined as a conjunction

$$RConf \geq 1.4 \wedge a_1 \geq 20 \wedge a_2 \geq 50,$$

and mode $\alpha \times \beta$ is used. Note that the ratio between minimal values for a_1 and a_2 corresponds to the ratio of the numbers of accidents with female and male drivers.

A run of the SD4ft-Miner with the above specified parameters resulted in 23 true SD4ft-patterns. 22 660 SD4ft-patterns were tested in 74 seconds, the same numbers as in the previous section. The SD4ft-pattern with the highest value of $RConf = \frac{Conf_{Female}}{Conf_{Male}}$ is available in Table 10.7 together with additional chosen interesting SD4ft-patterns. Here $Conf_{Female}$ denotes a confidence of rules

Table 10.7: Selected SD4ft-patterns – frequency of accidents higher for female drivers.

Segment	Severity	$Conf_{Female}$	$Conf_{Male}$	$\frac{Conf_{Female}}{Conf_{Male}}$
Hour(*12,13*) ∧ ∧ Maneouvre(*Parked*)	Fatal, Serious	0.211	0.104	2.018
Age(26-35) ∧ ∧ Maneouvre(*Parked*)	Fatal, Serious	0.200	0.101	1.978
Age(36-55) ∧ ∧ Type(*Goods 7.5 tonnes*)	Fatal, Serious	0.477	0.328	1.456

$S \approx$ Severity(ω)/Sex(*Female*) and $Conf_{Male}$ denotes a confidence of rules $S \approx$ Severity(ω)/Sex(*Male*). We write "Age(*36-55*)" instead of "Age(*36–45,46–55*)" in Table 10.7.

10.2.6 Similarities between male and female

The analytical question QM_F3 – "*For which segments S of accidents is a relative frequency of fatal or serious accidents for female and for male drivers approximately the same?*" is solved in this section by a run of the SD4ft-Miner procedure with the following parameters.

A set of relevant antecedents corresponding to a set of relevant segments S is specified by the $\mathscr{RC}$-expression AccidentsQM_F available in Table 10.5. A set of relevant succedents is defined by the $\mathscr{RP}$-expressions Fatal_Serious presented in Fig. 10.3. A set $\mathscr{S}_\alpha$ of relevant Boolean attributes α is specified by the $\mathscr{RP}$-expression Female and a set $\mathscr{S}_\beta$ of relevant Boolean attributes β is specified by the $\mathscr{RP}$-expression Male available in Fig. 10.9. We are interested in SD4ft-patterns

$$\text{Sex}(\textit{Female}) \bowtie \text{Sex}(\textit{Male}) : S \approx \text{Severity}(\omega)$$

where the SD4ft-quantifier $\approx$ is defined as a conjunction

$$RConf \geq 0.999 \wedge RConf \leq 0.1001 \wedge a_1 \geq 100 \wedge a_2 \geq 100,$$

and mode $\alpha \times \beta$ is used.

A run of the SD4ft-Miner with the above specified parameters resulted in 12 true SD4ft-patterns. 21 034 SD4ft-patterns were tested in 63 seconds. Six SD4ft-patterns with the value of $RConf = \frac{Conf_{Female}}{Conf_{Male}}$ as close as possible to 1 are available in Table 10.8. Here $Conf_{Female}$ denotes a confidence of rules $S \approx$ Severity(ω)/Sex(*Female*) and $Conf_{Male}$ denotes a confidence of rules $S \approx$ Severity(ω)/Sex(*Male*). A detailed output for the first SD4ft-pattern from Table 10.8 is available in Fig. 10.10.

There are two association rules in Fig. 10.10:

$$\mathscr{RMF}_1 : \text{Age}(\geq 66) \wedge \text{Month}(\textit{Nov, Dec}) \Rightarrow_{0.24302;444}$$
$$\text{Severity}(\textit{Fatal, Serious})/\textit{Female}$$

(10.15)

Table 10.8: Selected SD4ft-patterns – equal frequency of accidents for female and male drivers.

Segment	Severity	$Conf_{Female}$	$Conf_{Male}$	$\frac{Conf_{Female}}{Conf_{Male}}$
Age($\geq$ 66) $\wedge$ $\wedge$ Month(*Nov, Dec*)	*Fatal, Serious*	0.24302	0.24306	0.99983
Hour(*13*) $\wedge$ $\wedge$ Speed limit(*30*)	*Fatal, Serious*	0.19534	0.19538	0.99980
Age(*56-65*) $\wedge$ $\wedge$ Hour(*9,10*)	*Fatal, Serious*	0.20893	0.20899	0.99972
Age(*66-75*) $\wedge$ $\wedge$ Hour(*7,8,9*)	*Fatal, Serious*	0.21636	0.21630	1.00029
Day(*Saturday*) $\wedge$ $\wedge$ Maneouvre(*Turning right*)	*Fatal, Serious*	0.18959	0.18953	1.00031
Hour(*9,10,11*) $\wedge$ $\wedge$ Maneouvre(*Reversing*)	*Fatal, Serious*	0.19968	0.19961	1.00032

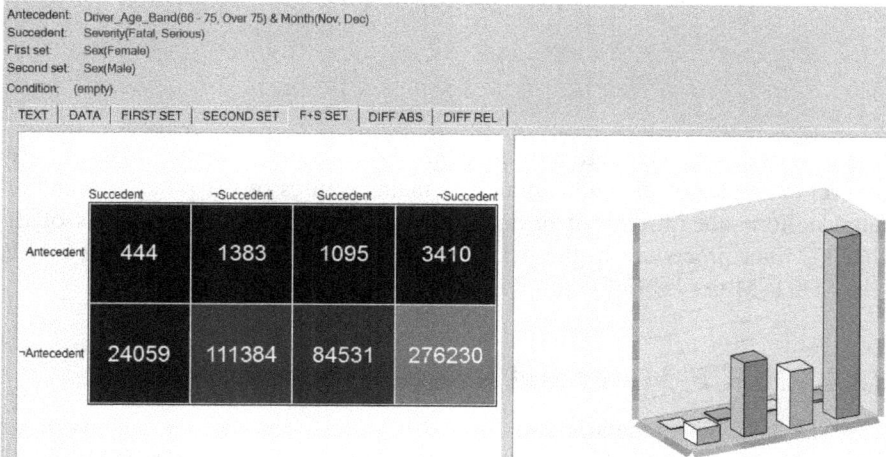

Figure 10.10: Detailed output for the first SD4ft-pattern from Table 10.8.

$$\mathscr{R}\mathscr{M}\mathscr{F}_2 : \text{Age}(\geq 66) \wedge \text{Month}(\textit{Nov,Dec}) \Rightarrow_{0.24306;1095}$$
$$\text{Severity}(\textit{Fatal,Serious})/\text{Male} \tag{10.16}$$

4ft-table of the rule $\mathscr{R}\mathscr{M}\mathscr{F}_1$ is $\langle 444, 1383, 24059, 111384 \rangle$ and 4ft-table of the rule $\mathscr{R}\mathscr{M}\mathscr{F}_2$ is $\langle 1095, 3410, 84531, 276230 \rangle$.

Chapter 11

SDCF-Miner—Couples of Histograms

The goal of this chapter is to introduce the SDCF-Miner procedure together with several examples of its application. The main features of the procedure and related notions are introduced in Section 11.1. Examples of applications of the SDCF-Miner procedure in the *Accidents* dataset are available in Section 11.2. The SDCF-Miner is briefly introduced also in [44, 128].

11.1 SDCF-Miner and Related Notions

The SDCF-Miner procedure and couples of association rules it deals with are informally introduced in Section 3.6. The procedure deals with SDCF-patterns in one of the forms $\alpha \bowtie \beta :\approx A/\chi, \alpha \bowtie \alpha \wedge \beta :\approx A/\chi, \alpha \bowtie \neg\alpha :\approx A/\chi$ mentioned in Section 3.6. A used form is determined by a parameter *Mode*.

Input and output of the SDCF-Miner procedure are described in Section 11.1.1. Necessary information on particular parameters are available in Sections 11.1.2–11.1.4.

11.1.1 SDCF-Miner input and output

We can say that the SDCF-Miner procedure generates and verifies relevant SDCF-patterns in the form $\kappa \bowtie \lambda : \approx A/\chi$, where κ and λ are Boolean attributes introduced below. An input of the SDCF-Miner procedure consists of a data matrix $\mathcal{M}$ and of the following parameters defining a set of relevant SDCF-patterns:

- a categorical attribute A with t categories

- an $\mathscr{RC}$-expression defining a set *Cond* of relevant conditions χ

- an $\mathscr{RC}$-expression defining a set $\mathscr{S}_\alpha$ of relevant Boolean attributes α

- an $\mathscr{RC}$-expression defining a set $\mathscr{S}_\beta$ of relevant Boolean attributes β

- *Mode* with possible values – strings $\alpha \bowtie \beta$, $\alpha \bowtie \alpha \wedge \beta$, and $\alpha \bowtie \neg \alpha$

- a list *ListSDCF* of basic SDCF-quantifiers relevant to the attributes A.

The SDCF-Miner procedure generates all patterns $\kappa \bowtie \lambda :\approx A/\chi$ such that

- $\chi \in Cond$

- κ and λ are created from Boolean attributes $\alpha \in \mathscr{S}_\alpha$ and $\beta \in \mathscr{S}_\beta$ in a way given by the parameter *Mode*, see Section 11.1.2

- SDCF-quantifier $\approx$ is defined as a conjunction of all basic SDCF-quantifiers from the list *ListSDCF*

- Boolean attributes κ and χ cannot have common attributes, the same is true for λ and χ; however, κ and λ can have common attributes.

Simple frequencies SDCF-quantifiers and SDCF-quantifiers concerning steps in the histogram are used as basic quantifiers in the below presented examples. These quantifiers are described in Sections 11.1.3 and 11.1.4. The conditions corresponding to the SDCF-quantifier $\approx$ are verified in SDCF-tables defined by the parameter *Mode*, see Section 11.1.2.

Each generated SDCF-pattern $\kappa \bowtie \lambda :\approx A/\chi$ is verified in the given data matrix $\mathscr{M}$. Output of the SDCF-Miner procedure consists of all true SDCF-patterns.

11.1.2 Modes of SDCF-Miner and SDCF-tables

SDCF-Miner deals with the SDCF-patterns of the form $\kappa \bowtie \lambda : \approx A/\chi$ where κ and λ are created from Boolean attributes $\alpha \in \mathscr{S}_\alpha$ and $\beta \in \mathscr{S}_\beta$. Possible forms of κ and λ are introduced below, they are shown already in Table 4.1 in Section 4.1.

SDCF-patterns are evaluated using couples of CF-tables introduced in Table 4.4 in Section 4.3. Such couples of CF-tables are called *SDCF-tables*. We deal here with SDCF-quantifiers requiring only a simplified form of SDCF-tables presented below. A form of the SDCF-table depends on a value of the *Mode* parameter. There are three possible values of this parameter, we use only first two of them.

If $Mode = \alpha \bowtie \beta$, then all $\kappa = \alpha$ and $\lambda = \beta$ where $\alpha \in \mathscr{S}_\alpha$ and $\beta \in \mathscr{S}_\beta$ are used in relevant SDCF-patterns. In this mode, we deal with SDCF-patterns $\alpha \bowtie \beta : \approx A/\chi$, thus histograms $A/\chi \wedge \alpha$ and $A/\chi \wedge \beta$ are compared.

If $Mode = \alpha \bowtie \alpha \wedge \beta$, then all $\kappa = \alpha$ and $\lambda = \alpha \wedge \beta$ where $\alpha \in \mathscr{S}_\alpha$ and $\beta \in \mathscr{S}_\beta$ are used in relevant SDCF-patterns. In this mode, we deal with SDCF-patterns $\alpha \bowtie \alpha \wedge \beta : \approx A/\chi$, thus histograms $A/\chi \wedge \alpha$ and $A/\chi \wedge \alpha \wedge \beta$ are compared.

SDCF-table $SDCF(A, \mathscr{M}, \chi, \kappa, \lambda)$ of $\kappa \bowtie \lambda : \approx A/\chi$ in a data matrix $\mathscr{M}$ is a $2t$-tuple

$$SDCF(A, \mathscr{M}, \chi, \kappa, \lambda) = \langle h_1^{(1)}, \ldots, h_t^{(1)}, h_1^{(2)}, \ldots, h_t^{(2)} \rangle \qquad (11.1)$$

of non-negative integers where t is the number of categories of the attribute A. $SDCF(A, \mathscr{M}, \chi, \kappa, \lambda)$ is often written in a form according to Fig. 11.1.

$\mathscr{M}$	a_1	$\ldots$	a_t
κ	$h_1^{(1)}$	$\ldots$	$h_t^{(1)}$
λ	$h_1^{(2)}$	$\ldots$	$h_t^{(2)}$

Figure 11.1: SDCF-table $(A, \mathscr{M}, \chi, \kappa, \lambda) = \langle h_1^{(1)}, \ldots, h_t^{(1)}, h_1^{(2)}, \ldots, h_t^{(2)} \rangle$.

We assume that the attribute A has categories $a_1, \ldots, a_t$. It holds for $i = 1, \ldots, t$:

■ $h_i^{(1)}$ is the number of rows of $\mathscr{M}$ satisfying both κ and $A(a_i)$

■ $h_i^{(2)}$ is the number of rows of $\mathscr{M}$ satisfying both λ and $A(a_i)$.

11.1.3 Simple frequencies SDCF-quantifiers

We deal with the simple frequencies SDCF-quantifiers in a form

$$\langle \text{SDCFSF: } Sum, u, v, Set \rangle \circledast Thr \qquad (11.2)$$

where

■ $\langle u, v \rangle$ is a range of a CF-quantifier for a histogram $\langle h_1, \ldots, h_t \rangle$, see Section 8.1.3

■ *Set* is one of the operation modes of SDCF-quantifier *Set*1, *Set*2

■ $\circledast$ is one of the relations $<, \leq, =, \geq, >$ concerning real numbers

■ *Thr* is a non-negative real number.

A quantifier $\langle$SDCFSF: $Sum, u, v, Set \rangle \circledast Thr$ defines a condition for an SDCF-table $SDCF(A, \mathcal{M}, \chi, \kappa, \lambda) = \langle h_1^{(1)}, \ldots, h_t^{(1)}, h_1^{(2)}, \ldots, h_t^{(2)} \rangle$ in the following way: If $Set = Set1$, then the relation (11.3) is used, if $Set = Set2$, then the relation (11.4) is used

$$Sum_1 \circledast Thr \quad \text{where} \quad Sum_1 = h_u^{(1)} + \cdots + h_v^{(1)} \tag{11.3}$$

$$Sum_2 \circledast Thr \quad \text{where} \quad Sum_2 = h_u^{(2)} + \cdots + h_v^{(2)} . \tag{11.4}$$

We say that an SDCF-quantifier $\langle$SDCFSF: $Sum, u, v, Set \rangle \circledast Thr$ is relevant for an attribute A with t categories if $v \leq t$.

Let us note that there are additional simple frequencies SDCF-quantifiers concerning conditions related to minimum, maximum and average values computed from SDCF-tables.

11.1.4   SDCF-quantifiers concerning steps in histogram

A step SDCF-quantifier is an expression

$$\langle \text{SDCFSt: } Dir, u, v, MinH, Cons, Set, \circledast, Thr \rangle \tag{11.5}$$

where

- *Dir* is one of the strings Up, *Down*

- $\langle u, v \rangle$ is a range of a CF-quantifier for a histogram $\langle h_1, \ldots, h_t \rangle$

- *MinH* is a real positive number

- *Cons* is one of the strings *Yes*, *No*

- *Set* is one of the operation modes of SDCF-quantifier *Set*1, *Set*2

- $\circledast$ is one of the relations $<, \leq, =, \geq, >$ concerning real numbers

- *Thr* is a non-negative real number.

We say that an SDCF-quantifier $\langle$SDCFSt: $Dir, u, v, MinH, Cons, Set, \circledast, Thr \rangle$ is relevant for an attribute A with t categories if $v \leq t$.

A quantifier $\langle$SDCFSt: $Dir, u, v, MinH, Cons, Set, \circledast, Thr \rangle$ relevant for an attribute A defines a condition for an SDCF-table $SDCF(A, \mathcal{M}, \chi, \kappa, \lambda) = \langle h_1^{(1)}, \ldots, h_t^{(1)}, h_1^{(2)}, \ldots, h_t^{(2)} \rangle$ by a relation

$$Steps(Dir, u, v, MinH, Cons, h_1, \ldots, h_t) \circledast Thr \tag{11.6}$$

depending on a value of operation mode *Set*. If $Set = Set1$, then $\langle h_1, \ldots, h_t \rangle = \langle h_1^{(1)}, \ldots, h_t^{(1)} \rangle$ and if $Set = Set2$ then $\langle h_1, \ldots, h_t \rangle = \langle h_1^{(2)}, \ldots, h_t^{(2)} \rangle$.

$Steps(Dir, u, v, MinH, Cons, h_1, \ldots, h_t)$ is a function assigning the number of steps in a sequence $h_u, \ldots, h_v$. It depends on parameters *Dir*, *MinH*, and *Cons*. It is defined using following notions.

- A couple $\langle h_i, h_{i+1} \rangle$ is a *step up of a minimal height MinH in a sequence* $h_u, \ldots, h_v$, if it holds both $h_{i+1} - h_i \geq MinH$ and $u \leq i \leq v - 1$.

- A couple $\langle h_i, h_{i+1} \rangle$ is a *step down of a minimal height MinH in a sequence* $h_u, \ldots, h_v$ if it holds both $h_i - h_{i+1} \geq MinH$ and $u \leq i \leq v - 1$.

- If $Seq = \langle \ \langle h_k, h_{k+1} \rangle, \langle h_{k+1}, h_{k+2} \rangle, \ldots, \langle h_{k+w}, h_{k+w+1} \rangle \ \rangle$ is a sequence of steps up or steps down of a minimal height $MinH$ in a sequence $h_u, \ldots, h_v$, then *w is a length of Seq.*

A value of $Steps(Dir, u, v, MinH, Cons, h_1, \ldots, h_t)$ is given in Fig. 11.2.

Dir	Cons	Value of $Steps(Dir, u, v, MinH, Cons, h_1, \ldots, h_t)$
Up	*No*	the number of steps up of a minimal height *MinH* in a sequence $h_u, \ldots, h_q$
Up	*Yes*	the length of the longest sequence of steps up of a minimal height *MinH* in a sequence $h_u, \ldots, h_q$
Down	*No*	the number of steps down of a minimal height *MinH* in a sequence $h_u, \ldots, h_q$
Down	*Yes*	the length of the longest sequence of steps down of a minimal height *MinH* in a sequence $h_u, \ldots, h_q$

Figure 11.2: Values of $Steps(Dir, u, v, MinH, Cons, h_1, \ldots, h_t)$.

11.2 Applying SDCF-Miner in *Accidents* Dataset

Two applications of the SDCF-Miner procedure to search for exceptions related to the general decreasing trend of the number of accidents in the period 2005–2015 are described. The first example concerns segments of accidents defined by vehicle age, driver's characteristics, and authorities such that the number of accidents in the whole segment is non-decreasing and the number of accidents in the segment belonging to an authority is at least partly decreasing. This goal leads to three related applications of the SDCF-Miner described in Section 11.2.1. A conclusion can be formulated that there are no clear interesting results concerning the whole period. However, the involvement of a domain expert can probably lead to interesting patterns.

Thus, an example of searching for interesting relations between trends in the last five years, i.e., in the period 2011–2015 is presented. We are interested in couples of police forces and types of vehicles such that there are both a non-decreasing trend of accidents of vehicles of a type belonging to a first police force and a non-increasing trend of accidents of vehicles of the same type belonging to a second police force. This example is presented in Section 11.2.2.

11.2.1 Exceptions to increasing trends and authorities

Recall that the general trend of the number of accidents in the period 2005–2015 is decreasing, see Fig. 8.17 in Section 8.3. The segments of accidents with an increasing trend in the whole period 2005–2015 are found in Section 8.3.2. Such segments are of course considered as exceptions to the general decreasing trend. Searching for exceptions to a concrete decreasing trend is described in Section 8.3.3. The CF-Miner procedure is used in both cases.

The goal of this section is to present possibilities of the SDCF-Miner procedure to search for an additional type of exceptions related to the general decreasing trend of the number of accidents in the period 2005–2015. We search for a segment of accidents defined by a Boolean attribute $\mathscr{S}$ (shortly segment $\mathscr{S}$) and an authority defined by a Boolean attribute $\mathscr{A}$ such that both the number of accidents from the segment defined by $\mathscr{S}$ is non-decreasing and the number of accidents from the segment $\mathscr{S}$ belonging to authority $\mathscr{A}$ is at least partly decreasing. Segment of accidents defined by a Boolean attribute $\mathscr{S}_{Exm}$

$$\mathscr{S}_{Exm} = \mathsf{Vehicle\_Age}(14,15) \wedge \mathsf{Age}(46\text{-}55) \wedge \mathsf{Sex}(Male) \qquad (11.7)$$

and authority Police(*Strathclyde*) satisfy these conditions, see histograms Year/$\mathscr{S}_{Exm}$ and Year/ ($\mathscr{S}_{Exm} \wedge$ Police(*Sussex*)) in Fig. 11.3.

Year/ $\mathscr{S}_{Exm}$

Year/ ($\mathscr{S}_{Exm} \wedge$Police(*Sussex*))

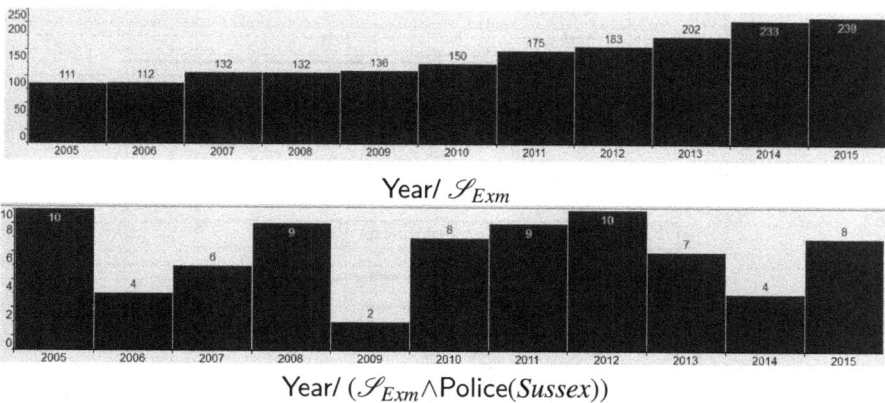

Figure 11.3: Histograms Year/$\mathscr{S}_{Exm}$ and Year/($\mathscr{S}_{Exm} \wedge$ Police(*Sussex*)).

A detailed definition of the above introduced conditions is given in the following way:

■ the requirement that the number of accidents from the segment defined by $\mathscr{S}$ is non-decreasing corresponds to a condition that there are 10 steps up with non-negative height in a histogram Year/$\mathscr{S}$

■ the requirement that the number of accidents from the segment $\mathscr{S}$ belonging to the authority $\mathscr{A}$ is at least partly decreasing corresponds to a condition that a histogram Year/$\mathscr{S} \wedge \mathscr{A}$ has at least four steps down.

We specify our task such that we are interested in segments $\mathscr{S}$ of accidents defined as $\mathscr{S} = \text{Vehicle\_Age}(\alpha) \wedge \mathscr{D}$, where $\mathscr{D}$ is a characteristic of drivers in one of forms

■ $\text{Sex}(\beta) \wedge \text{Age}(\gamma)$

■ $\text{Sex}(\beta) \wedge \text{Journey}(\delta)$.

The Boolean attributes Vehicle_Age(α) and $\mathscr{D}$ are specified in details by $\mathscr{RP}$-expressions Vehicle_Age and Driver available in Fig. 11.4. Segment $\mathscr{S}$ of accidents is defined by $\mathscr{RC}$-expression Accidents_S presented in Table 11.1.

Name: Vehicle_Age		Type: *Conjunction*		Length: $1-1$	
Attribute	Coefficients	Length	Gace	B/R	Class
Vehicle_Age	*Sequences*	$2-3$	*Pos*	B	$-$

Name: Driver		Type: *Conjunction*		Length: $1-2$	
Attribute	Coefficients	Length	Gace	B/R	Class
Sex	*Subsets*	$1-1$	*Pos*	B	$-$
Age	*Subsets*	$1-1$	*Pos*	R	$-$
Journey	*Subsets*	$1-1$	*Pos*	R	$-$

Figure 11.4: $\mathscr{RP}$-expressions Vehicle_Age and Driver.

Table 11.1: $\mathscr{RC}$-expression Accidents_S in a short tabular form.

Name: Vehicle_Driver_Conditions	Length: $2-3$
Set of partial cedents	Length
Vehicle_Age	$1-1$
Driver	$1-2$

A set of relevant authorities defined by a Boolean attribute is specified by an $\mathscr{RP}$-expression Authorities available in Fig. 11.5, see also Section 2.3.6.

We use the following input parameters for a run of the SDCF-Miner procedure: The procedure deals with the attribute Year. The set of relevant conditions contains only Boolean attribute *True* – it is defined by the $\mathscr{RP}$ expression True, see Fig. 10.6. A set $\mathscr{S}_\alpha$ of relevant Boolean attributes α is defined by the $\mathscr{RC}$-expression Accidents_S, see Table 11.1. A set $\mathscr{S}_\beta$ of relevant Boolean attributes β is defined by the $\mathscr{RP}$-expression Authorities, see Fig. 11.5. *Mode* $\alpha \bowtie \alpha \wedge \beta$ is used. *SDCF-quantifier* defined by the following basic SDCF-quantifiers is used:

Name: Authorities		Type: *Conjunction*		Length: $1-1$	
Attribute	Coefficients	Length	Gace	B/R	Class
District	*Subsets*	$1-1$	*Pos*	B	–
Highway	*Subsets*	$1-1$	*Pos*	B	–
Police	*Subsets*	$1-1$	*Pos*	B	–

Figure 11.5: $\mathscr{RP}$-expression Authorities.

- $Sum_1 \geq 500$ meaning that there are at least 500 accidents satisfying $\mathscr{S} = \text{Vehicle\_Age}(\alpha) \wedge \mathscr{D}$

- $Sum_2 \geq 40$ meaning that there are at least 40 accidents satisfying $\text{Vehicle\_Age}(\alpha) \wedge \mathscr{D} \wedge \mathscr{A}$ where $\mathscr{A}$ is a Boolean attribute defined by the $\mathscr{RP}$-expression Authorities

- SDCF-quantifier $\langle$SDCFSt: $Up, 1, 11, 0, Yes, Set1 \rangle = 10$ meaning that each histogram Year/$\mathscr{S}$ has 10 steps with non-negative height, i.e., the histogram is non-decreasing

- SDCF-quantifier $\langle$SDCFSt: $Down, 1, 11, 1, No, Set2 \rangle \geq 4$ meaning that each histogram Year/$(\mathscr{S} \wedge \mathscr{A})$ has at least four steps down with minimal height 1 and that these steps need not to be consecutive.

The task with the above described parameters resulted in 13 true SDCF-patterns. 122 388 SDCF-patterns were tested in 190 seconds. Overview of output patterns is available in Fig. 11.6.

```
2:Sum  Hypothesis
77.00  Year : Vehicle_Age(14,15) & Age(46 - 55) & Sex(Male) × Set1 & Police(Sussex)
69.00  Year : Vehicle_Age(14,15) & Age(46 - 55) & Sex(Male) × Set1 & Police(Devon and Cornwall)
61.00  Year : Vehicle_Age(14,15) & Age(46 - 55) & Sex(Male) × Set1 & Police(Kent)
54.00  Year : Vehicle_Age(14,15) & Age(46 - 55) & Sex(Male) × Set1 & Police(Strathclyde)
53.00  Year : Vehicle_Age(14,15) & Age(46 - 55) & Sex(Male) × Set1 & Highway 5000+(E10000016,Kent)
53.00  Year : Vehicle_Age(14,15) & Age(46 - 55) & Sex(Male) × Set1 & Police(Greater Manchester)
53.00  Year : Vehicle_Age(14,15) & Age(46 - 55) & Sex(Male) × Set1 & Police(Hampshire)
50.00  Year : Vehicle_Age(14,15) & Age(46 - 55) & Sex(Male) × Set1 & Police(Avon and Somerset)
49.00  Year : Vehicle_Age(14,15) & Age(46 - 55) & Sex(Male) × Set1 & Police(Essex)
49.00  Year : Vehicle_Age(14,15) & Age(46 - 55) & Sex(Male) × Set1 & Police(Lothian and Borders)
48.00  Year : Vehicle_Age(14,15) & Age(46 - 55) & Sex(Male) × Set1 & Police(Merseyside)
43.00  Year : Vehicle_Age(14,15) & Age(46 - 55) & Sex(Male) × Set1 & District2+(Birmingham)
43.00  Year : Vehicle_Age(14,15) & Age(46 - 55) & Sex(Male) × Set1 & Highway 5000+(E08000025,Birmingham)
```

Figure 11.6: 13 true SDCF-patterns with four steps down in histograms Year/$(\mathscr{S} \wedge \mathscr{A})$.

Let us note that all 13 patterns satisfy that $\mathscr{S} = \mathscr{S}_{Exm}$ where

$$\mathscr{S}_{Exm} = \text{Vehicle\_Age}(14,15) \wedge \text{Age}(46\text{-}55) \wedge \text{Sex}(Male) ,$$

see equation (11.7). The number of accidents satisfying $\mathscr{S} \wedge \mathscr{A}$ varies between 77 and 43. The first SDCF-pattern in Fig. 11.6 can be written as

$$\mathscr{S}_{Exm} \bowtie \mathscr{S}_{Exm} \wedge \text{Police}(Sussex) : \text{Year}/True . \qquad (11.8)$$

The corresponding histograms Year/$\mathscr{S}_{Exm}$ and Year/($\mathscr{S}_{Exm} \wedge$ Police($Sussex$)) are presented in Fig. 11.3. However, there is no clear trend down in the histogram Year/($\mathscr{S}_{Exm} \wedge$ Police($Sussex$)).

Thus we can ask if there is a histogram with four consecutive steps down. We will modify the above presented input parameters to answer this question. The fact that all 13 patterns satisfy that $\mathscr{S} = \mathscr{S}_{Exm}$ makes possible to use the the $\mathscr{R}\mathscr{P}$-expression Driver_Exm presented in Fig. 11.7 instead of $\mathscr{R}\mathscr{P}$-expression Driver introduced in Fig. 11.4. In addition, we use the SDCF-

Name: Driver_Exm		Type: *Conjunction*		Length: $2-2$	
Attribute	Coefficients	Length	Gace	B/R	Class
Sex	*One category*	*Male*	*Pos*	B	–
Age	*One category*	*46-55*	*Pos*	R	–

Figure 11.7: $\mathscr{R}\mathscr{P}$-expression Driver_Exm.

quantifier $\langle$SDCFSt: $Down, 1, 11, 1, Yes, Set2 \rangle \geq 4$ requiring that the histogram Year/ ($\mathscr{S} \wedge \mathscr{A}$) has at least four **consecutive steps** down instead of the SDCF-quantifier $\langle$SDCFSt: $Down, 1, 11, 1, No, Set2 \rangle \geq 4$. A run of the SDCF-Miner with the modified parameters resulted without any output. 4 371 SDCF-patterns were tested in 6 seconds.

Thus, a question arises if there is a histogram with at least five steps down. A run of the SDCF-Miner procedure with the SDCF-quantifier $\langle$SDCFSt: $Down, 1, 11, 1, No, Set2 \rangle \geq 5$ instead of $\langle$SDCFSt: $Down, 1, 11, 1, No, Set2 \rangle \geq 4$ and the $\mathscr{R}\mathscr{P}$-expression Driver_Exm instead of $\mathscr{R}\mathscr{P}$-expression Driver resulted in five true SDCF-patterns presented in Fig. 11.8. 4 371 SDCF-patterns were tested in 6 seconds.

2:Sum Hypothesis

48.00 Year : Vehicle_Age(*14, 15*) & Age(*46 - 55*) & Sex(*Male*) × Set1 & Police(*Merseyside*)
49.00 Year : Vehicle_Age(*14, 15*) & Age(*46 - 55*) & Sex(*Male*) × Set1 & Police(*Lothian and Borders*)
53.00 Year : Vehicle_Age(*14, 15*) & Age(*46 - 55*) & Sex(*Male*) × Set1 & Highway 5000+(*E10000016,Kent*)
53.00 Year : Vehicle_Age(*14, 15*) & Age(*46 - 55*) & Sex(*Male*) × Set1 & Police(*Greater Manchester*)
53.00 Year : Vehicle_Age(*14, 15*) & Age(*46 - 55*) & Sex(*Male*) × Set1 & Police(*Hampshire*)

Figure 11.8: Five true SDCF-patterns with five steps down in histograms Year/($\mathscr{S} \wedge \mathscr{A}$).

A run of the SDCF-Miner procedure with the SDCF-quantifier $\langle$SDCFSt: $Down, 1, 11, 1, No, Set2 \rangle \geq 6$ searching for SDCF-patterns with at least six steps down instead of SDCF-quantifier $\langle$SDCFSt: $Down,1,11,1,No,Set2\rangle \geq 5$ resulted in one true SDCF-pattern

$$\mathscr{S}_{Exm} \bowtie \mathscr{S}_{Exm} \wedge \text{Police}(Lothian\ and\ Borders) : \text{Year}/True . \qquad (11.9)$$

Again, 4 371 SDCF-patterns were tested in 6 seconds. There are 49 rows satisfying $\mathscr{S}_{Exm} \wedge \text{Police}(Lothian\ and\ Borders)$. The corresponding histograms Year/$\mathscr{S}_{Exm}$ and Year/($\mathscr{S}_{Exm}\wedge$ Police(Lothian and Borders)) are available in Fig. 11.9.

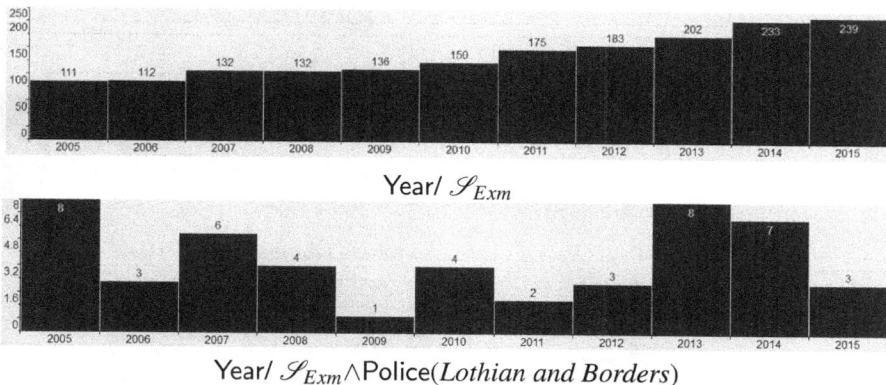

Year/ $\mathscr{S}_{Exm}$

Year/ $\mathscr{S}_{Exm}\wedge$Police(Lothian and Borders)

Figure 11.9: Histograms Year/$\mathscr{S}_{Exm}$ and Year/$\mathscr{S}_{Exm}\wedge$ Police(Lothian and Borders).

Finally, a run of the SDCF-Miner procedure with the SDCF-quantifier $\langle$SDCFSt: $Down,1,11,1,No,Set2\rangle \geq 7$ searching for SDCF-patterns with at least seven steps down finished without any output true SDCF-patterns.

We can conclude that there are only the above presented SDCF-patterns

$$\mathscr{S} \bowtie \mathscr{S} \wedge \mathscr{A} : \text{Year}/True .$$

Informally speaking, there are not important segments $\mathscr{S}$ of accidents defined by age of vehicles and characteristics of drivers and authorities $\mathscr{A}$ such that the trend of accidents of a segment $\mathscr{S}$ is increasing and the trend of such accidents in an authority $\mathscr{A}$ is systematically decreasing for the whole period 2005–2015. Results of particular runs of the SDCF-Miner procedure are very sensitive to parameters of the SDCF-Miner. The presented examples of applications of the SDCF-Miner only present possibilities of the procedure. Involvement of a domain expert is necessary to achieve practically interesting results.

11.2.2 Differences between police forces

We have not found an interesting relation between trends of accidents in the whole period 2005–2015 in the previous section. We show an example of searching for interesting relations between trends in the last five years, i.e., in the period 2011–2015. More precisely, we are interested in couples of Polices $\langle \text{Police}(P_1), \text{Police}(P_2) \rangle$ and types of vehicles $\text{Type}(T)$ such that there are both non-decreasing trends of accidents of vehicles of type $\text{Type}(T)$ belonging to Police(P_1) and non-increasing trend of accidents of vehicles of type $\text{Type}(T)$ belonging to Police(P_2) in period 2011–2015.

Polices Police(*Thames Valley*), Police(*Cheshire*) and type of vehicles Type(*Pedal cycle*) are examples of such Polices and a type of vehicle, see histograms Year/ (Type(*Pedal cycle*) ∧ Police(*Thames Valley*)) and Year/ (Type(*Pedal cycle*) ∧ Police(*Cheshire*)) shown in Fig. 11.10.

Year/ (Type(*Pedal cycle*) ∧ Police(*Thames Valley*))

Year/ (Type(*Pedal cycle*) ∧ Police(*Cheshire*))

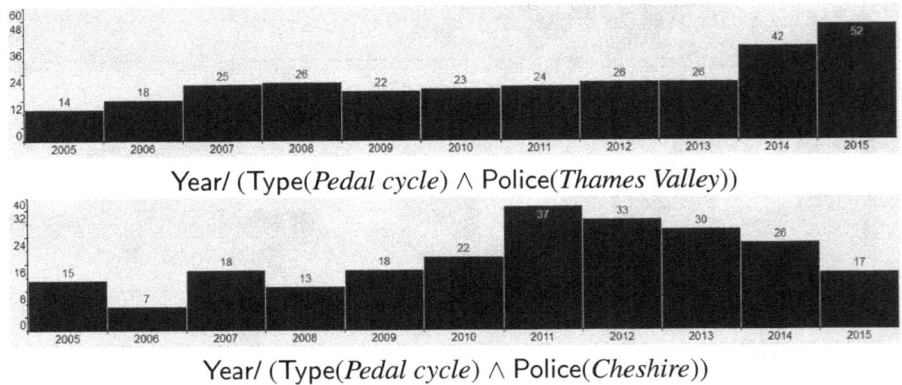

Figure 11.10: Histograms for Pedal cycle and Polices Thames Valley and Cheshire.

A detailed definition of the above introduced conditions is given in the following way:

■ the requirement that there is a non-decreasing trend of accidents of vehicles of type $\text{Type}(T)$ belonging to Police(P_1) in period 2011–2015 corresponds to a condition that there are four steps up with a non-negative height in range $\langle 7, 11 \rangle$

■ the requirement that there is a non-increasing trend of accidents of vehicles of type $\text{Type}(T)$ belonging to Police(P_2) in period 2011–2015 corresponds to a condition that there are four steps down with a non-negative height in range $\langle 7, 11 \rangle$

■ there are at least 200 accidents satisfying Police(P_1) ∧ Type(T) and at least 200 accidents satisfying Police(P_2) ∧ Type(T).

We specify our task such that we are interested in SDCF-patterns

$$\text{Police}(P_1) \bowtie \text{Police}(P_2) : \text{Year}/\text{Type}(T) . \tag{11.10}$$

We use $\mathscr{R}\mathscr{P}$-expressions Police and Type available in Fig. 11.11.

Name: Police		Type: *Conjunction*		Length: $1-1$	
Attribute	Coefficients	Length	Gace	B/R	Class
Police	*Subsets*	$1-1$	*Pos*	B	—

Name: Type		Type: *Conjunction*		Length: $1-2$	
Attribute	Coefficients	Length	Gace	B/R	Class
Type	*Subsets*	$1-1$	*Pos*	B	—

Figure 11.11: $\mathscr{R}\mathscr{P}$-expressions Police and Type.

We use the following input parameters for a run of the SDCF-Miner procedure: The procedure deals with the attribute Year. The set of relevant conditions is defined by the $\mathscr{R}\mathscr{P}$ expression Type, see Fig. 11.11. Both a set $\mathscr{S}_{\alpha}$ of relevant Boolean attributes α and a set $\mathscr{S}_{\beta}$ of relevant Boolean attributes β are defined by the $\mathscr{R}\mathscr{P}$-expression Police, see Fig. 11.11. *Mode* $\alpha \bowtie \beta$ is used, see Section 11.1.2. *SDCF-quantifier* defined by the following basic SDCF-quantifiers is used:

- $Sum_1 \geq 200$ and $Sum_2 \geq 200$ meaning that there are both at least 200 accidents satisfying $\text{Type}(T) \wedge \text{Police}(P_1)$ and at least 200 accidents satisfying $\text{Type}(T) \wedge \text{Police}(P_2)$

- SDCF-quantifier $\langle$SDCFSt: $Up, 7, 11, 0, Yes, Set1\rangle = 4$ meaning that histogram Year/ $(\text{Type}(T) \wedge \text{Police}(P_1))$ has four steps up with non-negative height in the range $\langle 7, 11\rangle$, i.e., the histogram is non-decreasing in this range

- SDCF-quantifier $\langle$SDCFSt: $Down, 7, 11, 0, Yes, Set2\rangle = 4$ meaning that histogram Year/ $\text{Type}(T) \wedge \text{Police}(P_2))$ has four steps down with non-negative height in the range $\langle 7, 11\rangle$, i.e., the histogram is non-increasing in this range.

The task with the above described parameters resulted in eight true SDCF-patterns; 41 560 SDCF-patterns were tested in 63 seconds. The overview of output patterns is available in Fig. 11.12.

Let us note that there are 4 232 accidents satisfying

Police(*Metropolitan Police*) $\wedge$ Type(*Motorcycle 125cc and under*) ,

this is relevant to the first four rows in Fig. 11.12 where only the expression "***" occurs in the column 1 : Sum. It also holds that there are 264 accidents satisfying

Police(*Greater Manchester*) $\wedge$ Type(*Motorcycle 125cc and under*) ,

see the column 2 : Sum. This is similar for additional rows.

1:Sum	2:Sum	Hypothesis
***	264.00	Year : Police(*Metropolitan Police*) × Police(*Greater Manchester*) / Type(*Motorcycle 125cc and under*)
***	248.00	Year : Police(*Metropolitan Police*) × Police(*Hertfordshire*) / Type(*Motorcycle 125cc and under*)
***	222.00	Year : Police(*Metropolitan Police*) × Police(*Humberside*) / Type(*Motorcycle 125cc and under*)
***	217.00	Year : Police(*Metropolitan Police*) × Police(*Derbyshire*) / Type(*Motorcycle 125cc and under*)
624.00	488.00	Year : Police(*Strathclyde*) × Police(*Northumbria*) / Type(*Van / Goods 3.5 tonnes mgw or under*)
624.00	344.00	Year : Police(*Strathclyde*) × Police(*Humberside*) / Type(*Van / Goods 3.5 tonnes mgw or under*)
298.00	236.00	Year : Police(*Thames Valley*) × Police(*Cheshire*) / Type(*Pedal cycle*)
214.00	219.00	Year : Police(*Essex*) × Police(*Humberside*) / Type(*Taxi/Private hire car*)

Figure 11.12: 8 true SDCF-patterns concerning range $\langle 7, 11 \rangle$.

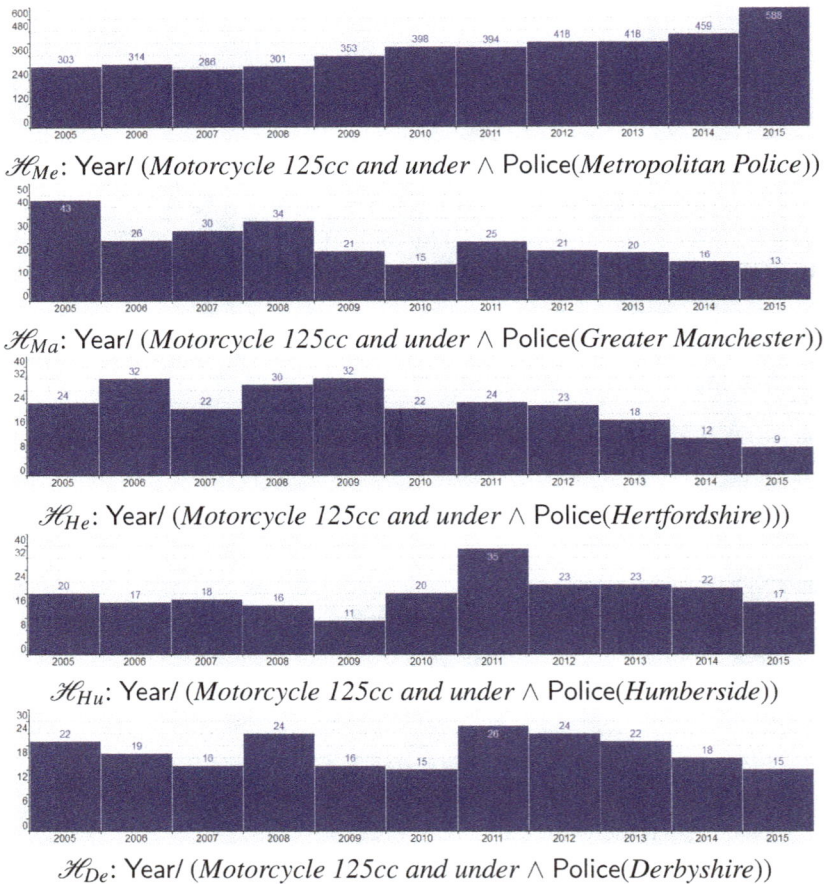

$\mathcal{H}_{Me}$: Year/ (*Motorcycle 125cc and under* $\wedge$ Police(*Metropolitan Police*))

$\mathcal{H}_{Ma}$: Year/ (*Motorcycle 125cc and under* $\wedge$ Police(*Greater Manchester*))

$\mathcal{H}_{He}$: Year/ (*Motorcycle 125cc and under* $\wedge$ Police(*Hertfordshire*)))

$\mathcal{H}_{Hu}$: Year/ (*Motorcycle 125cc and under* $\wedge$ Police(*Humberside*))

$\mathcal{H}_{De}$: Year/ (*Motorcycle 125cc and under* $\wedge$ Police(*Derbyshire*))

Figure 11.13: Histograms relevant to the first four rows of Fig. 11.12.

Histograms relevant to the first four rows of Fig. 11.12 are available in Fig. 11.13. We write "*Motorcycle 125cc and under*" instead of "Type(*Motorcycle 125cc and under*)".

We see that the histogram $\mathcal{H}_{Me}$ is non-decreasing in the range $\langle 7, 11 \rangle$ while the histograms $\mathcal{H}_{Ma}$, $\mathcal{H}_{He}$, $\mathcal{H}_{Hu}$, and $\mathcal{H}_{De}$ are non-increasing or even decreasing.

Histograms relevant to the fifth and sixth rows of Fig. 11.12 are available in Fig. 11.14. We write "*Van/Goods 3.5 tonnes mgw and under*" instead of "Type(*Van/Goods 3.5 tonnes mgw and under*)". The histogram $\mathcal{H}_{St}$ is non-decreasing in the range $\langle 7, 11 \rangle$ while the histograms $\mathcal{H}_{No}$ and $\mathcal{H}_{Hu\_Van}$ are non-increasing.

Histograms relevant to the seventh row of Fig. 11.12 are available in Fig. 11.10. Histograms relevant to the last row of Fig. 11.12 are available in Fig. 11.15. The histogram $\mathcal{H}_{Es}$ is non-decreasing in the range $\langle 7, 11 \rangle$ while the histograms $\mathcal{H}_{Hu\_Taxi}$ is non-increasing.

We can summarize that the shapes of particular histograms in the range $\langle 7, 11 \rangle$ satisfy the conditions given by the input parameters of the SDCF-Miner procedure. However, the shapes of these histograms in the range $\langle 1, 7 \rangle$ are very different. Let us again emphasize that the presented example of applications of the SDCF-Miner only points to possibilities of the procedure. Interpretation of the presented results and further analysis requires co-operation of a domain expert.

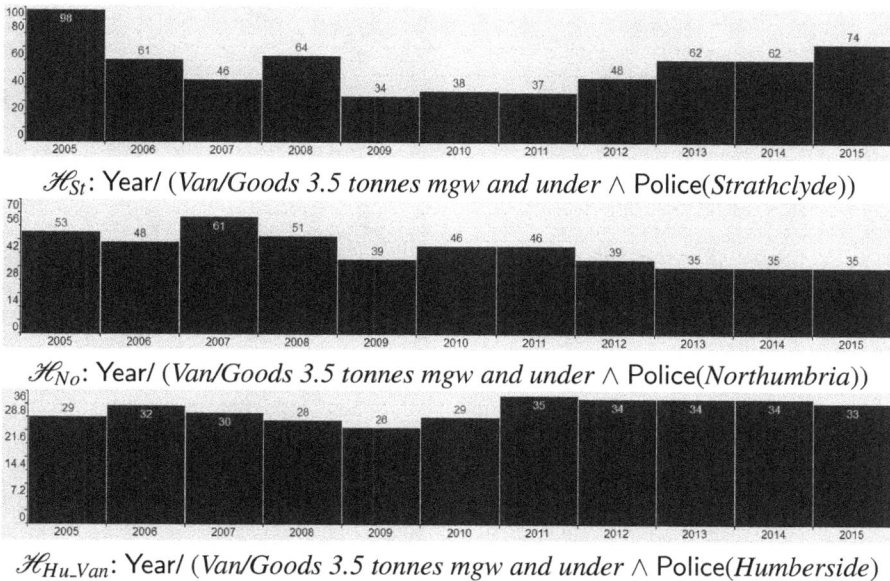

$\mathcal{H}_{St}$: Year/ (*Van/Goods 3.5 tonnes mgw and under* ∧ Police(*Strathclyde*))

$\mathcal{H}_{No}$: Year/ (*Van/Goods 3.5 tonnes mgw and under* ∧ Police(*Northumbria*))

$\mathcal{H}_{Hu\_Van}$: Year/ (*Van/Goods 3.5 tonnes mgw and under* ∧ Police(*Humberside*))

Figure 11.14: Histograms relevant to the fifth and sixth rows of Fig. 11.12.

$\mathcal{H}_{Es}$: Year/ (Type(*Taxi/Private hire car*) ∧ Police(*Essex*))

$\mathcal{H}_{Hu\_Taxi}$: Year/ (Type(*Taxi/Private hire car*) ∧ Police(*Humberside*))

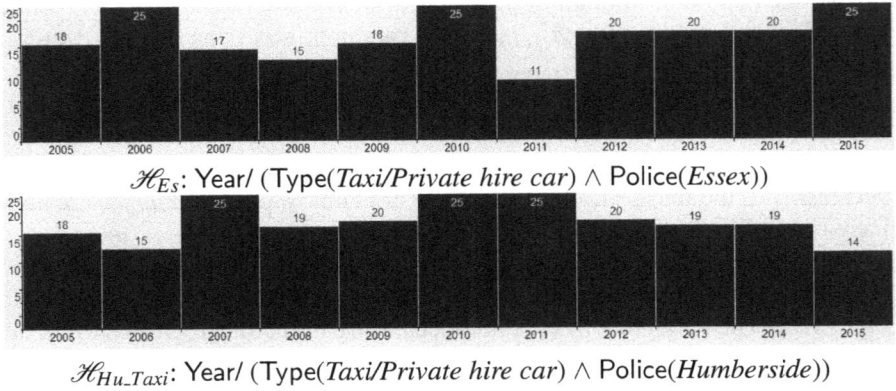

Figure 11.15: Histograms relevant to the last row of Fig. 11.12.

Chapter 12

SDKL-Miner—Couples of Pairs of Categorical Attributes

The goal of this chapter is to introduce the SDKL-Miner procedure and three examples of its application. The main features of the procedure are introduced in Section 12.1 together with related notions. Examples of application of the SDKL-Miner procedure in the *STULONG* dataset are described in Section 12.2. The SDKL-Miner is briefly introduced also in [44, 128].

12.1 SDKL-Miner and Related Notions

The SDKL-Miner procedure and couples of pairs of categorical attributes it deals with are informally introduced in Section 3.7. The procedure deals with SDKL-patterns in one of the forms $\alpha \bowtie \beta : R \approx C/\chi$, $\alpha \bowtie \alpha \wedge \beta : R \approx C/\chi$, $\alpha \bowtie \neg \alpha : R \approx C/\chi$ mentioned in Section 4.3. A used form is determined by a parameter *Mode*. The presented examples concern only the form $\alpha \bowtie \beta : R \approx C/\chi$ which corresponds to the value $\alpha \bowtie \beta$ of the parameter *Mode*. The additional used parameters and output of the procedure are introduced in Section 12.1.1. Generalised quantifiers that the SDKL-Miner procedure deals with are called *SDKL-quantifiers*. The SDKL-quantifiers used in the presented examples are introduced in Section 12.1.2.

12.1.1 SDKL-Miner input and output

There are the following parameters defining a set of relevant SDKL-patterns $\alpha \bowtie \beta : R \approx C/\chi$ to be generated and verified:

- a set *Row* of relevant row attributes R and a set *Col* of relevant column attributes C

- an $\mathscr{RC}$-expression defining a set *Cond* of relevant conditions χ

- an $\mathscr{RC}$-expression defining a set $\mathscr{S}_\alpha$ of relevant Boolean attributes α and an $\mathscr{RC}$-expression defining a set $\mathscr{S}_\beta$ of relevant Boolean attributes β

- a list *ListSDKL* of basic SDKL-quantifiers.

A set of relevant SDKL-patterns is defined as a set of all patterns $\alpha \bowtie \beta : R \approx C/\chi$ satisfying $R \in Row$, $C \in Col$, $\chi \in Cond$, $\alpha \in \mathscr{S}_\alpha$, $\beta \in \mathscr{S}_\beta$. SDKL-quantifier $\approx$ is defined as a conjunction of basic SDKL-quantifiers introduced in the list *ListSDKL*. In addition, it must hold that φ, ψ, χ have no common attributes and there is no attribute used both in χ and in α or β. However, α and β can have common attributes.

The SDKL-Miner procedure generates all relevant SDKL-patterns and verifies them in a given data matrix $\mathscr{M}$. Each generated SDKL-pattern is verified such that the SDKL-quantifier $\approx$ is applied to an SDKL-table given by the parameter *Mode* (i.e., $\alpha \bowtie \beta$ in examples below). SDKL-patterns satisfying the condition given by the SDKL-quantifier $\approx$ are listed in an output.

12.1.2 SDKL-quantifiers

An SDKL-pattern $\alpha \bowtie \beta : R \approx C/\chi$ is true in a data matrix $\mathscr{M}$ if a condition related to a SDKL-quantifier $\approx$ is satisfied for an SDKL-table $SDKL(R, C, \mathscr{M}, \chi, \alpha, \beta)$ introduced in Fig. 12.1.

$\mathscr{M}/\chi \wedge \alpha$	c_1	$\cdots$	c_L		$\mathscr{M}/\chi \wedge \beta$	c_1	$\cdots$	c_L	
r_1	$a_{1,1}$	$\cdots$	$a_{1,L}$	$a_{1,*}$	r_1	$b_{1,1}$	$\cdots$	$b_{1,L}$	$b_{1,*}$
$\vdots$	$\vdots$	$\ddots$	$\vdots$	$\vdots$	$\vdots$	$\vdots$	$\ddots$	$\vdots$	$\vdots$
r_K	$a_{K,1}$	$\cdots$	$a_{K,L}$	$a_{K,*}$	r_K	$b_{K,1}$	$\cdots$	$b_{K,L}$	$b_{K,*}$
	$a_{*,1}$	$\cdots$	$a_{*,L}$	a		$b_{*,1}$	$\cdots$	$b_{*,L}$	b

Figure 12.1: SDKL-table $SDKL(R, C, \mathscr{M}, \chi, \alpha, \beta) = \langle \{a_{i,j}\}, \{b_{i,j}\} \rangle$.

In the examples presented in this section we use two types of basic SDKL-quantifiers: simple frequencies SDKL-quantifiers and advanced SDKL-quantifiers based on Kendall's coefficient. We use simple frequencies SDKL-quantifiers in a form

$$\langle \text{SDKLSF: } Sum, Set \rangle \circledast Thr \qquad (12.1)$$

where *Set* is one of operation modes of SDKL-quantifier *Set*1, *Set*2, $\circledast$ is one of relations $<, \leq, =, \geq, >$ concerning real numbers, and *Thr* is a non-negative real number. A simple frequencies SDKL-quantifier $\langle \text{SDKLSF: } Sum, Set \rangle \circledast Thr$ defines a condition (12.2) if *Set* = *Set*1, and a condition (12.3) if *Set* = *Set*2.

$$a_{1,1} + \cdots + a_{1,L} + \cdots + a_{K,1} + \cdots + a_{K,L} \circledast Thr \qquad (12.2)$$

$$b_{1,1} + \cdots + b_{1,L} + \cdots + b_{K,1} + \cdots + b_{K,L} \circledast Thr \qquad (12.3)$$

Let us note that there are additional simple frequencies SDCF-quantifiers concerning conditions related to minimum, maximum and average values computed from SDKL-tables. Let us further note that we 'do not use here a range of SDKL-quantifiers which is analogous to the range of KL-quantifiers introduced in Section 9.1.3.

An advanced SDKL-quantifier is an expression

$$AdvKL \circledast Thr \qquad (12.4)$$

where *AdvKL* is a value derived from theoretical measures of dependencies of two categorical attributes. Kendall's coefficient and Cramer's V coefficient are examples of such theoretical measures. A value $Diff\tau_b$ defined as a difference of Kendall's coefficient is used in the examples presented in Section 12.2. It holds

$$Diff\tau_b = \tau_b^{(1)} - \tau_b^{(2)} \qquad (12.5)$$

where
$$\tau_b^{(1)} = \frac{2(P_a - Q_a)}{\sqrt{(a^2 - \sum_{k=1}^{K} a_{k,*}^2)(a^2 - \sum_{l=1}^{L} a_{*,l}^2)}}, \quad \tau_b^{(2)} = \frac{2(P_b - Q_b)}{\sqrt{(b^2 - \sum_{k=1}^{K} b_{k,*}^2)(b^2 - \sum_{l=1}^{L} b_{*,l}^2)}},$$
$P_a = \sum_{k=1}^{K} \sum_{l=1}^{L} a_{k,l} \sum_{i=k+1}^{K} \sum_{j=l+1}^{L} a_{i,j}, Q_a = \sum_{k=1}^{K} \sum_{l=1}^{L} a_{k,l} \sum_{i=k+1}^{K} \sum_{j=1}^{l-1} a_{i,j}, P_b = \sum_{k=1}^{K} \sum_{l=1}^{L} b_{k,l} \sum_{i=k+1}^{K} \sum_{j=l+1}^{L} b_{i,j}, Q_b = \sum_{k=1}^{K} \sum_{l=1}^{L} b_{k,l} \sum_{i=k+1}^{K} \sum_{j=1}^{l-1} b_{i,j}.$

Recall that the Kendall's coefficient τ_b takes values from $\langle 1, 1 \rangle$ with the following interpretation: $\tau_b > 0$ indicates positive ordinal dependence, $\tau_b < 0$ indicates negative ordinal dependence, $\tau_b = 0$ indicates ordinal independence, $|\tau_b = 1|$ indicates that a column attribute *C* in question is a function of row attribute *R*, see Section 9.2.

12.2 Applying SDKL-Miner in *STULONG* Dataset

We present three applications of the SDKL-Miner procedure. All of them concern the SDKL patterns $Alco(c_1) \bowtie Alco(c_2): R \approx C/\chi$ where

■ R, C are ordinal attributes chosen from the attributes BMI, Subscapularis, Triceps, Diastolic, Systolic, Cholesterol, Triglycerides, Beer, Wine, and Liquors

■ *Alco* is one of the attributes Beer, Wine, and Liquors; c_1 and c_2 are different categories of *Alco*

■ χ is a Boolean characteristic of the group Personal of attributes defined by the $\mathscr{RP}$-expression Personal introduced in Fig. 12.2.

Name: Personal		Type: *Conjunction*		Length: $0-4$	
Attribute	Coefficients	Length	Gace	B/R	Class
Education	*Subsets*	$1-1$	*Pos*	B	−
MaritalStatus	*Subsets*	$1-1$	*Pos*	B	−
Responsibility	*Subsets*	$1-1$	*Pos*	B	−
Age	*Subsets*	$1-1$	*Pos*	B	−

Figure 12.2: $\mathscr{RP}$-expression Personal.

The KL-pattern $Alco(c_1) \bowtie Alco(c_2) : R \approx C/\chi$ is evaluated using KL-tables $T_1 = KL(R,C,Entry/\chi \wedge Alco(c_1))$ and $T_2 = KL(R,C,Entry/\chi \wedge Alco(c_2))$. We are interested in R, C, χ, c_1, and c_2 for each *Alco* attribute such that the difference $\tau_B(T_1) - \tau_B(T_2)$ of Kendall's coefficients $\tau_B(T_1), \tau_B(T_2)$ of contingency tables T_1, T_2 is the highest possible. The highest difference $\tau_B(T_1) - \tau_B(T_2)$ corresponds to the highest difference between levels c_1 and c_2 of drinking alcohol *Alco* what concerns ordinal dependencies between attributes R and C for a group of patients given by the Boolean attribute χ. The following three tasks are solved below:

■ searching for groups of patients given by a Boolean attribute χ, attributes R and C and for levels c_1 a c_2 of drinking liquors such that $\tau_B(T_1) - \tau_B(T_2)$ is as high as possible, see Section 12.2.1

■ searching for groups of patients given by a Boolean attribute χ, attributes R and C and for levels c_1 a c_2 of drinking wine such that $\tau_B(T_1) - \tau_B(T_2)$ is as high as possible, see Section 12.2.2

■ searching for groups of patients given by a Boolean attribute χ, attributes R and C and for levels c_1 a c_2 of drinking beer such that $\tau_B(T_1) - \tau_B(T_2)$ is as high as possible, see Section 12.2.3.

In all three below described applications of the SDKL-Miner procedure, we use parameters according to the following principles.

■ A set *Row* of relevant row attributes R contains all attributes BMI, Subscapularis, Triceps, Diastolic, Systolic, Cholesterol, Triglycerides,

Beer, Wine, and Liquors except the *Alco* attribute is used to define sets $\mathscr{S}_\alpha$ and $\mathscr{S}_\beta$.

- A set *Col* of relevant column attributes C is defined in the same way as the *Row* set.

- Set of relevant conditions χ is given by the $\mathscr{R}\mathscr{P}$-expression Personal introduced in Fig. 12.2.

- Both a set $\mathscr{S}_\alpha$ of relevant Boolean attributes α and a set $\mathscr{S}_\beta$ of relevant Boolean attributes β are given as a set of all basic Boolean attributes $Alco(c)$ where c is a category of the *Alco* attribute.

- *SDKL-quantifier* $\approx$ is given by a list of three following simple SDKL-quantifiers:

  - $\langle$SDKLSF: *Sum, Set*$1\rangle \geq 20$ corresponding to at least 20 rows in a data sub-matrix $Entry/\chi \wedge Alco(c_1)$

  - $\langle$SDKLSF: *Sum, Set*$2\rangle \geq 20$ corresponding to at least 20 rows in a data sub-matrix $Entry/\chi \wedge Alco(c_2)$

  - $Diff\tau_b \geq 0.5$ corresponding to a condition that it holds at least $\tau_B(T_1) - \tau_B(T_2) \geq 0.5$.

Let us emphasize that the goal of these examples is not to present a serious statistical analysis. The goal is only to outline possibilities of the SDKL-Miner procedure.

12.2.1 Drinking liquors—groups with the highest τ_B difference

A run of the SDKL-Miner procedure with sets of relevant Boolean attributes α and a β defined by the $\mathscr{R}\mathscr{P}$-expression Liquors introduced in Fig. 12.3 and with sets *Row* and *Col* of relevant row and column attributes consisting of attributes BMI, Subscapularis, Triceps, Diastolic, Systolic, Cholesterol, Triglycerides, Beer, and Wine resulted in 74 true relevant SDKL-patterns. 77 328 SDKL-patterns were tested in two seconds. The SDKL-pattern

$$\text{Liquors}(\textit{up to 100 cc / day}) \bowtie \text{Liquors}(\textit{no}) : \text{Beer} \approx \text{Wine}/\chi_L \qquad (12.6)$$

Name: Liquors		Type: *Conjunction*		Length: $1-1$	
Attribute	Coefficients	Length	Gace	B/R	Class
Liquors	*Subsets*	$1-1$	*Pos*	*B*	—

Figure 12.3: $\mathscr{R}\mathscr{P}$-expression Liquors.

where $\chi_L = \text{Age}(\leq 45) \wedge manager \wedge married \wedge university$ is the KL-pattern with the highest $\tau_B(T_1) - \tau_B(T_2)$. The contingency tables
$TL_1 = KL(\text{Beer}, \text{Wine}, Entry/\chi_L \wedge \text{Liquors}(up\ to\ 100\ cc\ /\ day))$ and
$TL_2 = KL(\text{Beer}, \text{Wine}, Entry/\chi_L \wedge \text{Liquors}(no))$ are available in Fig. 12.4. It holds $\tau_B(TL_1) = 0.80$, $\tau_B(TL_2) = -0.18$, and $\tau_B(TL_1) - \tau_B(TL_2) = 0.98$.

Table TL_1		Wine	
Beer	no	$\leq \frac{1}{2}$ litre	$> \frac{1}{2}$ litre
no	2	1	0
≤ 1 litre	0	19	0
> 1 litre	0	0	0

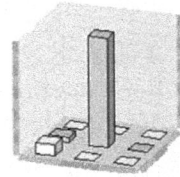

Table TL_2		Wine	
Beer	no	$\leq \frac{1}{2}$ litre	$> \frac{1}{2}$ litre
no	6	4	2
≤ 1 litre	5	4	0
> 1 litre	1	0	0

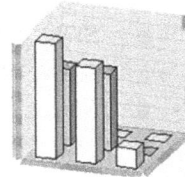

Figure 12.4: Contingency tables TL_1 and TL_2.

We write "*manager*" instead of "Responsibility(*manager*)", "*married*" instead of "MaritalStatus(*married*)", etc.

12.2.2 Drinking vine—groups with the highest τ_B difference

A run of the SDKL-Miner procedure with sets of relevant Boolean attributes α and β defined by the $\mathscr{RP}$-expression Wine introduced in Fig. 12.5 and with sets *Row* and *Col* of relevant row and column attributes consisting of attributes BMI, Subscapularis, Triceps, Diastolic, Systolic, Cholesterol, Triglycerides, Liquors and Beer resulted in 88 true relevant SDKL-patterns. 70 848 SDKL-patterns were tested in two seconds.

The SDKL-pattern

$$\text{Wine}(no) \bowtie \text{Wine}(up\ to\ half\ a\ litre\ /\ day) : \text{Beer} \approx \text{Triceps}/\chi_V \qquad (12.7)$$

Name: Wine		Type: *Conjunction*		Length: $1 - 1$	
Attribute	Coefficients	Length	Gace	B/R	Class
Wine	*Subsets*	$1 - 1$	*Pos*	*B*	—

Figure 12.5: $\mathscr{RP}$-expression Wine.

where $\chi_V = \mathrm{Age}(>45) \wedge manager \wedge married \wedge secondary$ is the KL-pattern with the highest $\tau_B(T_1) - \tau_B(T_2)$. We again write "*manager*" instead of "Responsibility(*manager*)", "*married*" instead of "MaritalStatus(*married*)" and "*secondary*" instead of "Education(*secondary*)".

The contingency tables
$TV_1 = KL(\mathrm{Beer}, \mathrm{Triceps}, Entry/\chi_V \wedge \mathrm{Vine}(no))$ and
$TV_2 = KL(\mathrm{Beer}, \mathrm{Triceps}, Entry/\chi_V \wedge \mathrm{Vine}(up\ to\ half\ a\ litre\ /\ day))$ are available in Fig. 12.6. We write "≤ 1 *litre*" instead of "*up to 1 litre / day*" etc. It holds $\tau_B(TV_1) = 0.60$, $\tau_B(TV_2) = -0.09$, and $\tau_B(TV_1) - \tau_B(TV_2) = 0.69$.

Table $T1_V$	Triceps		
Beer	low	average	high
no	7	1	0
≤ 1 litre	2	6	4
> 1 litre	0	1	0

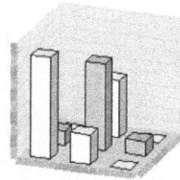

Table $T2_V$	Triceps		
Beer	low	average	high
no	4	3	5
≤ 1 litre	6	9	4
> 1 litre	1	0	1

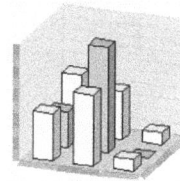

Figure 12.6: Contingency tables TV_1 and TV_2.

12.2.3 Drinking beer—groups with the highest τ_B difference

A run of the SDKL-Miner procedure with sets of relevant Boolean attributes α and β defined by the $\mathcal{RP}$-expression Beer introduced in Fig. 12.7 and with sets *Row* and *Col* of relevant rows and columns attributes consisting of attributes BMI, Subscapularis, Triceps, Diastolic, Systolic, Cholesterol, Triglycerides, Liquors, and Wine resulted in 78 true relevant SDKL-patterns. 83 088 SDKL-patterns were tested in two seconds.

The first three highest differences $\tau_B(T_1) - \tau_B(T_2)$ have values 0.80, 0.74, and 0.72. All of them belong to contingency tables with both three rows and

Name: Beer		Type: *Conjunction*	Length: $1-1$		
Attribute	Coefficients	Length	Gace	B/R	Class
Beer	*Subsets*	$1-1$	*Pos*	B	–

Figure 12.7: $\mathcal{RP}$-expression Beer.

three columns. An example of high difference belonging to a contingency table of a different size is the SDKL-pattern

$$\text{Beer}(no) \bowtie \text{Beer}(up\ to\ 1\ litre) : \text{Diastolic} \approx \text{Cholesterol}/\chi_B \qquad (12.8)$$

where $\chi_B = \text{Age}(\leq 45) \wedge partly\ independent \wedge married \wedge secondary$. We use "*partly independent*" instead of "Responsibility(*partly independent*)", etc. Contingency tables $T1_B = KL(\text{Diastolic}, \text{Cholesterol}, Entry/\chi_B \wedge \text{Beer}(no))$ and $T2_B = KL(\text{Diastolic}, \text{Cholesterol}, Entry/\chi_B \wedge \text{Beer}(upto1litre/day))$ are available in Fig. 12.8. It holds $\tau_B(T1_B) = 0.39$, $\tau_B(T2_B) = -0.26$, and $\tau_B(T1_B) - \tau_B(T2_B) = 0.65$.

Table $T1_B$	Cholesterol				
Diastolic	very low	lower	average	higher	very high
very low	2	1	3	1	0
lower	1	1	2	1	0
average	0	0	1	1	1
higher	0	0	0	1	0
very high	0	1	2	1	2

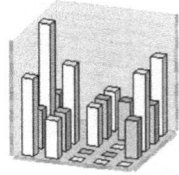

Table $T2_B$	Cholesterol				
Diastolic	very low	lower	average	higher	very high
very low	0	2	2	3	2
lower	1	1	3	0	2
average	0	2	1	3	1
higher	1	1	2	2	1
very high	4	0	2	1	0

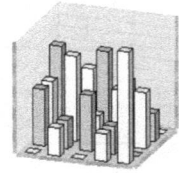

Figure 12.8: Contingency tables $T1_B$ and $T2_B$.

Chapter 13

Ac4ft-Miner—Action Rules

The Ac4ft-Miner procedure and action rules it deals with are informally introduced in Section 3.8. A more precise description of the Ac4ft-Miner procedure and related notions is available in Section 13.1. Examples of applications of the Ac4ft-Miner procedure in the *STULONG* dataset are presented in Section 13.2. Examples of applications of the Ac4ft-Miner procedure in the *Hotel* dataset are described in Section 13.3. Additional examples of applications of the Ac4ft-Miner procedures are available in [93, 121, 131].

13.1 Ac4ft-Miner and Related Notions

We start with notions of flexible attributes and of changes of Boolean attributes introduced in Sections 13.1.1 and 13.1.2. A definition of action rules is presented in Section 13.1.3. Ac4ft-quantifiers and truthfulness of action rules are defined in Section 13.1.4. A notion of relevant changes of the Boolean attribute defined in Section 13.1.5 is used to describe input and output of the Ac4ft-Miner procedure in Section 13.1.6.

13.1.1 Flexible and stable attributes

The first basic idea behind action rules is the existence of *flexible attributes*. Values of flexible attributes can be changed. The amount of interest can be changed by a bank for some clients; Body Mass Index (BMI) can be changed by a patient etc. On the contrary, the values of some attributes cannot be changed. Such attributes are called *stable attributes*. A place of birth is an example of a stable attribute.

The second basic idea behind the action rules is that there is available data about objects with the same values of stable attributes and different values of flexible attributes. This makes it possible to study the influence of changes of flexible attributes. An example of such a study is available in Section 3.8. BMI is used as a flexible attribute and the height of a patient is used as a stable attribute. The goal of the example is to study the influence of changes of BMI to a level of systolic blood pressure.

13.1.2 Changes of Boolean attributes

The change of a Boolean attribute is a crucial notion for action rules. Let A be an attribute with a set of possible values – categories $\{a_1, \ldots, a_t\}$. Then we define:

■ a *change of a basic Boolean attribute* is an expression $A(\alpha_1 \mapsto \alpha_2)$ where $\alpha_1 \subsetneq \{a_1, \ldots, a_t\}$, $\alpha_2 \subsetneq \{a_1, \ldots, a_t\}$, and $\alpha_1 \cap \alpha_2 = \emptyset$

■ a *before state* $\mathscr{B}(A(\alpha_1 \mapsto \alpha_2))$ of a change $A(\alpha_1 \mapsto \alpha_2)$ of basic Boolean attribute is a basic Boolean attribute $A(\alpha_1)$

■ an *after state* $\mathscr{A}(A(\alpha_1 \mapsto \alpha_2))$ of a change $A(\alpha_1 \mapsto \alpha_2)$ of basic Boolean attribute is a basic Boolean attribute $A(\alpha_2)$.

In addition, we define:

■ each change of a basic Boolean attribute is a *change of Boolean attribute*

■ if ω is a change of Boolean attribute, then $\neg\omega$ is a change of Boolean attribute, $\neg\mathscr{B}(\omega)$ is a *before state* of $\neg\omega$ and $\neg\mathscr{A}(\omega)$ is an *after state* of ω

■ if ω and τ are changes of Boolean attributes, then $\omega \wedge \tau$ is a change of a Boolean attribute, $\mathscr{B}(\omega) \wedge \mathscr{B}(\tau)$ is a *before state* $\mathscr{B}(\omega \wedge \tau)$ of $\omega \wedge \tau$, and $\mathscr{A}(\omega) \wedge \mathscr{A}(\tau)$ is an *after state* $\mathscr{A}(\omega \wedge \tau)$ of $\omega \wedge \tau$

■ if ω and τ are changes of Boolean attributes, then $\omega \vee \tau$ is a change of a Boolean attribute, $\mathscr{B}(\omega) \vee \mathscr{B}(\tau)$ is a *before state* $\mathscr{B}(\omega \vee \tau)$ of $\omega \vee \tau$, and $\mathscr{A}(\omega) \vee \mathscr{A}(\tau)$ is an *after state* $\mathscr{A}(\omega \vee \tau)$ of $\omega \vee \tau$.

13.1.3 Action rules

An *action rule* is one of the following expressions

$$\varphi \wedge \Phi \approx \psi \wedge \Psi \qquad\qquad \varphi \wedge \Phi \approx \psi \wedge \Psi / \chi, \qquad (13.1)$$

where φ, ψ, χ are Boolean attributes and Φ, Ψ are changes of Boolean attributes. Here, φ is a *stable antecedent*, ψ is a *stable succedent*, χ is a *condition*, Φ is a *change of antecedent,* Ψ is a *change of succedent* and a symbol $\approx$ is an Ac4ft-quantifier introduced in Section 13.1.4. There are no common attributes in φ, Φ, ψ, Ψ and χ. The expression $\varphi \wedge \Phi \approx \psi \wedge \Psi / \chi$ is a *conditional action rule.*

Each action rule $\varphi \wedge \Phi \approx \psi \wedge \Psi$ can be seen as a couple of association rules

$$\varphi \wedge \mathcal{B}(\Phi) \approx \psi \wedge \mathcal{B}(\Psi) \qquad\qquad \varphi \wedge \mathcal{A}(\Phi) \approx \psi \wedge \mathcal{A}(\Psi), \qquad (13.2)$$

and each conditional action rule action rule $\varphi \wedge \Phi \approx \psi \wedge \Psi / \chi$ can be seen as a couple of conditional association rules

$$\varphi \wedge \mathcal{B}(\Phi) \approx \psi \wedge \mathcal{B}(\Psi)/\chi \qquad\qquad \varphi \wedge \mathcal{A}(\Phi) \approx \psi \wedge \mathcal{A}(\Psi)/\chi. \qquad (13.3)$$

13.1.4 Ac4ft-quantifiers and truthfulness of action rules

The symbol $\approx$ used in action rules is an Ac4ft-quantifier. It corresponds to a condition concerning a couple of 4ft-tables. An action rule $\varphi \wedge \mathcal{B}(\Phi) \approx \psi \wedge \mathcal{B}(\Psi)$ is true in a data matrix $\mathcal{M}$ if a condition related to a given Ac4ft-quantifier $\approx$ is satisfied for a couple of 4ft-tables

■ $T_B = \langle a_B, b_B, c_B, d_B \rangle = 4ft(\varphi \wedge \mathcal{B}(\Phi), \psi \wedge \mathcal{B}(\Psi), \mathcal{M})$

■ $T_A = \langle a_A, b_A, c_A, d_A \rangle = 4ft(\varphi \wedge \mathcal{A}(\Phi), \psi \wedge \mathcal{A}(\Psi), \mathcal{M})$.

A conditional action rule $\varphi \wedge \mathcal{B}(\Phi) \approx \psi \wedge \mathcal{B}(\Psi)/\chi$ is true in a data matrix $\mathcal{M}$ if a condition related to a given Ac4ft-quantifier $\approx$ is satisfied for 4ft-tables

■ $T_B = \langle a_B, b_B, c_B, d_B \rangle = 4ft(\varphi \wedge \mathcal{B}(\Phi), \psi \wedge \mathcal{B}(\Psi), \mathcal{M}/\chi)$

■ $T_A = \langle a_A, b_A, c_A, d_A \rangle = 4ft(\varphi \wedge \mathcal{A}(\Phi), \psi \wedge \mathcal{A}(\Psi), \mathcal{M}/\chi)$.

Ac4ft-quantifier is usually a conjunction of several basic Ac4ft-quantifiers. Seven examples of basic Ac4ft-quantifiers are introduced in Table 13.1.

Table 13.1: Simple SD4ft-quantifiers.

#	Name	Symbolic notation	$\approx (a,b,c,d) = 1$ if and only if
1	$Base_B$	–	$a_B \circledast B_B$
2	$Base_A$	–	$a_A \circledast B_A$
3	$Conf_B$	$Conf_B \circledast p_B$	$\frac{a_B}{a_B+b_B} \circledast p_B \wedge a_B + b_B > 0$
4	$Conf_A$	$Conf_A \circledast p_A$	$\frac{a_A}{a_A+b_A} \circledast p_A \wedge a_A + b_A > 0$
5	$RConf$	$RConf \circledast q$	$\frac{a_B}{a_B+b_B} / \frac{a_A}{a_A+b_A} \circledast q \wedge a_B + b_B > 0 \wedge a_A + b_A > 0$
6	$Supp_B$	$Supp_B \circledast s_B$	$\frac{a_B}{a_B+b_B+c_B+d_B} \circledast s_B \wedge a_B + b_B + c_B + d_B > 0$
7	$Supp_A$	$Supp_A \circledast s_A$	$\frac{a_A}{a_A+b_A+c_A+d_A} \circledast s_A \wedge a_A + b_A + c_A + d_A > 0$

It holds:

- $B_B > 0, B_A > 0$ are integer numbers

- $0 \leq p_B \leq 1, 0 \leq p_B \leq 1, 0 \leq s_B \leq 1, 0 \leq s_A \leq 1$ are real numbers

- $q \geq 0$ is a real number and

- ⊛ is one of the relations $<, \leq, =, \geq, >$ concerning real numbers.

Conditions in the last column of Table 13.1 with the symbol ⊛ are satisfied if relations given by ⊛ are satisfied. A conjunction

$$a_B \geq 100 \wedge a_A \geq 100 \wedge RConf \geq 2$$

is an example of Ac4ft-quantifier.

13.1.5   Relevant changes of Boolean attribute

A definition of a set of relevant changes of a Boolean attribute is an important part of the input parameters of the Ac4ft-Miner procedure. A definition of a set of relevant cedents is used to define a set of relevant changes of a Boolean attribute. Recall that a set of relevant cedents is given by an $\mathscr{RC}$-expression

$$\langle Name_C, Min_C, Max_C; NameP_1, \ldots, NameP_r \rangle \,,$$

see Section 4.4.5. Here $Name_C$ is a name of the set of relevant cedents, Min_P is a minimal length, Max_P is a maximal length of relevant cedents, it holds $0 \leq Min_C \leq Max_C$. $NameP_1, \ldots, NameP_r$ are names of sets of relevant partial cedents. For details see Section 4.4.5.

A set of relevant changes of a Boolean attribute given by a set $Name_C$ of relevant cedents is defined as a set of changes Λ of Boolean attributes satisfying both $\mathscr{B}(\Lambda) \in Name_C$ and $\mathscr{A}(\Lambda) \in Name_C$.

13.1.6   Ac4ft-Miner procedure input and output

The Ac4ft-Miner procedure generates and verifies action rules $\varphi \wedge \Phi \approx \psi \wedge \Psi$ or conditional action rules $\varphi \wedge \Phi \approx \psi \wedge \Psi / \chi$. An input of the procedure consists of a data matrix $\mathscr{M}$ to be analysed and of the following parameters defining a set of relevant Ac4ft-action rules:

- an $\mathscr{RC}$-expression defining a set *StabAnt* of relevant stable antecedents

- an $\mathscr{RC}$-expression defining a set *ChangeAnt* of relevant changes of antecedent

- an $\mathscr{RC}$-expression defining a set *StabSucc* of relevant stable succedents

- an $\mathscr{RC}$-expression defining a set *ChangeSucc* of relevant changes of succedent

- an $\mathscr{RC}$-expression defining a set *Cond* of relevant conditions

- a list *ListAc4ft* of basic Ac4ft-quantifiers.

The $\mathscr{RC}$-expressions defining sets *StabAnt, ChangeAnt, StabSucc, ChangeSucc* and *Cond* can contain common attributes.

If the set *Cond* is empty, then all action rules $\varphi \wedge \Phi \approx \psi \wedge \Psi$ such that $\varphi \in StabAnt$, $\Phi \in ChangeAnt$, $\psi \in StabSucc$, and $\Psi \in ChangeSucc$ where φ, Φ, ψ, and Ψ have no common attributes are generated.

If the set *Cond* is not empty, then all action rules $\varphi \wedge \Phi \approx \psi \wedge \Psi / \chi$ such that $\varphi \in StabAnt$, $\Phi \in ChangeAnt$, $\psi \in StabSucc$, $\Psi \in ChangeSucc$ satisfying that φ, Φ, ψ, Ψ, and χ have no common attributes generated.

In both cases, Ac4ft-quantifier $\approx$ is defined as a conjunction of all basic Ac4ft-quantifiers from *ListAc4ft*. Each generated action rule is verified in the given data matrix $\mathscr{M}$. The output of the Ac4ft-Miner procedure consists of all true action rules.

13.2   Applying Ac4ft-Miner in *STULONG* Dataset

We present two simple examples of applications of the Ac4ft-Miner procedure concerning generally known relations of BMI and blood pressure, see also Section 3.8. Common features of both examples are introduced in Section 13.2.1. The examples are described in Sections 13.2.2 and 13.2.3.

13.2.1   Two analytical questions–common features

We deal with attributes Age, Education, MaritalStatus, and Responsibility from the group Personal described in Section 2.4.2. We use also the attributes Diastolic and Systolic from the group Blood pressure described in Section 2.4.7. Finally, we use the attribute BMI which belongs to the group Measurement, see Section 2.4.5.

It is a known fact that decreasing BMI results in decreasing probability of high blood pressure. Let us consider two association rules

$$\text{Age}(> 45) \wedge \text{BMI}(\textit{overweight}) \rightarrow_{0.128} \text{Diastolic}(\textit{higher}) \qquad (13.4)$$

$$\text{Age}(> 45) \wedge \text{BMI}(\textit{normal}) \rightarrow_{0.105} \text{Diastolic}(\textit{higher}). \qquad (13.5)$$

The confidence of the rule (13.4) is 0.128 and the confidence of the rule (13.5) is 0.105. This can be seen such that a change of BMI from overweight to normal for patients satisfying Age(> 45) causes decreasing probability Diastolic(*higher*) from 0.128 to 0.105, i.e., $\frac{0.128}{0.105} = 1.22$ times.

The couple of the rules (13.4) and (13.5) is understood as an action rule

$$\text{Age}(> 45) \wedge \text{BMI}(overweight \mapsto normal) \downarrow_{1.22} \text{Diastolic}(higher). \quad (13.6)$$

The expression BMI(*overweight* $\mapsto$ *normal*) is a change of the flexible attribute BMI, the attribute Age is an example of stable attribute.

We consider all the attributes of the Personal group as stable attributes. We are going to solve two related tasks:

■ *Which changes of the attribute BMI decrease a probability of high blood pressure?* This task is solved in Section 13.2.2.

■ *Which changes of the attribute BMI increase a probability of average blood pressure?* This task is solved in Section 13.2.3.

13.2.2 BMI and decreasing probability of high blood pressure

We search for changes of the attribute BMI decreasing at least two times a probability of high blood pressure $\mathscr{H}\mathscr{P}$. We are interested in groups of patients described by the Boolean attribute $\mathscr{P}$ derived from the group Personal. Thus, we are interested in action rules

$$\mathscr{P} \wedge \text{BMI}(Lev_B \mapsto Lev_A) \downarrow_{2.0,30,30} \mathscr{H}\mathscr{P} \quad (13.7)$$

equivalent to a couple of association rules

$$\mathscr{P} \wedge \text{BMI}(Lev_B) \rightarrow_{p_B,30} \mathscr{H}\mathscr{P} \quad (13.8)$$

and

$$\mathscr{P} \wedge \text{BMI}(Lev_A) \rightarrow_{p_A,30} \mathscr{H}\mathscr{P} . \quad (13.9)$$

4ft-tables $4ft(\mathscr{P} \wedge \text{BMI}(Lev_B), \mathscr{H}\mathscr{P}, Entry)$ and $4ft(\mathscr{P} \wedge \text{BMI}(Lev_A), \mathscr{H}\mathscr{P}, Entry)$ corresponding to these rules are available in Fig. 13.1.

Entry	$\mathscr{H}\mathscr{P}$	$\neg\mathscr{H}\mathscr{P}$		*Entry*	$\mathscr{H}\mathscr{P}$	$\neg\mathscr{H}\mathscr{P}$
$\mathscr{P} \wedge \text{BMI}(Lev_B)$	a_B	b_B		$\mathscr{P} \wedge \text{BMI}(Lev_A)$	a_A	b_A
$\neg(\mathscr{P} \wedge \text{BMI}(Lev_B))$	c_B	d_B		$\neg(\mathscr{P} \wedge \text{BMI}(Lev_A))$	c_A	d_A

Figure 13.1: $4ft(\mathscr{P} \wedge \text{BMI}(Lev_B), \mathscr{H}\mathscr{P}, Entry)$ and $4ft(\mathscr{P} \wedge \text{BMI}(Lev_A), \mathscr{H}\mathscr{P}, Entry)$.

Ac4ft quantifier $\downarrow_{2.0,30,30}$ used in (13.7) corresponds to the fact that actions suggested by the action rules decrease the confidence. It is defined by the condition

$$\frac{p_B}{p_A} \geq 2 \ \wedge a_B \geq 30 \ \wedge a_A \geq 30 \quad (13.10)$$

where $p_B = \frac{a_B}{a_B+b_B}$ and $p_A = \frac{a_A}{a_A+b_A}$. This is the same as

$$RConf \geq 2 \ \wedge a_B \geq 30 \ \wedge a_A \geq 30 , \qquad (13.11)$$

since it holds $RConf = \frac{p_B}{p_A}$, see row 5 in Table 13.1. We can conclude that we use three basic Ac4ft-quantifiers $Base_B$, $Base_A$, and $RConf$ specified in equation (13.11).

$\mathscr{P}$ is a Boolean characteristic of the group Personal given by the $\mathscr{RP}$-expression Personal introduced in Fig. 13.2, see also Fig. 7.12.

Name: Personal		Type: *Conjunction*		Length: $1-4$	
Attribute	Coefficients	Length	Gace	B/R	Class
Education	*Subsets*	$1-1$	*Pos*	B	−
MaritalStatus	*Subsets*	$1-1$	*Pos*	B	−
Responsibility	*Subsets*	$1-1$	*Pos*	B	−
Age	*Subsets*	$1-1$	*Pos*	B	−

Figure 13.2: $\mathscr{RP}$-expression Personal.

BMI($Lev_B \mapsto Lev_A$) is a change of a Boolean attribute BMI. Both Lev_B (level before) and Lev_A (level after) belong to the set {underweight, normal, overweight, obese class I, obese class II, obese class III} of categories of the attribute BMI, see Section 2.4.5. This means that a set of relevant changes of the attribute BMI is given by an $\mathscr{RP}$-expression Changes of BMI available in Fig. 13.3.

Name: Changes of BMI		Type: *Conjunction*		Length: $1-1$	
Attribute	Coefficients	Length	Gace	B/R	Class
BMI	*Subsets*	$1-1$	*Pos*	B	−

Figure 13.3: $\mathscr{RP}$-expression Changes of BMI.

$\mathscr{HP}$ used in (13.7), (13.8), and (13.9) is one of the Boolean attributes Diastolic(*very high*), Diastolic(*high, very high*), Systolic(*very high*), Systolic(*high, very high*), Diastolic(*very high*) ∧ Systolic(*very high*), ..., Diastolic(*high, very high*) ∧ Systolic(*high, very high*). They are given by the $\mathscr{RP}$-expression High Pressure introduced in Fig. 13.4.

A run of the Ac4ft-Miner procedure with the above specified parameters resulted in 40 true relevant action rules. 1 648 action rules were tested in less than one second. A detailed output of the action rule with the highest ratio $RConf$ is available in Fig. 13.5.

This rule can be written as

married ∧ BMI(*obese class I* $\mapsto$ *normal*) $\downarrow_{3.137,47,61}$ Systolic(*very high*) (13.12)

Name: High Pressure		Type: *Conjunction*		Length: $1-2$	
Attribute	Coefficients	Length	Gace	B/R	Class
Diastolic	*Right cuts*	$1-2$	*Pos*	B	—
Systolic	*Right cuts*	$1-2$	*Pos*	B	—

Figure 13.4: $\mathscr{RP}$-expression High Pressure.

Antecedent: MaritalStatus(married) : (BMI(obese class I) -> BMI(normal))
Succedent: Systolic(very high)
State before: MaritalStatus(married) && BMI(obese class I) >-< Systolic(very high)
State after: MaritalStatus(married) && BMI(normal) >+< Systolic(very high)
Condition: (empty)

TEXT | DATA | STATE BEFORE | STATE AFTER | B+A | DIFF ABS | DIFF REL |

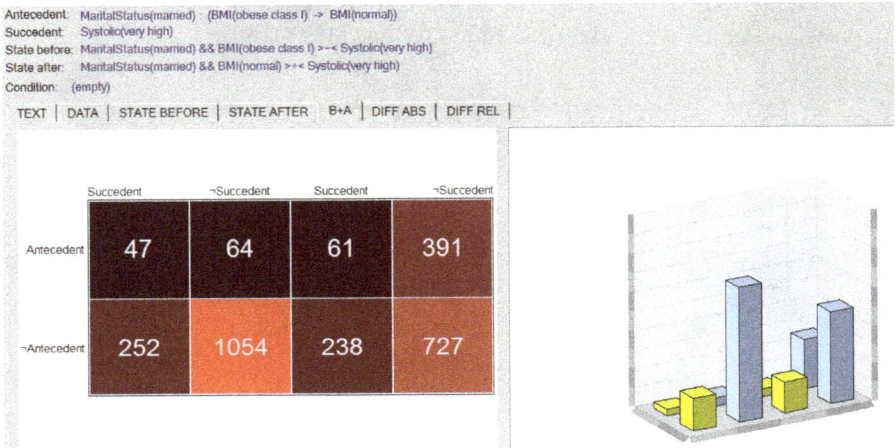

Figure 13.5: A detailed output of the action rule with the highest ratio *RConf*.

where we use "*married*" instead of "MaritalStatus(*married*)". This rule is equivalent to a couple of association rules (13.13) and (13.14)

$$married \wedge \mathsf{BMI}(obese\ class\ I) \rightarrow_{0.423,47} \mathsf{Systolic}(very\ high) \qquad (13.13)$$

$$married \wedge \mathsf{BMI}(normal) \rightarrow_{0.135,61} \mathsf{Systolic}(very\ high). \qquad (13.14)$$

4ft-tables corresponding to these rules are available in Fig. 13.6.

Entry	Systolic(*very high*)	¬ Systolic(*very high*)
married $\wedge$ BMI(*obese class I*)	47	64
¬(*married* $\wedge$ BMI(*obese class I*))	252	1054

Entry	Systolic(*very high*)	¬ Systolic(*very high*)
married $\wedge$ BMI(*normal*)	61	391
¬(*married* $\wedge$ BMI(*normal*))	238	727

Figure 13.6: 4ft-tables corresponding to rules (13.13) and (13.14).

Note that it holds $0.423 = \frac{47}{47+64}$, $0.135 = \frac{61}{61+391}$ and $3.137 = \frac{\frac{47}{47+64}}{\frac{61}{61+391}}$.

Let us emphasize that the action rule (13.13) corresponds to a well known fact. The same is true for the additional output rules. Our goal is not to present a new medical knowledge; our goal is only to demonstrate possibilities of the Ac4ft-Miner procedure.

13.2.3 *Increasing probability of average blood pressure*

The goal of the example presented in this section is similar to that presented in the previous section. However, we are now trying to *increase* the probability of average blood pressure. We again deal with groups of patients described by the attributes from the group Personal. We are interested in action rules

$$\mathscr{P} \wedge \mathsf{BMI}(Lev_B \mapsto Lev_A) \uparrow_{q,30,30} \mathscr{A}\mathscr{P} \qquad (13.15)$$

equivalent to a couple of association rules (13.16) and (13.17)

$$\mathscr{P} \wedge \mathsf{BMI}(Lev_B) \to_{p_B,30} \mathscr{A}\mathscr{P} \qquad (13.16)$$

$$\mathscr{P} \wedge \mathsf{BMI}(Lev_A) \to_{p_A,30} \mathscr{A}\mathscr{P} \qquad (13.17)$$

satisfying $q = \frac{p_A}{p_B} > 1$.

4ft-tables $4ft(\mathscr{P} \wedge \mathsf{BMI}(Lev_B), \mathscr{A}\mathscr{P}, Entry)$ and $4ft(\mathscr{P} \wedge \mathsf{BMI}(Lev_A), \mathscr{A}\mathscr{P}, Entry)$ corresponding to these rules are available in Fig. 13.7.

$Entry$	$\mathscr{H}\mathscr{P}$	$\neg\mathscr{A}\mathscr{P}$		$Entry$	$\mathscr{H}\mathscr{P}$	$\neg\mathscr{A}\mathscr{P}$
$\mathscr{P} \wedge \mathsf{BMI}(Lev_B)$	a_B	b_B		$\mathscr{P} \wedge \mathsf{BMI}(Lev_A)$	a_A	b_A
$\neg(\mathscr{P} \wedge \mathsf{BMI}(Lev_B))$	c_B	d_B		$\neg(\mathscr{P} \wedge \mathsf{BMI}(Lev_A))$	c_A	d_A

Figure 13.7: $4ft(\mathscr{P} \wedge \mathsf{BMI}(Lev_B), \mathscr{A}\mathscr{P}, Entry)$ and $4ft(\mathscr{P} \wedge \mathsf{BMI}(Lev_A), \mathscr{A}\mathscr{P}, Entry)$.

Our requirements can be formulated in the following way: Ac4ft quantifier $\uparrow_{q,30,30}$ is defined by the condition

$$\frac{p_A}{p_B} \geq q \,\wedge a_B \geq 30 \,\wedge a_A \geq 30 \qquad (13.18)$$

where $p_B = \frac{a_B}{a_B+b_B}$ and $p_A = \frac{a_A}{a_A+b_A}$. This is the same as

$$RConf \leq \frac{1}{q} \,\wedge\, a_B \geq 30 \,\wedge a_A \geq 30 \,, \qquad (13.19)$$

since it holds $RConf = \frac{p_B}{p_A} = \frac{1}{\frac{p_A}{p_B}}$, see row 5 in Table 13.1. We can conclude that we use three basic Ac4ft-quantifiers $Base_B$, $Base_A$, and $RConf$ specified in equation (13.19). The arrow $\uparrow_{q,30,30}$ corresponds to the fact that actions suggested by the action rules increase the confidence.

$\mathscr{P}$ is a Boolean characteristic of the group Personal given by the $\mathscr{RP}$-expression Personal introduced in Fig. 13.2. BMI($Lev_B \mapsto Lev_A$) is a change of a Boolean attribute BMI defined by a $\mathscr{RP}$-expression Changes of BMI in Fig. 13.3. $\mathscr{AP}$ is one of the Boolean attributes Diastolic(*average*), Systolic(*average*), Diastolic(*average*) $\wedge$ Systolic(*average*) given by the $\mathscr{RP}$-expression Average Pressure available in Fig. 13.8.

Name: Average Pressure		Type: *Conjunction*		Length: 1–2	
Attribute	Coefficients	Length	Gace	B/R	Class
Diastolic	*One category*	*average*	*Pos*	B	—
Systolic	*One category*	*average*	*Pos*	B	—

Figure 13.8: $\mathscr{RP}$-expression Average Pressure.

An experimental run of the Ac4ft-Miner procedure with these parameters and with the Ac4ft-quantifier $\uparrow_{2.0,30,30}$ resulted in no true action rule. A run with the Ac4ft-quantifier $\uparrow_{1.25,30,30}$ resulted in six true relevant action rules. 618 action rules were tested in less than one second. Note that the Ac4ft-quantifier $\uparrow_{1.25,30,30}$ corresponds to a condition

$$RConf \leq 0.8 \,\wedge\, a_B \geq 30 \,\wedge a_A \geq 30 \cdot \qquad (13.20)$$

A detailed output of the action rule with the lowest ratio $RConf$ (i.e., with the highest q in $\uparrow_{q,30,30}$) is available in Fig. 13.9.

This rule can be written as

Age(≤ 45) $\wedge$ *married* $\wedge$ BMI(*overweight* $\mapsto$ *normal*)

$$\uparrow_{1.59,42,61} \text{Diastolic}(average) \quad (13.21)$$

where we write "*married*" instead of "MaritalStatus(*married*)". It is equivalent to a couple of association rules
(13.22) and (13.23)

Age(≤ 45) $\wedge$ *married* $\wedge$ BMI(*overweight*) $\rightarrow_{0.188,42}$ Diastolic(*average*)

$$(13.22)$$

Age(≤ 45) $\wedge$ *married* $\wedge$ BMI(*normal*) $\rightarrow_{0.299,61}$ Diastolic(*average*). (13.23)

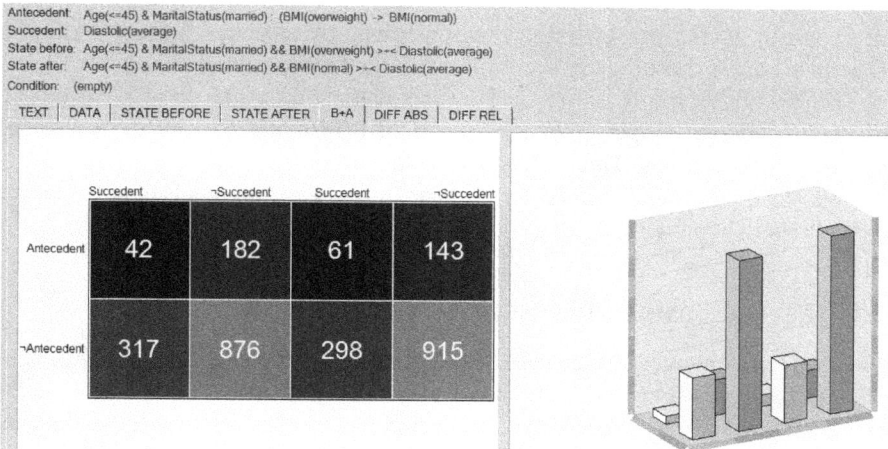

Antecedent: Age(<=45) & MaritalStatus(married) : (BMI(overweight) -> BMI(normal))
Succedent: Diastolic(average)
State before: Age(<=45) & MaritalStatus(married) && BMI(overweight) >-< Diastolic(average)
State after: Age(<=45) & MaritalStatus(married) && BMI(normal) >-< Diastolic(average)
Condition: (empty)

TEXT | DATA | STATE BEFORE | STATE AFTER | B+A | DIFF ABS | DIFF REL

Figure 13.9: A detailed output of the action rule with the lowest ratio *RConf*.

	Entry	Diastolic(*average*)	¬ Diastolic(*average*)
Age(≤ 45) $\wedge$ *married* $\wedge$ $\wedge$ BMI(*overweight*)		42	182
¬(Age(≤ 45) $\wedge$ *married* $\wedge$ $\wedge$ BMI(*overweight*))		317	876

	Entry	Diastolic(*average*)	¬ Diastolic(*average*)
Age(≤ 45) $\wedge$ *married* $\wedge$ $\wedge$ BMI(*normal*)		61	143
¬(Age(≤ 45) $\wedge$ *married* $\wedge$ $\wedge$ BMI(*normal*))		298	915

Figure 13.10: 4ft-tables corresponding to rules (13.22) and (13.23).

4ft-tables corresponding to these rules are available in Fig. 13.10. Note that it holds $0.188 = \frac{42}{42+182}$, $0.299 = \frac{61}{61+143}$ and $1.59 = \frac{\frac{61}{61+143}}{\frac{42}{42+182}}$. All the output action rules correspond to well known facts. However, let us again emphasize that our goal is not to present a new medical knowledge; our goal is only to demonstrate possibilities of the Ac4ft-Miner procedure.

13.3   Applying Ac4ft-Miner in *Hotel* Dataset

We present two examples of applications of the Ac4ft-Miner procedure concerning the artificial dataset *Hotel* introduced in Section 2.5. The first example concerns possibilities of increasing relative frequency of high overall satisfaction of guests by increasing partial evaluations given by attributes QRoom, QFood, QService, and QCulture, see Section 2.5.5. This task is solved in Section 13.3.1.

The second example is similar. However, we are also interested in the consequences of increasing partial evaluations. This is formalised by an application of changes of succedents. This task is solved in Section 13.3.2.

We use the artificial data *Hotel* because the data mining goals and results are easily understandable.

13.3.1   *Increasing guest satisfaction*

We search for changes of the attributes QRoom, QFood, QService, and QCulture leading to at least two times higher relative frequency of stays in a hotel satisfying QSatisfaction(*high*). We are interested in segments of stays described by attributes from the groups Guest (see Section 2.5.2), Meteo (see Section 2.5.4), and Check-in (see Section 2.5.7). This means that we are interested in action rules

$$\mathscr{G} \wedge \mathscr{ME} \wedge \mathscr{C} \wedge Ch(\mathscr{Q}) \uparrow_{2.0,30,30} \text{QSatisfaction}(high) \qquad (13.24)$$

equivalent to a couple of association rules (13.25) and (13.26)

$$\mathscr{G} \wedge \mathscr{ME} \wedge \mathscr{C} \wedge \mathscr{B}(Ch(\mathscr{Q})) \rightarrow_{p_B,30} \text{QSatisfaction}(high) \qquad (13.25)$$

$$\mathscr{G} \wedge \mathscr{ME} \wedge \mathscr{C} \wedge \mathscr{A}(Ch(\mathscr{Q})) \rightarrow_{p_A,30} \text{QSatisfaction}(high) \; . \qquad (13.26)$$

Here $\mathscr{G}$, $\mathscr{ME}$, and $\mathscr{C}$ are Boolean attributes derived from the groups Guest, Meteo, and Check-in respectively. $Ch(\mathscr{Q})$ denotes a change of a Boolean attribute derived from attributes QRoom, QFood, QService, and QCulture. Our requirements can be formulated in the following way:

Ac4ft quantifier $\uparrow_{2,60,60}$ is defined by the condition

$$\frac{p_A}{p_B} \geq 2 \; \wedge a_B \geq 60 \; \wedge a_A \geq 60 \; . \qquad (13.27)$$

This is the same as

$$RConf \leq \frac{1}{2} \; \wedge \; a_B \geq 60 \; \wedge a_A \geq 60 \; , \qquad (13.28)$$

it holds $RConf = \frac{p_B}{p_A} = \frac{1}{\frac{p_A}{p_B}}$, see row 5 in Table 13.1. We can conclude that we use three basic Ac4ft-quantifiers $Base_B$, $Base_A$, and $RConf$ specified in equation (13.28).

Relevant Boolean attributes $\mathscr{G}$, $\mathscr{ME}$, and $\mathscr{C}$ derived from the groups Guest, Meteo, and Check-in respectively are specified by the $\mathscr{RP}$-expressions Guest, Meteo, and Check-in introduced in Fig. 13.11. Set of relevant stable antecedents is defined by the $\mathscr{RC}$-expression Stable antecedent increasing satisfaction shown in Table 13.2.

Name: Guest		Type: *Conjunction*		Length: $0-2$	
Attribute	Coefficients	Length	Gace	B/R	Class
Age	*Subsets*	$1-1$	*Pos*	B	—
Sex	*Subsets*	$1-1$	*Pos*	B	—

Name: Meteo		Type: *Conjunction*		Length: $0-2$	
Attribute	Coefficients	Length	Gace	B/R	Class
Sky	*Subsets*	$1-1$	*Pos*	B	—
Temperature	*Subsets*	$1-1$	*Pos*	B	—

Name: Check-in		Type: *Conjunction*		Length: $0-4$	
Attribute	Coefficients	Length	Gace	B/R	Class
DayOfWeek	*Cyclical*	$1-3$	*Pos*	B	*W-End_Day-W*
WeekEnd	*Subsets*	$1-1$	*Pos*	B	*W-End_Day-W*
Month	*Cyclical*	$1-3$	*Pos*	B	*Season_Month*
Season	*Subsets*	$1-1$	*Pos*	B	*Season_Month*

Figure 13.11: $\mathscr{R}\mathscr{P}$-expressions Guest, Meteo, and Check-in.

Table 13.2: $\mathscr{R}\mathscr{C}$-expression Stable antecedent increasing satisfaction.

Name: Stable antecedent increasing satisfaction	Length: $1-8$
Set of partial cedents	Length
Guest	$0-2$
Meteo	$0-2$
Check-in	$0-4$

Note that the class of equivalence *W-End_Day-W* used in $\mathscr{R}\mathscr{P}$-expression Check-in ensures that no relevant Boolean attribute defined by the $\mathscr{R}\mathscr{P}$-expression Check-in contains both DayOfWeek and WeekEnd attributes, see also Section 4.4.4. This is the same for the class of equivalence *Season_Month* and attributes Month and Season.

A set of relevant changes $Ch(\mathscr{Q})$ of Boolean attributes derived from attributes QRoom, QFood, QService, and QCulture is defined by the $\mathscr{R}\mathscr{P}$-expression Questionnaire introduced in Fig. 13.12.

Name: Questionnaire		Type: *Conjunction*		Length: $1-4$	
Attribute	Coefficients	Length	Gace	B/R	Class
QRoom	*Subsets*	$1-1$	*Pos*	B	—
QFood	*Subsets*	$1-1$	*Pos*	B	—
QService	*Subsets*	$1-1$	*Pos*	B	—
QCulture	*Subsets*	$1-1$	*Pos*	B	—

Figure 13.12: $\mathscr{R}\mathscr{P}$-expression Questionnaire.

Succedent QSatisfaction(*high*) is of course defined by the $\mathscr{R}\mathscr{P}$-expression QSatisfaction available in Fig. 13.13.

Name: QSatisfaction		Type: *Conjunction*		Length: $1 - 1$	
Attribute	Coefficients	Length	Gace	B/R	Class
QSatisfaction	*One category*	high	*Pos*	B	—

Figure 13.13: $\mathscr{R}\mathscr{P}$-expression QSatisfaction.

A run of the Ac4ft-Miner procedure with the above specified parameters resulted in two true relevant action rules. 538 258 action rules were tested in 109 seconds. A detailed output of the action rule with the lowest ratio $RConf$ (i.e., with the highest ratio $\frac{p_A}{p_B}$) is available in Fig. 13.14. This rule can be written as

$$\text{DayOfWeek}(\textit{Thu, Fri, Sat}) \wedge \text{QCulture}(\textit{very low} \mapsto \textit{higher})$$
$$\uparrow_{2.24,60,147} \text{QSatisfaction}(\textit{high}) \,. \quad (13.29)$$

This is equivalent to a couple of association rules (13.30) and (13.31)

$$\text{DayOfWeek}(\textit{Thu, Fri, Sat}) \wedge \text{QCulture}(\textit{very low})$$
$$\rightarrow_{0.23,60} \text{QSatisfaction}(\textit{high}) \quad (13.30)$$

$$\text{DayOfWeek}(\textit{Thu, Fri, Sat}) \wedge \text{QCulture}(\textit{higher}) \rightarrow_{0.51,147} \text{QSatisfaction}(\textit{high}) \,.$$
$$(13.31)$$

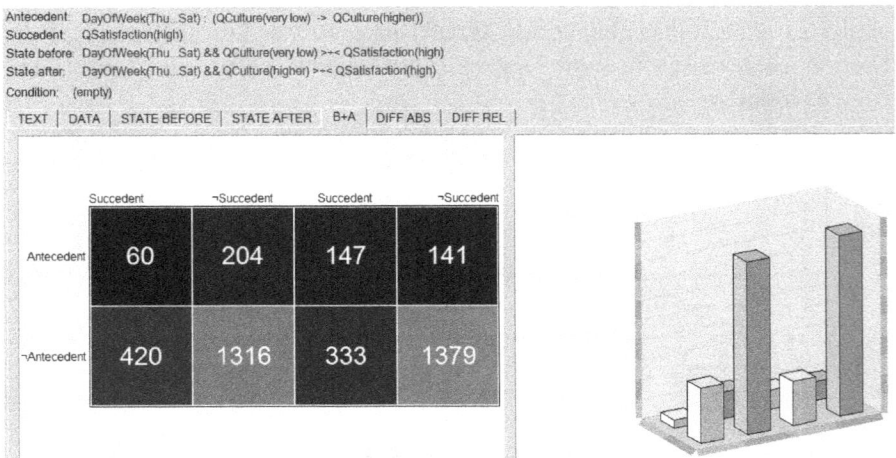

Figure 13.14: A detailed output of the action rule with the highest ratio $\frac{p_A}{p_B}$.

We can conclude that increasing evaluation of culture from *very low* to *higher* increases 2.24 times the probability of overall satisfaction at level *high* for a segment of stays satisfying DayOfWeek(*Wed, Thu, Fri*) (i.e., check-in in Wed, Thu, or Fri).

13.3.2 *Increasing guest satisfaction and consequences*

We have seen how the overall satisfaction can be increased at level *high*. A natural induced question is a question about the consequences of increasing a particular evaluation. We are especially interested in consequences concerning price of stays.

This means that we are interested in action rules

$$\mathcal{G} \wedge \mathcal{ME} \wedge \mathcal{C} \wedge Ch(\mathcal{Q}) \uparrow_{2.0,60,60} \text{QSatisfaction}(high) \wedge Ch(\mathcal{P}) \quad (13.32)$$

equivalent to a couple of association rules (13.25) and (13.26)

$$\mathcal{G} \wedge \mathcal{ME} \wedge \mathcal{C} \wedge \mathcal{B}(Ch(\mathcal{Q})) \rightarrow_{p_B,60} \text{QSatisfaction}(high) \wedge Ch(\mathcal{P}) \quad (13.33)$$

$$\mathcal{G} \wedge \mathcal{ME} \wedge \mathcal{C} \wedge \mathcal{A}(Ch(\mathcal{Q})) \rightarrow_{p_A,60} \text{QSatisfaction}(high) \wedge Ch(\mathcal{P}) \ . \quad (13.34)$$

Here $\mathcal{G}$, $\mathcal{ME}$, and $\mathcal{C}$ are Boolean attributes derived from the groups Guest, Meteo, and Check-in respectively. $Ch(\mathcal{Q})$ denotes a change of the Boolean attribute derived from attributes QRoom, QFood, QService, and QCulture. $Ch(\mathcal{P})$ denotes a change of the Boolean attribute derived from attributes PriceRoom and PriceFood, see Section 2.5.8.

Our requirements can be formulated in the following way:

Ac4ft quantifier $\uparrow_{2,60,60}$ is defined by the condition

$$\frac{p_A}{p_B} \geq 2 \ \wedge a_B \geq 60 \ \wedge a_A \geq 60 \ , \quad (13.35)$$

which is equivalent to

$$RConf \leq \frac{1}{2} \ \wedge \ a_B \geq 60 \ \wedge a_A \geq 60 \ . \quad (13.36)$$

We can conclude that we use three basic Ac4ft-quantifiers $Base_B$, $Base_A$, and $RConf$ specified in equation (13.36).

Relevant Boolean attributes $\mathcal{G}$, $\mathcal{ME}$, and $\mathcal{C}$ derived from the groups Guest, Meteo, and Check-in respectively are specified by the $\mathcal{RP}$-expressions Guest, Meteo, and Check-in, defined in the same as in the previous section, see Fig. 13.11. However, the set of relevant stable antecedents was defined by the $\mathcal{RC}$-expression Stable antecedent increasing satisfaction – short, see Table 13.3.

A set of relevant changes $Ch(\mathcal{Q})$ of Boolean attributes derived from attributes QRoom, QFood, QService, and QCulture and succedent QSatisfaction(high) are

Table 13.3: $\mathscr{RC}$-expression Stable antecedent increasing satisfaction – short.

Name: Stable antecedent increasing satisfaction – short	Length: $1-2$
Set of partial cedents	Length
Guest	$0-2$
Meteo	$0-2$
Check-in	$0-4$

Name: Prices		Type: *Conjunction*		Length: $1-2$	
Attribute	Coefficients	Length	Gace	B/R	Class
PriceRoom	*Subsets*	$1-1$	*Pos*	B	–
PriceFood	*Subsets*	$1-1$	*Pos*	B	–

Figure 13.15: $\mathscr{RP}$-expression Prices.

also defined in the same way as in the previous section, see Figs. 13.12 and 13.13. A set of relevant changes $Ch(\mathscr{P})$ derived from attributes PriceRoom and PriceFood is defined by the $\mathscr{RP}$-expression Prices introduced in Fig. 13.15.

A run of the Ac4ft-Miner procedure with the above specified parameters resulted in no true relevant action rules. After several attempts with various Ac4ft-quantifiers we get two true relevant action rules for Ac4ft quantifier $\uparrow_{1.4,60,60}$ defined by the condition

$$\frac{p_A}{p_B} \geq 1.4 \ \wedge a_B \geq 30 \ \wedge a_A \geq 30 \,, \tag{13.37}$$

which is equivalent to

$$RConf \leq \frac{1}{1.4} = 0.71 \ \wedge \ a_B \geq 30 \ \wedge a_A \geq 30 \,. \tag{13.38}$$

6 578 264 action rules were tested in 8 minutes and 42 seconds. A detailed output of the action rule with the lowest ratio $RConf$ (i.e., with the highest ratio $\frac{p_A}{p_B}$) is available in Fig. 13.16. This rule can be written as

DayOfWeek(*Thu, Fri, Sat*) $\wedge$ QFood(*higher* $\mapsto$ *very high*) $\uparrow_{1.53,37,57}$
QSatisfaction(*high*) $\wedge$ QPriceFood(*very low* $\mapsto$ *higher*) . (13.39)

This is equivalent to a couple of association rules (13.40) and (13.41)

DayOfWeek(*Thu, Fri, Sat*) $\wedge$ QFood(*higher*)
$\rightarrow_{0.12,37}$ QSatisfaction(*high*) $\wedge$ QPriceFood(*very low*) (13.40)

DayOfWeek(*Wed, Thu, Fri*) $\wedge$ QFood(*very high*)
$\rightarrow_{0.19,57}$ QSatisfaction(*high*) $\wedge$ QPriceFood(*higher*) (13.41)

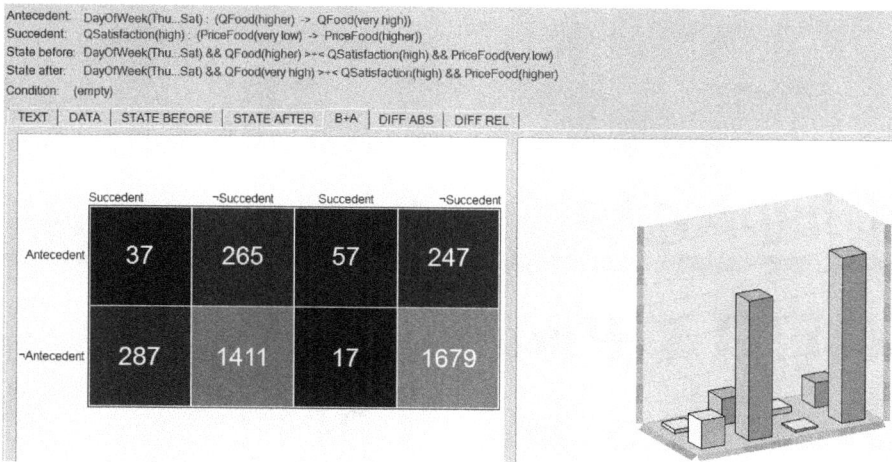

Figure 13.16: A detailed output of the action rule with the highest ratio $\frac{p_A}{p_B}$.

We can conclude that increasing evaluation of food from *higher* to *very high* increases 1.53 times the probability of overall satisfaction at level *high* for a segment of stays satisfying DayOfWeek(*Thu, Fri, Sat*) (i.e., check-in in Thu, Fri or Sat). However, this also means that price of food increases from *very low* to *higher*.

Let us note that the run of the Ac4ft-Miner procedure with the set of relevant stable antecedents defined by the $\mathcal{RC}$-expression Stable antecedent increasing satisfaction, see Table 13.2, resulted in the same two action rules. However, 59 224 704 action rules were tested in 89 minutes.

Let us again emphasize that we deal with artificial data and that an application of the Ac4ft-Miner in a right way requires a close co-operation with a domain expert.

Chapter 14

GUHA Procedures and Business Intelligence

The reason that this chapter is presented in the book is to show that business intelligence can benefit from mechanized hypotheses formation, represented by GUHA method. The goal of this chapter is also to show the benefits and possibilities of interconnecting these methods. The chapter is based on the dissertation thesis of its author [11]. One of the goals of the thesis was proposing and testing scenarios of complementary usage of OLAP and association rules mining. This chapter extends this goal by focusing not only on association rules represented by 4ft-Miner procedure, but also another GUHA procedures, especially CF-Miner. In contrast to the dissertation where a simple tool for creating OLAP cubes were used[1], in this chapter, business intelligence is represented by one of the most favorite commercial tool, PowerBI.

The chapter is organized as follows. The term business intelligence and related term self service business intelligence are presented in Section 14.1. The process of analysis, its common parts and differences using self service business intelligence tool and LISp-Miner procedure are presented in Section 14.2. Section 14.3 introduces two scenarios of possible complementary usage of self service business intelligence and the ability to generate and verify a large set of relevant patterns by GUHA procedures. The usage of these scenarios is demonstrated on practical examples based on the Accidents dataset presented in Section 14.4. Section 14.5 concludes this chapter.

---

[1] http://olapcube.com.

14.1 Business Intelligence and Self Service BI

Business intelligence is a set of approaches and ICT applications that support the analytical and planning activities of the business. We can find many definitions, some of which are listed.

"Business Intelligence can be defined as having the right access to the right data or information needed to make the right business decisions at the right time. The data might be raw or might have been analysed in some way. Having access to such information enables management of the business by fact instead of by primarily relying on intuition" [148].

"We define Business Intelligence as providing decision makers with valuable information and knowledge by levering a variety of sources of data as well as structured and unstructured information" [144].

"Business intelligence is a set of technologies and processes that allow people at all levels of an organization to access and analyze data. Without people to interpret the information and act on it, business intelligence achieves nothing. For this reason, business intelligence is less about technology than about culture, creativity, and whether people view data as a critical asset." ([69]).

Business intelligence is used to present data to the management so they can use it to gain knowledge and take appropriate action. It enables delivery of information to business users. It is a visible part of the corporate data system.

In large companies, business intelligence is based on technologies like ETL, data warehouses, data marts, multidimensional modelling, OLAP databases, etc. The description of these technologies is out of the scope of this book and can be found in the "bible" of multidimensional modelling, The Data Warehouse Toolkit: The Definitive Guide to Multidimensional Modelling [77].

The sub-area of business intelligence, called self service business intelligence, is more relevant for this book. Self service business intelligence is used for small groups of users or individual users. Although the main principles such as ETL or multidimensional modelling remain the same, the main difference from the standard business intelligence is usually the use of a smaller volume of data, the lesser system complexity, and lower costs. The typical representatives of self service business intelligence tools are Tableau, PowerBI, Qlik Sense, PowerPivot and these tools try to be as user-friendly as possible.

PowerBI was chosen as a tool for demonstrating business intelligence possibilities. The main reason is the overall popularity and usability of the tool which is expressed by the good location in Gartners' Magic Quadrant[2]. In the basic version, the tool is available for free[3] and its user interface is very well known from Excel spreadsheet editor.

---

[2]https://www.sisense.com/gartner-magic-quadrant-business-intelligence/.
[3]at the time of writing this book.

14.2 Comparing Analysis Performed by Self Service BI and GUHA

This section compares the process of GUHA method data mining and analysis performed by self service business intelligence tool (see Fig. 14.1). The common part is the first phase of data preprocessing. Both techniques require to load the data in the appropriate format. Data mining tools usually require a flat table, self service business intelligence tools support many forms and formats, large databases composed of many tables are usually not a problem. These tools also support the transformation of complex data into one flat file.

At this point it should be noted that in the practical examples in the following section only two GUHA procedures are going to be used, 4ft-Miner and CF-Miner. The reason is that other procedures offer too complex results. The simulation of these results in self service BI tool would be too artificial and difficult for interpretation.

The second phase of preprocessing varies for individual techniques. When working with self service business intelligence tools it is necessary to determine which attributes create dimensions, their hierarchical levels, what the hierarchy for particular dimensions is, and which attributes create measures. In GUHA based data mining represented by the LISp-Miner system, the preprocessing phase consists of dealing mainly with numeric attributes which need to be discretized into categories and with nominal attributes which can be grouped. A very important feature of the LISp-Miner is the support of dealing with groups of attributes that correspond to partial cedents. Group of attributes helps the user to deal with domain knowledge and to create corresponding analytical questions, see Section 1.5.1.

The analytical phase in self service BI tools consists of trying to manually find some interesting and useful information about the data, and display them in the clear form of dashboards. It is done through creating measures, aggregating and grouping data and interfacing it with the data presentation area. The summary of indicators and dashboards is used to create constant data surveillance with the possibility of automatic reporting. In GUHA based data mining, analytical phases can be divided into three parts. The first part is setting the task in data mining template screen where the required attributes for the analysis together with proper quantifiers with the threshold parameters are selected. The second phase is the search of the hypotheses performed by the proper algorithm and the third phase is the postprocessing of obtained hypotheses, which can be sorted, filtered and the interesting ones can be interpreted and further used.

The last phase for both techniques is the phase of transforming the results into a clearer and understandable form, resulting in new knowledge from the dataset. As mentioned before, self service BI tools presents data in the form of individual visuals arranged in the dashboard. A dashboard is a visual display, summarizing two or more reports and (or) graphs on a single well-arranged screen. Dashboards

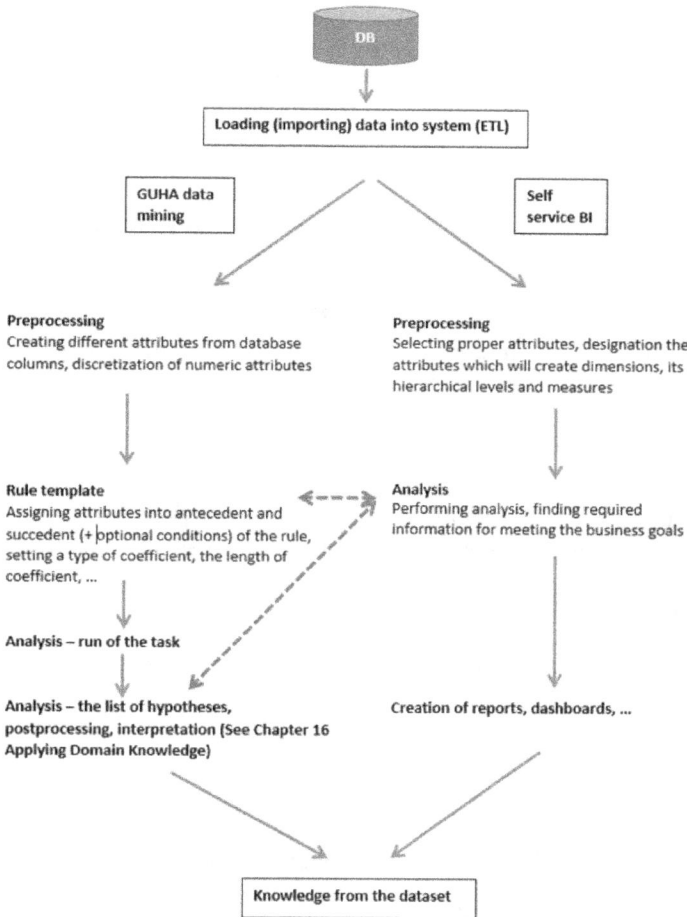

Figure 14.1: Schema of analytical process.

are used by all levels of management to obtain an overall, clear and simple picture of various aspects of the business. Gaining knowledge of data mining results is trickier than in the case of self service BI tools.

14.3 Scenarios of Complementary Usage of BI and GUHA

It is possible to identify two scenarios of using a self service business intelligence tool and GUHA data mining in one dataset. In the first scenario, the analyst identifies an interesting part of a dataset by BI analysis. This part is then examined more closely with GUHA method data mining. In the second scenario, GUHA method data mining is used to recommend interesting parts of a dataset, where the observed measure behaves differently from the general trend in the data. These parts of the dataset can then be explored and visualized by self service BI tool, summarized and reported to the data owner.

14.3.1 Gaining insight into specific (interesting) part of the dataset

This scenario, which can also be titled: "BI first, then GUHA data mining", uses BI analysis to find some interesting or preferable part of the dataset and then uses GUHA data mining for further examination.

A typical process of this scenario follows.

1. Identify an interesting part of the dataset with BI analysis.

2. In the data mining template screen of LISp-Miner, choose an appropriate combination of cedents (possibly partial cedents), literals, coefficient types, coefficient lengths, and/or conditions to mine rules from the specified part of the dataset.

3. Check the resulting rules with the chosen part of the dataset. Repeat point 2 until the rules from specified parts of the dataset do not prevail.

4. Use the obtained rules to gain additional knowledge of the selected part of the dataset.

14.3.2 Automatic BI analysis using GUHA data mining

This scenario, called Automatic BI analysis by GUHA data mining, is focused on completely bypassing the ineffective and time-consuming work of an analyst in the manual finding of interesting patterns in data. This work is performed by the implementation of GUHA procedures. The resulting interesting rules can then be manually verified by BI analysis by specifying and exploring the proper part of the dataset.

A typical process of this scenario is as follows.

1. Set the GUHA data mining task as generally as possible

2. Rules' postprocessing.

3. Subjective evaluation of the resulting rules. If the rules are uninteresting, run the task again with different (more specific) settings.

4. Verification of the results by BI analysis.

5. Use the obtained rules to gain additional knowledge of the dataset.

14.4　Examples on Accidents Dataset

Individual scenarios on the Accidents dataset have been introduced in detail in Chapter 2.3.

14.4.1　Automatic BI analysis using GUHA data mining

Let us consider the basic frequencies of severity of accidents (Fig. 14.2) and the number of accidents (Fig. 14.3) during the years obtained from the PowerBI tool.

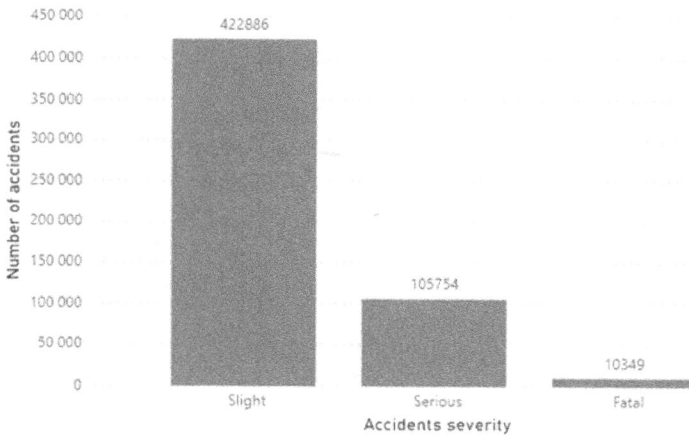

Figure 14.2: Histogram of total accidents severity.

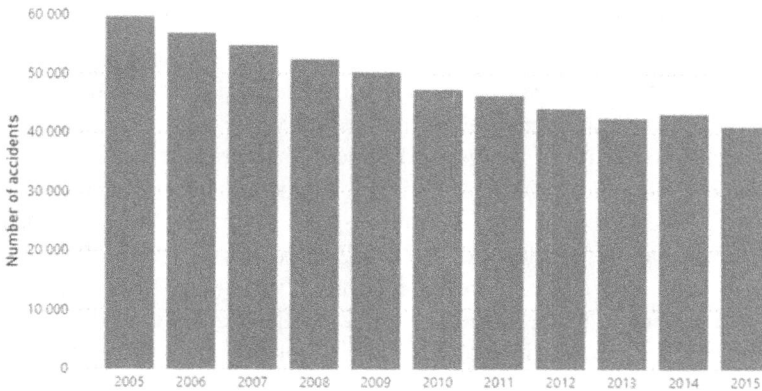

Figure 14.3: Total number of accidents per years.

We are trying to find some exceptions from these general trends, whether there is any subset of data that behaves differently. Starting with the severity of accidents we know that the relative distribution of accidents severity looks as follows (Fig. 14.4).

Accident severity	Relative count
Slight	78,46%
Serious	19,62%
Fatal	1,92%

Figure 14.4: Relative distribution of accidents in total.

Using 4ft-Miner in Section 7.4.1 we know that the relative number of serious accidents for motorcycles over 500 cc is much higher (particular rule (7.12) is limited to Aberdeenshire district, but we can generalize it to all motorcycles over 500 cc category). This can be verified by creating a relative number of accidents just for the motorcycles over the 500cc category. The resulting relative number of accidents can be seen in Fig. 14.5.

Accident severity	Relative count
Slight	54,29%
Serious	41,42%
Fatal	4,29%

Figure 14.5: Accidents severity for Motorcycles over 500 cc category.

The result can be summarized as in the event of an accident on big motorcycles there is significantly more probability to suffer serious or even fatal injury.

In the case of a number of accidents during years we are trying to find the trend which is opposite of the general trend. Whether there is a subset of data for which the number of accidents is rising. Again, using GUHA procedure CF-Miner we found that such a subgroup exists (Section 8.3.2, Fig. 8.23). It is a group of cars having accidents on a 20 MPH speed limit. In the PowerBI tool we can find this subgroup with a combination of filters (Fig. 14.6), focusing only on vehicle type car and the speed limit of 20 MPH. As we can see in Fig. 14.7 the number of accidents is increasing. This is probably caused due to the overall reduction of the speed limits in rural areas[4]. The same result could be found using

---

[4]https://www.rospa.com/rospaweb/docs/advice-services/road-safety/drivers/20-mph-zone-factsheet.pdf.

Vehicle_Type		Speed_limit	
☐ Agricultural vehicle		■ 20 MPH	
☐ Bus or coach (17 or more pass seats)		☐ 30 MPH	
■ Car		☐ 40 MPH	
☐ Goods 7.5 tonnes mgw and over		☐ 50 MPH	
☐ Goods over 3.5t. and under 7.5t		☐ 60 MPH	
☐ Goods vehicle - unknown weight		☐ 70 MPH	
☐ Minibus (8 - 16 passenger seats)			
☐ Mobility scooter			
☐ Motorcycle 125cc and under			
☐ Motorcycle 50cc and under			
☐ Motorcycle over 125cc and up to 500cc			
☐ Motorcycle over 500cc			
☐ Other vehicle			
☐ Pedal cycle			
☐ Taxi/Private hire car			
☐ Tram			
☐ Van / Goods 3.5 tonnes mgw or under			

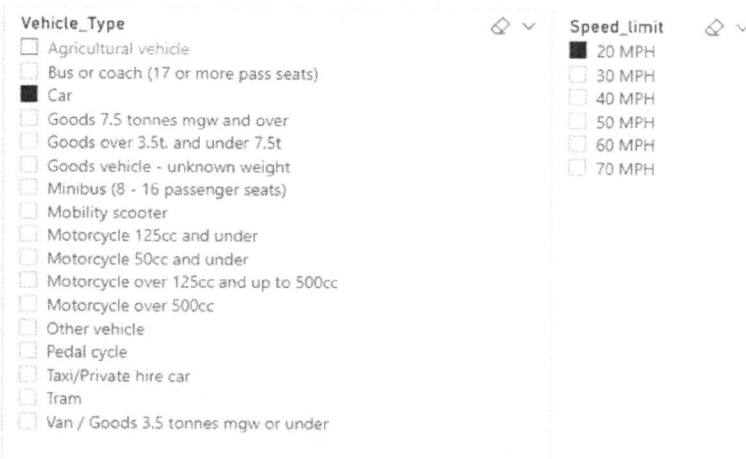

Figure 14.6: Using filter in PowerBI.

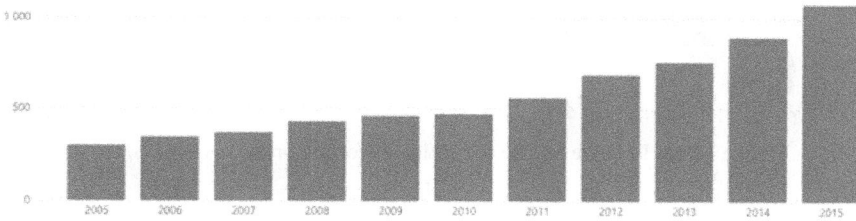

Figure 14.7: Exception of the trend, increasing number of accidents for 20 MPH speed limit.

PowerBI, but it would mean examining (all) the combinations given by the two filters. In our case it would be $17 \times 6 = 102$ combinations.

This example showed the possibility of presenting interesting insights found by GUHA data mining through BI analysis.

14.4.2 Gaining inside into specific parts of dataset

Let us consider that we are interested in the accidents happened on dual carriageway. From the counting of relative number of accidents on dual carriageway in PowerBI we can see that the number of more serious accidents is greater than in general (Fig. 14.8).

Using filters we can easily find what types of vehicles have the highest relative number of fatal accidents on dual carriageways and these are truck over 7,5 tons (Fig. 14.9) and motorcycles over 500 cc (Fig. 14.10).

Now we will be interested in whether there is some subgroup of data for which applies that the relative number of fatal accidents is higher and vice versa

Accident severity	Relative count
Slight	76,83%
Serious	19,98%
Fatal	3,19%

Figure 14.8: Accidents severity on dual carriegeway.

Accident severity	Relative count
Slight	64,87%
Serious	22,60%
Fatal	12,53%

Figure 14.9: Relative number of accidents for trucks over 7,5 tonnes.

Accident severity	Relative count
Slight	52,38%
Serious	41,44%
Fatal	6,18%

Figure 14.10: Relative number of accidents for motorcycles over 500 cc.

the subgroups for which the number of slight accidents is higher. For this purpose we will use the 4ft-Miner procedure with above average dependency quantifiers. As a threshold value for above average quantifier we are going to use the value of p=0.15 meaning that we are interested in subgroups of accidents happening by 15% more often than average.

In the first case we are trying to find subgroup of accidents on the dual carriageway where trucks over 7.5 tonnes are involved. The resulted accidents are fatal. The task is set as follows in Fig. 14.11 and Fig. 14.12.

With this task setting no results were found.

The second case is the subgroup of accidents on the dual carriageway where truck over 7,5 tonnes are involved, where resulted accidents are slight. This task setting brings 20 rules, Fig. 14.13 show the 6 shortest rules for good readibility.

From the resulted rules it is possible to determine that slight accidents on the dual carriageway where trucks over 7,5 tonnes are involved have a slightly better chance of being less serious in case of a 70 MPH speed limit, in January, daylight and rural areas.

The third case is the subgroup of accidents on the dual carriageway where motorcycles over 500 cc are involved. The resulted accidents are fatal. With this task setting one rule was found. This rule has only minimum informational value,

ANTECEDENT

Vehicle	Con, 0 - 5
» Vehicle_Age (subset), 1 - 1	B, pos
Conditions	Con, 0 - 5
» Area (subset), 1 - 1	B, pos
» Light (subset), 1 - 1	B, pos
» Speed_limit (subset), 1 - 1	B, pos
Driver	Con, 0 - 5
» Driver_Age_Band (subset), 1 - 1	B, pos
» Driver_Sex (subset), 1 - 1	B, pos
» Child (subset), 1 - 1	B, pos
» Journey (subset), 1 - 1	B, pos
Date / time	Con, 0 - 5
» Month (subset), 1 - 1	B, pos
» WeekDay (subset), 1 - 1	B, pos

Figure 14.11: The setting of the antecedent.

CONDITION

Default Partial Cedent	Con, 2 - 5
» Vehicle_Type(21,Goods 7.5 tonnes +)	B, pos
» Road_Type(3,Dual carriageway)	B, pos

Figure 14.12: The settings of the condition.

1	7	0.156	**Speed_limit**(*70*) & **Month**(*Jan*) >+< **Severity**(*Slight*) / **Type**(*21,Goods 7.5 tonnes* +) & **Road_Type**(*3,Dual carriageway*)
2	9	0.169	**Speed_limit**(*70*) & **Sex**(*1,Male*) & **Month**(*Jan*) >+< **Severity**(*Slight*) / **Type**(*21,Goods 7.5 tonnes* +) & **Road_Type**(*3,Dual carriageway*)
3	8	0.166	**Speed_limit**(*70*) & **Child**(*No*) & **Month**(*Jan*) >+< **Severity**(*Slight*) / **Type**(*21,Goods 7.5 tonnes* +) & **Road_Type**(*3,Dual carriageway*)
4	14	0.156	**Area**(*2,Rural*) & **Light**(*1,Daylight*) & **Age**(*46 - 55*) >+< **Severity**(*Slight*) / **Type**(*21,Goods 7.5 tonnes* +) & **Road_Type**(*3,Dual carriageway*)
5	2	0.156	**Area**(*2,Rural*) & **Sex**(*1,Male*) & **Month**(*Jan*) >+< **Severity**(*Slight*) / **Type**(*21,Goods 7.5 tonnes* +) & **Road_Type**(*3,Dual carriageway*)
6	1	0.154	**Area**(*2,Rural*) & **Child**(*No*) & **Month**(*Jan*) >+< **Severity**(*Slight*) / **Type**(*21,Goods 7.5 tonnes* +) & **Road_Type**(*3,Dual carriageway*)

Figure 14.13: Resulted rules for slight accidents with trucks over 7,5 tonnes.

stating that there is a 15,8% greater chance of fatal accidents for motorcycles over 500 cc on the dual carriageway when the driver is male, has no child with him and the purpose of journey is other or not known.

The fourth case is the subgroup of accidents on the dual carriageway where motorcycles over 500 cc are involved. The resulted accidents are slight. With this task setting 76 rules were found (Fig. 14.14, shows the shortest 12 rules).

From these rules we can conclude that urban area, daylight, and a low speed of limit (30 MPH) contribute to the slight severity of accidents.

The following Table 14.1 summarizes the number of resulting rules for identified subgroups of accidents.

1	11	0.269 **Speed_limit**(*30*) >+< **Severity**(*Slight*) / **Type**(*5.Motorcycle over 500cc*) & **Road_Type**(*3.Dual carriageway*)
2	1	0.197 **Journey**(*Part of work*) >+< **Severity**(*Slight*) / **Type**(*5.Motorcycle over 500cc*) & **Road_Type**(*3.Dual carriageway*)
3	15	0.336 **Speed_limit**(*30*) & **Age**(*26 - 35*) >+< **Severity**(*Slight*) / **Type**(*5.Motorcycle over 500cc*) & **Road_Type**(*3.Dual carriageway*)
4	62	0.330 **Light**(*1.Daylight*) & **Speed_limit**(*30*) >+< **Severity**(*Slight*) / **Type**(*5.Motorcycle over 500cc*) & **Road_Type**(*3.Dual carriageway*)
5	48	0.288 **Area**(*1.Urban*) & **Speed_limit**(*30*) >+< **Severity**(*Slight*) / **Type**(*5.Motorcycle over 500cc*) & **Road_Type**(*3.Dual carriageway*)
6	12	0.270 **Speed_limit**(*30*) & **Child**(*No*) >+< **Severity**(*Slight*) / **Type**(*5.Motorcycle over 500cc*) & **Road_Type**(*3.Dual carriageway*)
7	14	0.263 **Speed_limit**(*30*) & **Sex**(*1.Male*) >+< **Severity**(*Slight*) / **Type**(*5.Motorcycle over 500cc*) & **Road_Type**(*3.Dual carriageway*)
8	13	0.253 **Speed_limit**(*30*) & **Journey**(*5 Other/Not known*) >+< **Severity**(*Slight*) / **Type**(*5.Motorcycle over 500cc*) & **Road_Type**(*3.Dual carriageway*)
9	28	0.240 **Area**(*1.Urban*) & **Day**(*Tuesday*) >+< **Severity**(*Slight*) / **Type**(*5.Motorcycle over 500cc*) & **Road_Type**(*3.Dual carriageway*)
10	16	0.236 **Speed_limit**(*30*) & **Age**(*36 - 45*) >+< **Severity**(*Slight*) / **Type**(*5.Motorcycle over 500cc*) & **Road_Type**(*3.Dual carriageway*)
11	5	0.198 **Age**(*36 - 45*) & **Journey**(*2.Commuting to/from work*) >+< **Severity**(*Slight*) / **Type**(*5.Motorcycle over 500cc*) & **Road_Type**(*3.Dual carriageway*)
12	31	0.197 **Area**(*1.Urban*) & **Journey**(*2.Commuting to/from work*) >+< **Severity**(*Slight*) / **Type**(*5.Motorcycle over 500cc*) & **Road_Type**(*3.Dual carriageway*)

Figure 14.14: Slight accidents on dual carriageway for 500 cc motorcycles.

Table 14.1: Summarization of the results.

	Task	Number of rules
1	Truck over 7.5 tonnes, fatal severity	0 rules
2	Truck over 7.5 tonnes, slight severity	20 rules
3	Motorcycles over 500 cc, fatal severity	1 rule
4	Motorcycles over 500 cc, slight severity	76 rules

14.5 Possible Extension of the Work

There are several possible ways to extend the work. One option is to compare the results of the remaining GUHA procedures with the outputs of the self service business intelligence tool. However, the outputs of other procedures are complex and more difficult to interpret. In many cases, it might not be possible to show a direct relationship between outputs. The practical use of this comparison would probably not be very beneficial.

Another possibility is to analyze a real business problem assigned by a data owner or domain expert. In this case, it would be useful to perform an analysis independently by both methods. Then present the results to a domain expert or data owner, who would evaluate the interest of the outputs. Subsequently, together with data analyst, he could define a task combining both approaches according to the scenarios defined in Section 14.3.

Chapter 15

CleverMiner—GUHA and Python

CleverMiner [89] is a project with a goal to transfer GUHA procedures into state of the art data science and machine learning language. After initial research, Python language [96] had been selected. This chapter describes why we chose Python (15.1), what are goals of Python implementation (15.2), what is the required format of data input to Cleverminer procedure (15.3), what functionality is currently implemented in Cleverminer (15.4). Example of Cleverminer call, including results, is shown in 15.5. Future plans with Cleverminer are in 15.6.

Note that in this Chapter, when we refer to any GUHA procedure, we mean CleverMiner implementation of this procedure in version 0.0.84 unless implicitly stated otherwise.

15.1 Why GUHA in Python

GUHA has currently very advanced implementation with many workbench functions suitable for preparation of categorical data (see description of the LISp-Miner system in Chapter 5). As many functions may be called from external software, it is based on Visual C and Microsoft Access.

Currently, many data scientists use the state of the art software like Python and R and asked us to make these procedures available also for them. As data preparation (load, transformation, categorization) is in the box as well as visual evaluation, we decided to make these procedures available in such language.

There are several languages that may be suitable for the implementation of machine learning procedures. There are two groups of languages—low level languages (that currently include C/Java), where absolute control and efficient algorithms may be fluently implemented, and advanced statistics/data science languages, as R/Python, that are high level languages (no type declaration needed, interpreted and not compiled code is executed). For end user language, we definitely decided for a high level language.

As statisticians use mainly *R* [97] and data scientists use mainly *Python* [96], Python better supports the entire CRISP-DM [10] process (see, e.g., [154]), and many SW vendors (incl. SAP) move from R to Python, we decided on Python. We discussed whether implementation will be in low level language or native Python. We decided to compromise. Implementation is in native Python but it includes low-level function and tie-to data type implementation and its properties to reach efficiency. Note that this approach is not used by 99+% of Python end users – data scientists.

Let us summarize the reasons for high-level language (Python). Main reasons are:

data load Python allows (typically with pandas and/or data connectors) to load any data type or connect to the database quickly and load data including metadata (data types, names, etc.)

data preparation allows easy data transformations (like connecting tables, calculations is almost native form)

categorization allows easy data categorization via cut and qcut functions for equidistance and equifrequency categorization

experience of Python users most Python users are able to use scripting, join data and process them; data scientists typically use pandas to work with data

easily to get, display and show results there is no need to work on procedures that ensure hypothesis displaying and charting as most of Python users are able to do more – to sort results given by dictionary and display it easily, with many options to display it in charts. It extends direct possibilities of using GUHA results compared to LISp-Miner.

Python with its common libraries covers in the box the first 3 phases of the CRISP-DM process so there is no need to reimplement this.

We need to emphasise that there are extensions like Cython [4] that is C for Python and allows low-level efficient implementations, but we decided to use native Python functions that are very efficient and override disadvantages of the use of high level functions. We compared six approaches and decided to use the method that uses native data types and extremely optimized functions in Python.

15.2   Goals of Python Implementation of GUHA

The main goals to implement GUHA in Python are:

allowing data scientists to use GUHA procedures — the main goal of this implementation is making GUHA available to current GUHA users that typically also use other software for data preparation, in many cases Python. They are able to prepare data, do any part of data preparation and are able to read semitechnical results like JSONs. So making GUHA procedures available in Python in a native way may help to combine GUHA procedures with Python directly so they may find advantages of GUHA and Python.

teaching in current standard — currently, GUHA is taught in several courses at the Prague University of Economics and Business so that students learn to work with categorical data. Many students have asked us to work in Python as they use it in work for data analytics. So it will be a seamless fit to teach working with categorical data as well as to go through the entire CRISP-DM process in one language.

way to disseminate GUHA — even though GUHA is a natural extension of the Apriori algorithm with much more possibilities (see, e.g., [138]), it is not so widespreadly used in the data science community. Making it available as a community package may be a way to disseminate GUHA to the data science community. There is also a good theoretical background of the GUHA method. This may be interesting for theoretically oriented data scientists, see Introduction, Chapter 18 and [41, 121]. Also, research related to dealing with domain knowledge in data mining related to the GUHA method (see Chapter 17) can be interesting for some data scientists.

15.3   Data Requirements and Representation on Analyzed Data Matrix

15.3.1   Requirements on input matrix and how to achieve it

CleverMiner requires to have a panda data matrix with nominal or categorical data as input. In general, it allows any data type, for non-categorical types (incl. continuous numbers), the nominal data type is assigned and each value is considered as an individual category.

As Python has many data manipulation libraries available, it is very easy to prepare data in the desired form. In the following example, we will show how to prepare Accidents dataset (Accidents dataset has been introduced in Section 2.3) to a form suitable for CleverMiner.

Simple data load of the *Accidents* dataset with selection of attributes Age, DayOfWeek, Area, Engine_Capacity, Month, Pedestrian_Control, Road_Type, Severity, Year can be done with the following code:

```
#load data from csv
df = pd.read_csv ('c:\\path_to_file\\Accidents.txt ',
                 encoding='cp1250', sep='\t')

#replace -1 values by 1
df=df.replace(-1,1)

#choose only several attributes
dfres=df[['Age','DayOfWeek','Area',
        'Engine_Capacity', 'Month', 'Pedestrian_Control',
        'Road_Type','Severity','Year']]
```

As all attributes have already been binned, there was no need for categorization. When we would need to prepare categorized attribute, e.g., from year, the code would be

```
#do categorization, 3 categories, equifrequent
dfa =pd.qcut(df['Year'],3)

#merge all categorized data into original dataset
df=pd.concat([df,dfa],axis=1)
```

15.3.2 Internal representation of data matrix in CleverMiner

As input, we use the panda dataset. For internal implementation, where a data matrix is heavily used for calculations of similar pattern, it may exist a better solution. Therefore, we considered several ways on how to implement a data matrix for GUHA procedure, from native data structures as pandas or NumPy as well as to binary strings. Python has typically many procedures extremely well optimized so that reimplementing them into a non-native way (like Cython) leads to extremely poor performance. But there are many alternatives and our task is not to make one calculation but many calculations in similar way so there are some optimizations available. For thousands of verifications, there is negligible total time, for millions or tens of millions of verifications (similar calculations) that may worth to using some advanced technique.

In Python, for data science, there are two main ways how to process data - pandas, where high level data operations are available, and NumPy, where optimized numeric matrix operations are available. We choose the third way, to implement binary chains (see also Section 5.5.3) in other available Python native concept so we were able to achieve four times better performance than NumPy implementation with possibility to achieve even better results when using Python 3.10, that is planned for second half of 2021.

15.4 CleverMiner Procedures

15.4.1 General parameters and calling

CleverMiner is designed as single space search engine for all GUHA procedures. So any extension like conjunction/disjunction, cedent and attribute length, and attribute set type (subsets, sequences) are implemented at one single place so if there is no immediate blocker (like nature of individual procedure), the feature is available for all procedures.

CleverMiner currently supports the following GUHA procedures:

1. CF-Miner, which is a GUHA procedure implemented in Python and inspired by the CF-Miner procedure described in Chapter 8.

2. 4ft-Miner, which is a GUHA procedure implemented in Python and inspired by the 4ft-Miner procedure described in Chapter 7.

Parameters of CleverMiner procedure are shown in Table 15.1. For each cedent, there are parameters that can control the minimal and maximal cedent length (`minlen`, `maxlen`) and cedent type – con for conjunction and dis for disjuction. For individual attributes, CleverMiner implements the following types of coefficients with the usual minlen-maxlen limitation. They are shown in Table 15.2, see also Table 4.6 and Section 4.4.1.

Table 15.1: Parameters of CleverMiner procedure.

Parameter name	Description
df	panda dataset
proc	GUHA procedure name (`CF-Miner`, `4ft-Miner`)
quantifiers	list of quantifiers for called procedure
ante, succ, cond, ...	cedents relevant to procedure
target, ...	procedure specific parameters (`target` for CF-Miner)

Table 15.2: Cleverminer type of coefficients selection.

Coefficient type	Description
subset	coefficients of type *Subsets*
sequence	coefficients of type *Sequences*
lcut	coefficients of type *Left cuts*
rcut	coefficients of type *Right cuts*

15.4.2 Quantifiers for individual GUHA procedures

Quantifiers are dependent on procedure type. Cleverminer implements following quantifiers for individual procedures.

CF-Miner procedure dealing with conditional histograms $A/[\chi, TypeHist]$ described in Section 8.1.1 provides quantifiers:

- selected simple frequencies CF-quantifiers described in Section 8.1.4, (informally speaking minimal sum of absolute or relative frequencies of particular categories)

- maximal value for maximum of absolute or relative frequencies of particular categories

- minimal value for maximum of absolute or relative frequencies of particular categories

- quantifiers concerning steps in histogram described in Section 8.1.5.

4ft-Miner procedure dealing with association rules $\varphi \approx \psi$ described in Section 7.1.1 provides three of the quantifiers introduced in Section 7.1.2:

- Base $\oplus_{Base}$ corresponding to minimal number of rows satisfying both φ and ψ

- support $\odot_s$ corresponding to minimal relative number of rows satisfying both φ and ψ

- confidence $\rightarrow_p$.

15.5 Calling CleverMiner Procedures

CleverMiner contains all procedures in one function call. The procedure name is given by a parameter. Other settings are given by other variable parameters. We will show this in examples.

In our examples we will use the *Accidents* dataset. First example is a *CF-Miner* procedure. We will look for histogram of attribute Severity with conditions $Base \geq 200000$ and $S_{Up} \geq 2$. Condition is given by following attributes (conjunction, total number of attributes used 1 to 2):

Attribute	Category selection	Minimal length	Maximal length
Area	subset	1	2
Engine_Capacity	subset	1	2
Road_Type	subset	1	2
Age	subset	1	2

Python code for CF-Miner procedure call is

```
hypo = CleverMiner(df=df, target='Severity', proc='CF-Miner',
        quantifiers= {'S_Up':2, 'Base':200000},
        cond ={
            'attributes':[

                {'name': 'Area', 'type': 'subset',
                 'minlen': 1, 'maxlen': 2},

                {'name': 'Engine_Capacity', 'type': 'subset',
                 'minlen': 1, 'maxlen': 2},

                {'name': 'Road_Type', 'type': 'subset',
                 'minlen': 1, 'maxlen': 2},

                {'name': 'Age', 'type': 'subset',
                 'minlen': 1, 'maxlen': 2}

            ],
            'minlen':1, 'maxlen':2, 'type':'con'}
    )
```

Procedure `CleverMiner` proceeds CF-Miner space search and generates list of hypotheses including task information into dictionary `hypo`. The resulting dictionary `result` contains the following items:

taskinfo copy of key input parameters/task information

summary statistics summary task processing information like processing time, number of verifications, number of hypotheses, ...

hypotheses list of resulting hypotheses, containing list of quantifiers values, human readable form of cedents and internal format information about traces of cedent and attributes with reference ids of columns that can be used for additional processing if needed

datalabels information about original variable and category names to be able to convert traces of cedents and attributes to human readable form or advanced post processing

Example of result is shown below (narrated version, deleted text has been replaced by three dots):

```
{
    'taskinfo': {
        'task_type': 'CF-Miner',
        'target': 'Severity',
        'quantifiers': {
            'S_Up': 2,
            'Base': 200000
        },
        'cond': {
            'attributes': [{
                    'name': 'Area',
                    'type': 'subset',
                    'minlen': 1,
                    'maxlen': 2
                }, {
                    'name': 'Engine_Capacity',
                    'type': 'subset',
                    'minlen': 1,
                    'maxlen': 2
                }, {
                    'name': 'Road_Type',
                    'type': 'subset',
                    'minlen': 1,
                    'maxlen': 2
                }, {
                    'name': 'Age',
                    'type': 'subset',
                    'minlen': 1,
                    'maxlen': 2
                }
            ],
            'minlen': 1,
            'maxlen': 2,
            'type': 'con'
        }
    },
    'summary_statistics': {
        'total_verifications': 5037,
        'valid_hypotheses': 17,
        'time_prep': 183.2076919078827,
        'time_processing': 47.60326957702637,
```

```
        'time_total': 230.81096148490906
},
'hypotheses': [{
        'hypo_id': 1,
        'cedents': {
            'cond': 'Area(1,Urban )'
        },
        'params': {
            'base': 340200,
            'rel_base': 0.6311817124282685,
            's_up': 2,
            's_down': 0,
            's_any_up': 2,
            's_any_down': 0,
            'max': 269606,
            'min': 4747,
            'rel_max': 0.5002068687858193,
            'rel_min': 0.008807229832148707,
            'hist': [4747, 65847, 269606]
        },
        'trace_cedent': [],
        'traces': []
    }, {

    ...

        'hypo_id': 17,
        'cedents': {
            'cond': 'Road_Type(6,Single carriageway
                            7,Slip road )'
        },
        'params': {
            'base': 432745,
            'rel_base': 0.802882804658351,
            's_up': 2,
            's_down': 0,
            's_any_up': 2,
            's_any_down': 0,
            'max': 339380,
            'min': 7819,
            'rel_max': 0.6296603455729152,
            'rel_min': 0.014506789563423373,
            'hist': [7819, 85546, 339380]
```

```
        },
        'trace_cedent': [],
        'traces': []
      }
    ],
    'datalabels': {
        'varname': ['Age', 'DayOfWeek', ...],
        'catnames': [['16 - 20', '21 - 25', '26 - 35',  ...]
    }
}
```

In the previous result, we have 17 hypotheses. Note we have compared the results with the LISp-Miner system and results are the same. Timing has not been compared due to LISp-Miner's architecture (Access/data prep written to DB) and the early optimalization state of Cleverminer (see 15.6).

We will look at the last hypothesis, i.e., hypothesis with id 17. This hypothesis has condition Road_Type(*6,Single carriageway7,Slip road*), i.e., the road type is either a single carriageway or a slip road. Other information is self explanatory. Histogram of Severity for the given condition has 3 values – 7819, 85546 and 339380. It can be plotted by *seaborn* or *matplotlib* library - Fig. 15.1.

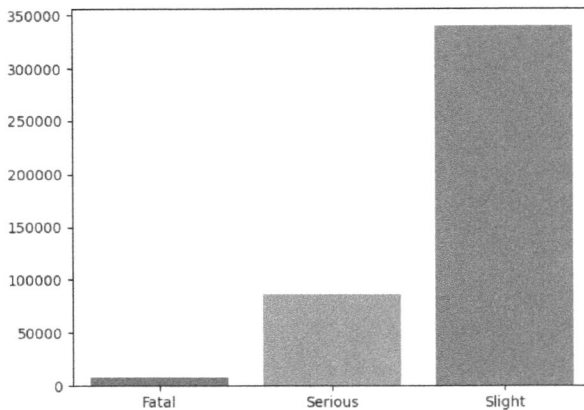

Figure 15.1: Displaying histogram results using Python's seaborn library.

Code to display is very simple.

```
sb.barplot(x=varlabels,y=histvalues)
plt.show()
```

Variable *varlabels* contains a list of labels for target variable, in our case *['Fatal', 'Serious', 'Slight']*, variable *histvalues* contains a list of histogram values extracted from the output, in our case [7819, 85546, 339380].

15.6   Future Plans with CleverMiner

Currently, CleverMiner is in the early stage of development. It implements two GUHA procedures with most common options except *partial cedents* and any optimizations. CleverMiner plans include to cover

optimization – optimization of space search as well as code timing

partial cedents – partial cedents may be helpful for advanced GUHA procedures

documentation – documentation as well as tuturials; examples are essential for bigger publicity

input checks – input checks, both formal and practical (like checking the number of categories, limiting the number of resulting hypotheses, or space search limits may be extremely helpful)

other procedures – implementation of other procedures is planned, starting with SD4ft-Miner inspired by the SD4ft-Miner procedure described in Chapter 10.

III

RELATED RESEARCH AND THEORY

A goal of Section III is to introduce important research directions related to the GUHA method. The first direction concerns generating artificial data. It is introduced in Chapter 16.

The second research direction deals with applications of domain knowledge in data mining. Special attention is devoted to the GUHA procedures 4ft-Miner and CF-Miner dealing with association rules and histograms. Three examples of applications of items of domain knowledge are presented in Chapter 17. An expert deduction is defined as incorrect deduction rules supported by indisputable facts concerning the domain of application of the used procedures.

Observational calculi are introduced in Chapter 18 as a tool for the study of expert deduction rules. An overview of relevant results concerning correct logical deduction rules are summarized. Correct and incorrect expert deduction rules are defined and their properties are presented.

Chapter 16

Artificial Data Generation and LM ReverseMiner Module

There are circumstances where we need several datasets with well known properties and simultaneously which are suitable for analysis by a given data-mining technique. A good example is teaching lessons where we need both suitable datasets for demonstration and several others for students to improve their skills. Data has to be simple enough to be wholly understandable during teaching lessons and has to be carefully prepared (or thoroughly analysed before) to contain hidden patterns already known to the teacher. Another example is a development of new data-mining tools or features that need to be tested for a proper functionality.

Using real data is not so easy. Firstly, they are hard to obtain. If so, anonymized data only are usually available, so very difficult to understand and interpret results. Moreover, every real data require time-consuming cleaning before they could be analysed and their properties have to be well-understood still before they could be used for demonstration or teaching. Finally, not all data are suitable for teaching or testing a given data-mining technique. Therefore artificially prepared data, with all the necessary properties incorporated, but still with a realistic look could be of a great benefit. The importance of synthetic data for testing and evaluation of data mining tools is stressed in [145].

A manual preparation of each artificial dataset is very laborious, with the hard part being to incorporate desired properties into them and simultaneously to hide their artificial origin. A hand-made synthetic data suffer often from clearly visible repeating patterns and usually, from small sizes of datasets, not suitable for all kinds of data mining techniques. On the other hand, preparing demonstrational data for teaching has some specifics – there is usually quite a lot of time for preparation (e.g., during summer holidays). And no life-critical activity will be based on them, so a sub-optimal solution is acceptable. Thus, evolutionary algorithms could be utilized. Our approach tries to address all the mentioned problems and to automatically generate near-to-real-like looking data while allowing for a precise definition of values and their distributions and mainly, of patterns hidden into data. The synthetically prepared data could be analysed afterward in any data mining tool. The presented approach could be used moreover with any other data mining tool, provided that it supports a sufficient range of types of patterns it could mine for and it has automated means for data-mining tasks creations and computations.

We treat an artificial data preparation as a reverse process to the classical data mining analysis where given data is mined for hidden patterns. Here, we define desired patterns first and then try to prepare data where such patterns are hidden. Our approach utilizes an evolutionary algorithm together with several types of syntactically rich patterns the LISp-Miner system modules are already capable of mining for. Compliance of data with the desired patterns is used to guide the evolution.

In this chapter, a proposed approach is introduced, followed by details of evolutionary operations and fitness value computation. Then, the ReverseMiner module that implements the proposed approach is described with its main features. A comprehensive demonstration concerning the preparation of data from a fictional hotel (also used in this book) is presented. Experiences, advantages, and limitations of the evolutionary approach are summarized.

16.1   Evolutionary Approach

Two main points characterize the proposed approach – evolutionary algorithms and syntactically rich patterns that could be mined for by all the already implemented modules of the LISp-Miner system (see Section 4.1).

There is a description of the target database table – of its columns, their datatypes and the number of rows, as the first part of the input. An individual represents one possible variant of the generated data. At the beginning of evolution, a population of individuals is created at random. There could be constraints defined for values generated into each column (valid range of values or enumeration of possible values). Frequencies of values could conform to one of the pre-defined distributions (uniform, Gaussian) or user-defined frequencies for enumerations.

Let suppose we want to generate a simple bank dataset of clients and their loans. In this demo, we will have only one table with just three columns: Loan

Status with possible values of A/B/C/D (meaning successfully paid, written-off, paid on time, and paid with problems respectively), District of clients domicile, and their Salary. The target number of rows will be just 1 000. We define initial probabilities for Loan Status values to be 30%, 4%, 58%, and 8% respectively based on the Czech Statistical Office report on non-performing loans. Initial randomly generated values for this column will have this distribution. We download a list of possible values for column District from the ten-year census, together with the number of people living in each district. The numbers are automatically proportionally recalculated to sum-up to 1 000 (the target number of rows in our demo table). Finally, we define constraints for column Salary to be within the interval from 8 100 EUR to 12 541 EUR and having Gaussian distribution with a mean value of 9 000 EUR and standard deviation of 738 EUR. The mean value shifted towards the lower limit assures a proper shape of the distribution for salaries (an average salary is higher than the median due to a long tail of few people with higher salaries). Initial randomly generated values for Salary will be within defined limits and will have the proper distribution.

The second important part of the input is a user-defined set of data mining tasks looking for different kinds of patterns (and their strengths) which should be present in the synthetically generated data. This set of tasks is solved for each individual in population and the fitness of an individual is calculated as a sum of weighted differences between the actual and the desired number of patterns found for each data mining task.

In our demo, we would like to generate data where loan delinquency is two times higher in the economically depressed district of Louny than it is in the whole dataset. To logically support this pattern we also want to skew the distribution of salaries in this district (and two other districts based again on data from the Czech Statistical Office) so more than 50% of clients from district Louny, Sumperk, and Ceska Lipa have a low salary (less than 8 678 EUR). It is also important to prevent an above-average number of bad loans to appear somewhere else apart from Louny district.

A fuzzy concept is implemented of computing the fitness not only from the binary possibilities (pattern found × pattern not found) but also directly from the strength of the pattern (e.g., confidence). Thus, if we are looking for data with, e.g., a pattern A $\Rightarrow$ B with confidence of 90%, an individual's fitness is improved even if such a pattern was found with confidence of 60% only. This individual is then preferred to another one supporting the same rule with confidence of 50% only.

There is a randomly generated population of individual datasets at the beginning of evolution. The fitness value is computed for each of them, such that all the defined data-mining tasks are solved on these datasets and found patterns are assessed. More perspective individuals (with a higher fitness value – i.e., with the more desired patterns found in their variant of data) are favored to propagate their properties into the next generation by evolutionary operations of crossover, mutation and reproduction (see later). After a sufficient number of evolutionary steps, data with all (or nearly the all) desired patterns hidden in them are prepared.

An important property of the evolution algorithms is their suboptimal nature. There is no certainty in reaching perfect results. So the number of iteration is limited to a maximal number defined by the user, so is the maximal time. The evolutionary algorithm is also implemented interactively so it is clear in each step how the fitness of the best individual has evolved and what patterns are already incorporated in the current version of data—see Section 16.3 later.

16.2 Evolutionary Operations

Implementation of the evolutionary algorithm was kept intentionally as simple as possible. It is based on the description in [162] with several user-parameters available to tweak the evolution process accordingly to the size of data generated and its properties.

A whole new population of individuals is created in each evolutionary step. New individuals are created by one of the three possible evolutionary operations – crossover, mutation, and reproduction. A truly new individual emerges from crossover or mutation, the reproduction copies an already existing individual from the previous evolutionary step without any change. Individuals are included into the new population until its size matches the user-defined number. A type of evolutionary operation is chosen at random for each newly created individual with proportions given again by user-defined parameters. Parameters are available also to define the size of the elite and whether to deny entry into the population if a new individual is worse than the worst one from the previous step.

The most promising individuals to become parents are selected by a tournament. A user-defined number of picks is done and the individual with the highest fitness value is the tournament winner. So only the order of individuals is taken into account, not the absolute difference of fitness among individuals. The bigger the number of picks the smaller is probability of a worse individual being the winner of the tournament. Thus, larger tournaments lead to a decrease in the diversity of population.

Two parents are necessary to create a new individual by a crossover. Both parents are selected in the tournament (with a check that two different individuals were chosen). Two new individuals are created in a crossover evolutionary operation (with complementary properties of theirs parents). Two possible direction of crossover are:

a) **By data columns** – the first new individual takes the whole data column from the first parent and the second new individual takes this column from the second parent (or vice versa – it is chosen at random for each column) and so on for every column. This type of the crossover is always allowed, regardless of user-settings.

b) By data rows – the first new individual takes the first half of row values beginning at row n in the given column from the first parent and the second half from the second parent; the second new individual takes complementary row values from its parents. The value of *n* is chosen at random from a range of *one* to *number-of-rows*. Rows in a column are treated cyclically – after the last, row it goes to the top. This type of crossover is allowed only for columns where the distribution of values could change. (There are user parameters to preserve the distribution of values in some columns and then this type of crossover is not possible).

Both new individuals emerge gradually, column by column. A type of crossover is chosen for each column (if both types are possible). The proportion of both types is influenced by user-settings. The same value *n* of the starting row is used for each column where a crossover in sense of rows is used to preserve possible patterns between values in different columns.

Mutation means a random change of a value in a column. There is only one parent (a winner of the tournament) and only a single individual newly created with (slightly) different properties. There are once again two types of mutation possible:

a) Values swap – a consecutive sequence of values beginning at row *n* and length of *l* is swapped with another consecutive sequence with the same length. The length is chosen as a random number from *one* to the maximal possible length given by the user (not greater than *number-of-rows/2*). The rows in a column are treated again as cyclical. It is assured that there is no overlap between swapped regions. A user-parameter is available to control how many times a swap is repeated for different *n* but with the same *l*. This type of mutation is allowed always, regardless of user-settings.

b) Values change – a consecutive sequence of values beginning at row *n* and length of *l* is set to a randomly chosen value (maintaining given data-type and value range or value enumeration for this column). The length is chosen as a random number from *one* to the maximal possible length given by the user. Again, it is possible to repeat value change several times. This type of mutation is allowed only for columns where values distribution could change.

The proportion between both types of mutation is user-controlled (if both are possible for a given column).

Reproduction is an exact copy of an individual into a new generation. This individual is selected as a tournament winner. Several reproductions are derived from the user-defined percentage number of crossovers and mutations to be theirs complement to 100%. The best individual from the previous round is always copied into the next generation. Size of the elite could be defined as a user-parameter, meaning the number of top individuals to be copied too.

16.2.1 Evolutionary Fitness

Fitness of an individual (i.e., the dataset it represents) is evaluated as a number of desired patterns that are successfully mined from this dataset.

There are nine possible GUHA procedures and corresponding types of patterns available currently in the LISp-Miner system (see Chapter 5). Any of them could be used to define a data mining task trying to lookup a specific pattern (or group of patterns) in data. When the fitness of an individual needs to be computed, all the data mining tasks are solved on this data as if it were any real data to be analysed. It is just a matter of current combination values in the dataset and the definition of data mining tasks, whether desired patterns are found or not. The more such patterns hold in data, the better is the fitness of this individual (dataset).

An important feature is, that the presence of a pattern in data is not evaluated in binary form of yes or no, but using fuzzy concept and range of $\langle 0; 1 \rangle$. Therefore, a dataset where an association rule pattern has a confidence of 60% is better than a dataset where it holds just for 30%, even if we are aiming at a confidence of 90%. The first individual has better fitness values and therefore a higher chance to propagate its properties during evolution.

16.3 ReverseMiner Module

ReverseMiner is a module implemented into the LISp-Miner system. It utilizes the above mentioned approach and allows users to setup several evolutionary tasks each of them specifies all the parameters for synthetic data preparation. It is available through the LISp-Miner system home-page (`https://lispminer.vse.cz`) and there are examples included to present its functionality.

16.3.1 Evolutionary Task Definition

To define an evolutionary task means to describe the format of a target dataset (its columns, their data-types and number of rows), to define desired patterns in terms of data-mining tasks and expected patterns to be found and to define control parameters for the evolutionary process. There are pre-defined values for each input with an option to alter them and simultaneously to compare them to values in another evolutionary process – see Fig. 16.1.

A new evolutionary task is possible to create also by cloning an existing task with all the settings copied, so they could be slightly tweaked based on previous

Name
Demo F

| Compare To | Demo C |

DataSet
Database Table: Loans

Target data rows: 10000   × 1000

Initial Data Values
• Newly created
○ From sample data

Processing Mode
• TaskPooler
○ Grid

RandSeed Init
119719430
× 119719430

Evolutionary algorithm

Fitness Tolerance	Population size	Crossover prob	Mutation prob	Reproduction prob	Mutation parameters
0 × 0	50 × 50	40 × 50	50 × 40	10 % × 10 %	

Max Iterations	Initial population	Crossover parameters		Swap prob	Fill prob
1000 × 100	100 × 50			50 × 50	50 % × 50 %

Cross Col prob   Cross Row prob
70 × 50   30 % × 50 %

Swap Len Max   Swap Count Max
100 × 100   2 × 2

Max Time (mins)	Elite size
600 × 60	1 × 1

Crossover Row parameters
Cross Rand prob   Cross Half prob
0 × 0   100 % × 100 %

Fill Len Max   Fill Count Max
5 × 5   4 × 4

Tournament size
5 × 5

☐ Include the best for each task

Figure 16.1: Evolution parameters for Demo F and comparison with Demo C (see highlight).

experiences. A complete history of evolutionary tasks definitions is archived by using this function.

16.3.2 Evolutionary process

After an evolutionary task is defined, it is possible to start the evolutionary process – see Fig. 16.2. The evolution is interactive, so we could see which requirements are met, i.e., which patterns are already present in data – see the plus (+) sign in the list of tasks in the column Ok. There is a history of evolution shown in a graph at the bottom of the screen, in terms of fitness value for the best and the worst individual at the moment. Therefore, both the speed of convergence to a desired solution and the diversity of the population are available. State of evolution, iteration number, and elapsed time are shown in the top-left corner.

There is no capability in the ReverseMiner module to solve data-mining tasks. A purpose-built LISp-Miner module called *LMTaskPooler* dedicated for batch processing of tasks is periodically called during evolution. *LMTaskPooler* module solves every data mining task in the background as fast as possible thanks to all the optimizations in the LISp-Miner system computation core. Alternatively, another module (*LMGridPooler*) could be called to accommodate parallel solving of tasks on a distributed grid/cloud to speed-up solution times (see [161]).

After all the data-mining tasks are solved, ReverseMiner computes the overall fitness of a given individual dataset and evaluates its standing in the whole population. Evolution stops if an individual meets all the requirements, mainly the found patterns. This individual dataset is presented to the user and could be exported for further use. The exported data could be used then as a basis for a data analysis as if they were real. Even in the midst of evolution, the process

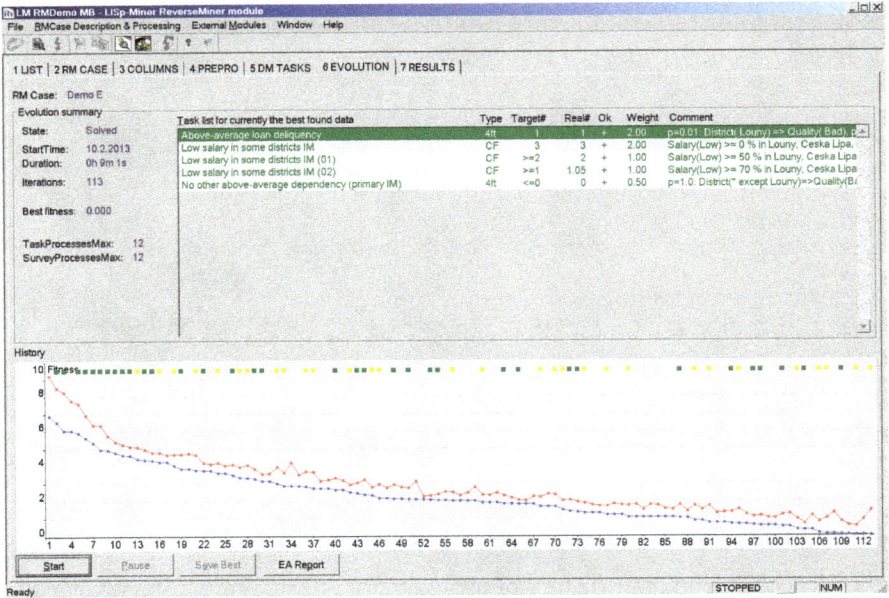

Figure 16.2: Evolution progress with list of requirements.

could be paused and the currently best found data could be exported or be used as a basis for another task.

16.3.3   Repeatibility of evolution

An important feature of the *ReverseMiner* implementation of evolution is its repeatability. Although evolutionary algorithms are theoretically based on random number generators, there are only pseudo-random number generators available in today's computers. They are initialized by a so called *initial seed number* that completely determines the whole random sequence. So if the *initial seed number* is set to the same value again, the whole evolution is repeated in exactly the same way as before (provided, that no other parameter was changed). If we change the *initial seed number*, the evolution takes a different path with potentially completely different data as its result. This behavior is successfully exploited both in teaching courses, in demonstrational examples, and in testing of the *ReverseMiner* functionality.

16.4   Evolution Helpers

After a first version of the ReverseMiner module was implemented and some experiments with the evolutionary approach were undertaken, it became clear that a purely random evolution is a significantly time-consuming process and sometimes even frustrating for users to watch the last of the desired patterns met only at 99% and still thereafter another 1000 evolution steps.

Another problem was a deterioration of datasets in the population into a degenerated state, for example with just a few highly frequent values appearing in a given column. Although a flawless solution from the point of view of desired patterns included, such data had plainly visible artificial artifacts and were useless for the intended demonstrational purposes.

First of all, we dealt with vast changes from initial frequency distributions of values in columns. So called frequency guiding rules were introduced as a simple associational rule limiting frequency change of every single value using upper and lower bounds. They became a part of the set of desired patterns from which the fitness function is computed. If a frequency of a particular value drifts outside given bounds, the confidence of corresponding frequency guiding rules decreases and so does the fitness. This individual thus risks being eliminated from the population through natural selection. But meanwhile, it has some non-zero chance of propagating some of its (otherwise) desirable properties onto other individuals by crossover and mutation operations.

Another new important feature implemented was a possibility to divide the complex evolution process of creating artificial data (meeting all the requirements) into several steps, each solved by a partial evolution. Moreover, each partial evolution could build upon results found in the previous steps. An individual representing the found solution of a particular step could be included in the population of the next partial evolution. The first benefit is an option to divide requirements on data into groups each concerning only a (non-overlapping) subset of columns. After each partial evolution is solved the complex solution is reached just by combining the columns together into a single dataset. The second benefit was found in a speed-up of evolution process because each subsequent evolution doesn't start from zero, but with an individual (or with individuals) present in the initial population that is the result of some of the previous partial evolutions and thus do already met some of requirements imposed on generated data. All the requirements from previous steps are included also in this partial evolution fitness function calculation. This ensures that already found desired properties are not lost during further evolution process.

Another problem faced arose from the random nature of evolution. For complex and simultaneously strong patterns especially to be incorporated into data, it was very improbable that some random changes to data would result in just the right combinations of values. So the evolution process was very time consuming. A special type of individual called mutant was devised to steer evolution in the right direction. The mutant dataset consists of just columns the complex pattern is related to. Being a dataset as any other individual in the population, it has the same number of rows also. But there are only values in each row, which support the desired complex pattern. For example, if an associational rule pattern of satisfaction of guests coming from Prague is below average should be present in data, the mutant dataset will consist from columns City and Satisfaction only (for column detail see Section 2.5), and all the rows will have the value of Prague in the first column and a value of low in the second. Such an individual has a very good fitness value for the particular pattern it is made for, but very poor for others. Albeit, its presence in the population during the evolution improves other individuals' fitness as the mutant still could happen to be the second parent in a crossover operation.

The last improvement of the evolutionary process regarded better options for initializing values in columns and mainly for the definition of dependant values. The initialization of values could be random or based on the uniform-, Gaussian- or even a user-defined distribution or the values even could be initialized by a (possibly) complex (arithmetic) formula with even date-related functions (DayOfWeek, DayOfYear, WeekOfYear...) available. It is also possible to lookup a value from a code-table based on its primary key. The same formula definition could be used also for computing derived columns based on values in an other column or columns. So, it is possible to compute a total price from partial prices for accommodation, food and other services or it is possible to look-up State for actual value in the City column. Such dependant values are automatically updated during the evolution process each time the original value changed due to an evolutionary operation of mutation or crossover.

16.5 Artifical Data Hotel

Some examples in this publication are based on a dataset about guests in a fictitious hotel, see Section 2.5. The whole process of the preparation of this dataset is described below demonstrating important features of the evolutionary approach to generating artificial data using the ReverseMiner module.

16.5.1 Data Specifications

It is very important to consider beforehand, what data we want to generate and which patterns we want to hide in them. A precise definition of goals and a careful consideration of desired patterns are necessary prerequisites for obtaining good results in a reasonable time frame.

Data specifications consist of:

■ aim of data generation (for what use we want to prepare data);

■ selection of business domain, that will be described by our data;

■ size of data – number of records or time periods covered by data;

■ structure of data, including "external" data – columns and their data types, valid values and value ranges;

- initial distributions of frequencies and tolerated deviations;

- domain knowledge;

- specific patterns to be hidden into data, so they could be later mined for;

The **goal** of this demo process is to generate artificial data, with a realistic look, but easy to understand. Properties of data and patterns hidden in them should allow for demonstration of a wide range of features the LISp-Miner system is capable of, including all the implemented GUHA procedures and in each phase of the data mining process as described by the CRIP-DM methodology (see Section 1.2.2).

Generated data will be used mainly for educational purposes in basic- and advanced-level courses of data mining and in this publication. ReverseMiner module (part of the LISp-Miner system) will be used.

A chosen **business domain** must be familiar for everybody, with easy to understand columns and values, with comprehensible domain knowledge and later found patterns. As nearly everyone has an experience of being accommodated in a hotel, so data about guests checking in, staying a few nights and then checking out are good examples of such domain. It is easy to understand such data and values, easy to understand motivations of a hotel owner for formulation given analytical questions. An add-on data in form of results from a guest-satisfaction questionnaire is also something everyone is familiar with.

Hotel domain is also sufficiently structured to allow for demonstration of available functions and GUHA procedures implemented into the LISp-Miner system.

Size of data was decided to be 2000 records – not trivially small, but still easily to browse manually through all records. Enlargement is still possible later using the Randomization function of ReverseMiner module for multiplication of rows and adding given amount of noise while simultaneously preserving all the desired patterns in data.

Time period of data will be two years (from Jan 2012 to Dec 2013), so at least two seasons are covered. It is also convenient to include the fall of 2013 where a monetary intervention of the Czech central bank led to a depreciation of the Czech koruna, which resulted in a price reduction for foreigners.

Structure of main data is implied from the chosen business domain, but educational purposes must include all supported data types – text, integer number, decimal number, data/time and a Boolean value of true/false. It is desired also to include nominal, ordinal and cardinal values. Still, the number of columns should stay as low as possible for simplicity reasons.

Apart from the main data, there should be some external data added to be connected with the main data. We have two sets of them – weather measurements and foreign exchange rates for the whole time period 2012–2013. We have real values that could be connected with main data by date (we use the day of the start of the visit to keep things simple). The relationship between main data and external data is *n:1*.

A detailed description of the data structure is in Section 2.5.

Well known **domain facts** have to be included in generated data, so they look realistic.

First of data properties are frequencies of particular values. They should be similar to reality and natural for a proposed hotel. So for each column, an initial distribution of frequencies of possible values was preset. For example, guests from the Czech Republic and Austria are the most frequent, and guests from far away countries are less so. Pre-defined distributions (mainly Gaussian distribution) are available for numeric values (e.g., Age) with user-defined mean value and variance. Distribution of frequencies could diverge from initial settings during evolution, but not too far outside pre-defined boundaries.

Specific threshold values are other examples of domain knowledge. The most important for a hotel could the beginning and the end of travel season. Our fictional hotel is situated in Sumava Mountains, so we should define both summer and winter season – June to August for summer season and December to February for winter.

Accommodation for tourists groups tends to be in turns, usually from Saturday and for a week, two or three in length. Some discount vouchers could be given to guests. A guest could obtain no voucher, a small voucher (200 CZK) or a large voucher (above 200 CZK).

There are some mutual influences among attributes. Recreational visits starts usually on Fridays (weekend only visits) or on Saturdays (group turns for 7, 14 or 21 days). Business travelers visits starts usually on working days (Monday to Thursday) and stay just one night only. There are also some long-term business visits for four weeks (e.g., specialists on shifts in a factory nearby) starting on Mondays with four people in one room.

Price of accommodation is calculated from a list price for a single person, multiplied by the number of nights and the actual number of persons. So the price is positively correlated with both the number of nights and the number of persons. Price of food (breakfast and dinner) is also calculated from a list price multiplied by the number of people and days present. But only visits in the main season have food included and just 70% of them only. A discount voucher is provided only for visitors staying over Saturday night (200 CZK) and for groups outside the main season (400 CZK per week).

The total price is computed from the above mentioned partial prices as: VPriceTotal = VPriceRoom + VPriceFood − VPriceDiscount.

Apart from relationships induced from domain knowledge, some other **patterns** have to be incorporated into data, so they could be later mined for. We choose to create patterns based on the home address of guests. They were defined in terms of association rules with State and City of guests' origin on the left side and several other attributes on the right side:

- Poles come mainly outside the main season.

- Recreational visits of Germans in the year 2013 were mainly in turns of groups.

- Visits from Bratislava have started significantly only after depreciation of Czech koruna.

- Guests from near-by towns of Linz (Austria) and České Budějovice (Czech Republic) come for weekend visits, but only in good weather.

- Satisfaction of guests from Brno is above-average.

Mainly to support the demonstration of clustering and decision tree methods, a statistical model was created to compute an overall satisfaction from partial satisfaction with accommodation, food and entertainment.

16.5.2 Requirements Checklist

Data specifications were validated against a checklist of necessary properties required for teaching materials and for each phase of the data mining process, see Table 16.1.

16.5.3 Evolution setup

Each step from Data Specification has to be converted into an appropriate setup of parameters in the ReverMiner module. The most important parts are (1) definition of attributes (columns) with their data types, allowed values and initial distribution of frequencies and (2) set up of data mining tasks to guide evolution towards datasets with desired patterns incorporated into them.

We are trying to keep the target size of datasets as small as possible – large enough to be able to hold all the patterns but not any larger to avoid a lengthy evolution process. An exact size of data is specific case by case and is often a matter of trial and error. But with a growing experience with the evolutionary algorithm it could be identified fairly quickly from data specifications, mainly from patterns meant to be implanted into data.

For our fictional hotel, only 200 records were enough during evolution. All the patterns could be successfully hidden into data and the time necessary for evolution to find a possible solution was acceptably short. After a solution is found for 200 records, it is easy to enlarge the dataset to 2000 (or more) records (see below).

Table 16.1: Hotel data requirements checklist.

Phase	Property	Check
Domain Understanding	Domain knowledge incorporated	Yes
Data Understanding	Several data types used for columns	Yes
	Joining of database tables	Yes
Data Preparation	Calculation of derived values	Yes
	Handling of complex Date/Time values, including day of week...	Yes
	Groups of attributes	Yes
	Enumeration values	Yes
	Numerical values for discretization	Yes
	Joining of multiple values or range of values	Yes
Interactive Analysis	At least two numerical attributes for XY plot	Yes
	Suitable numerical attributes for Principal component analysis (PCA)	Yes
	Geographical coordinates for geodata analysis	Yes
Modelling	Suitable categorical attributes for rich-syntax 4ft association rules (4ft-Miner)	Yes
	Suitable categorical attribute for conditional frequency analysis (CF-Miner)	Yes
	At least two suitable categorical attributes for conditional contingency analysis (KL-Miner) using Kendall τ_b and Cramer's V coefficients	Yes
	Target class for decision tree classification	Yes
	Suitable attributes for clustering	Yes
	Suitable attributes for set-difference (SD) GUHA procedures	Yes Yes
	Suitable attributes for 4ft-action rules	Yes
Evaluation	Easy to understand interpretation of found patterns	Yes
Deployment	Simple-to-understand domain so students could reasonably propose solutions based on evaluation	Yes Yes

16.5.4   *Data Generation*

In this step, we use evolution to find out data where the requirements from data specifications are met. According to number and types of requirements, we could either try one single evolution looking for all requirements at once or use several partial evolutions, each looking for a single requirement or for a subgroup of requirements (e.g., concerning only some columns only). For simple tasks, the first option is more straightforward and faster. For more complex tasks the second option helps to speed up evolution by looking for possible partial solutions one by one and use them afterwards as breeding pools from which the global solution is cooked up.

We choose the second option for our hotel – not only are requirements more than trivial, this option allows also to demonstrate the entire evolution better, in three steps:

- frequencies of values are initialized

- domain knowledge is incorporated

- patterns are hidden into data.

The goal of the first evolution is no more than to prepare data of a given size where the frequencies of values in all the columns correspond to the specification of the **initial distributions of frequencies**. To achieve this we have to setup all columns of our dataset, their data types and dependencies among columns as defined above (e.g., between total price and price of accommodation, food and discount voucher or that the GState value must correspond to value in the GCity column). As a bonus, such definition need to be done just once and is reused in every other step.

Possible values for each column and their initial frequencies were copied from the data specification. A mathematical formula was prepared for columns, on which values should be computed. Weather and foreign exchange values were looked-up according to the actual date of the beginning of the visit (VFrom column).

There are no patterns considered yet. This step is necessary to prove that the initial distributions of frequencies are feasible, with a no contradiction due to dependencies among columns. The dataset generated in this step serves also as a fallback source of values for all follow-up partial evolutions to fill values in columns not targeted by a given partial evolution.

Because of no contradiction in data specifications, an acceptable solution was found immediately.

Incorporating influence among attributes specified as **domain knowledge** could be separated into two steps—(1) type of visit and its beginning and (2) type of visit and its length.

It is easier to solve the first step beforehand and only after a dataset with a convenient subset of recreational visits and business trips is prepared we use it as a base for the second step. We use two CF-Miner data mining tasks (see Chapter 8) to guide evolution towards our demand – recreational visits starting mainly on Fridays and Saturdays, while business trips starting on working days only.

The share of recreational visits is large and we want the majority of them to start in just two of seven possible days of week. It would take an unnecessary long time for the evolution to slowly converge to such a skewed shape of data. So we provide a helping hand to nudge evolution in the desired direction by preparing a random dataset with only the desired combination of values (with VTypeOfViset preset to recreational and VFrom preset to any date as long as it is Friday or Saturday). This dataset is used during evolution for crossovers with other datasets and it de facto injects desired properties into any individual it cross-bred with. Such a dataset is easily defined using the ReverseMiner feature called mutant (see Evolutionary helpers above) and it is guaranteed that it stays present in the population during the whole evolution process. There could be more than one mutant dataset prepared and their existence in the population significantly improves solution times.

Only six minutes were needed for evolution to find an acceptable solution in terms of a dataset that meets our conditions on both the recreational visits and business trips and theirs beginning.

Now, it is time to solve the second step – conditions on the lengths of visits according to theirs type. Recreational visits of groups should be those that start on Saturdays and with the length in whole weeks (7, 14 or 21 days). Individual recreational visits should be for weekends starting either on Friday (two nights) or on Saturday (one night only). Four weeks of length is reserved for longer business stages starting on Monday.

A 4ft-Miner data mining task (see Chapter 7) was prepared for each of the conditions. Another three mutant datasets were prepared and included into the population to speed up evolution. An important feature is that the dataset found as a solution of the first step is also included into the population, so the already achieved solution is not lost and properties of conditions from this step are preserved and propagated into the best individual via crossovers.

It took 21 minutes for the evolution to come up with a dataset with properties matching all the conditions implied from the domain knowledge (from both the first and the second step).

A total of five **patterns** have to be hidden into the data according to specifications. A single evolution task was set-up to look up a dataset where all the patterns hold. The dataset with frequencies and domain knowledge prepared in the previous step is used as a starting point for this evolution. While the evolution is trying to tweak data values to incorporate patterns into them, it is continuously guarded that the already found dependencies are kept intact.

Group of 4ft-Miner data mining tasks was created to define desired patterns concerning guest domicile and the properties influenced by it. There had to be two tasks created for some requirements – one to fulfill strength of the pattern (confidence) and another to fulfill minimal number of cases for which it holds (support).

Once again, a mutant dataset was prepared with an exact combination of values to help speed up solution times.

It was the most complex evolution so far, but after about 40 minutes, a dataset was found with combination of values in its columns that match all the requirements from the data specification in terms of frequencies of values, domain knowledge and the stored patterns.

Only a minor problem was encountered, concerning the pattern of Poles coming mainly outside the main season. The found dataset fulfills this patterns with only slightly lower confidence than requested. After consideration, the evolution process was paused using the interactive nature of the ReverseMiner, tolerance for confidence levels for this pattern was loosened a little bit and the evolution restarted. It finished immediately, because now all the requirements have been met (in given tolerance bound).

We have now a dataset with all the desired properties incorporated, but of very small size of 200 records only. Therefore, we use a special type of evolutionary task to **enlarge data** by multiplying records and adding a noise to avoid plainly visible duplicities among rows.

We could create a dataset with 2000, 5000, 10000... records from an original of 200 records easily by duplicating every record n-times. To avoid an artificial look, the evolution operations of mutation are used to randomly change some values of some records. The noise addition must bear in mind all the required properties previously laboriously incorporated into data. So every change is evaluated using all the conditions from the previous evolution processes and a new fitness value is computed. Only datasets still matching all the requirements are allowed to live on to the next evolution step.

We choose to enlarge our dataset to 2000 rows. Some columns could be randomized freely, because they are not used in any domain knowledge or a pattern – GAge and GSex. The only restriction is a condition on frequencies distribution. We have randomized those two columns separately and immediately we have a dataset with 27% values changed, but preserving frequencies distributes tolerance and preserving also all the other conditions on data.

Now we randomize values in the VFrom column. There are several restrictions and influences regarding this column, so adding a noise must proceed with care. The majority of requirements are based on the day of the week. Not to mess things too far, we decided to add noise in sense of $+/- 7$ days (or multiplications). Values in VFrom column could change, but the day of the week stays the same. Once again, all the other requirements are still in place too during this evolution.

Slowly, number of changes to the VFrom column and thus noise added is increased. Once again, only changes that do not break the desired properties are allowed to propagate into population. After 19 minutes, a dataset with 6.7% changes was found. Evolution was paused and after a manual evaluation of the data appearance we decided that it is acceptable and evolution was stopped.

A city, a guests comes from, affects directly the state. So, only values in the GCity column are changed (and GState is automatically derived). Once again, many requirements are based on those two columns. There were no constraints on how a value of city could change, but still only such changes that do not disrupt dependencies already stored into data have chances to propagate.

It took 16 minutes to change 4.24% of values and we found this difference to be acceptable.

Now, there are artificial artefacts visible in the VPersons column only. This column has fewer requirements associated with so adding a noise was not so difficult for evolution. After 3 minutes it reached 4.12% of different values, again acceptable amount after a visual check.

We had therefore achieved a final dataset with all the desired properties as specified above and with size of 2 000 records which is a convenient size for educational data.

16.5.5 Experiences

Several experiences were already acquired with the *ReverseMiner* module through experiments, students' seminars and diploma theses and while preparing the above mentioned Hotel data and demo Loan cases (they are a part of the ReverseMiner installation file). Solution times are good (in minutes) for simple and well designed medium size evolutionary tasks, allowing for interactive work. The times of solution are linearly dependant on the number of iterations needed to reach a solution. The number of iterations is more than linearly dependant on the target number of rows. Further experiments show that even 100 thousand rows tables could be prepared within hours. So we could prepare synthetic data up to medium sized datasets in acceptable solution times given the non-critical nature of the synthetic-data-generating task.

The most time-consuming phase is the computing of data mining tasks in the *LMTaskPooler* module. Some time is spent in disk-related operations (communication with database, saving data to disk), so some improvements could be achieved with a faster *hard-disk* or even *solid-state disk*. Time spent in evolutionary operations is negligible.

Good solution times are possible thanks to both the fast solving of data mining tasks by GUHA procedures which is linearly dependant on the number of data rows, see [127], and thanks to multi-core processors in today's computers, where each individual's data mining tasks are solved in parallel up to maximum number of cores.

A (nearly unlimited) scale-up of processing power is possible and already implemented through the deployment of distributed-solution of tasks using the *Techila Grid*. It allows both for a grid consisting of unused PCs on the net (a similar design to the *SETI@home*) and for a computer cloud (using *Windows Azure*). So even if the above-mentioned rich syntax and growing user needs lead to the growing complexity of tasks and therefore growing solution times, they could be solved quickly in a distributed environment. For details see [161].

Further analysis of the evolutionary process is possible through a detailed report generated by the *ReverseMiner module*. There are four ways how an individual could be created – randomly (while initializing the population at the beginning), by one of two types of crossovers or by two possible types of mutations. The findings support the general notion that crossovers are more prevailing in the first part of evolution when a diversified population is exploited. Mutations are best at looking for slightly-different versions of the already good individuals in later stages.

Some hints are already available for the set-up of cases. First, it is important to choose a suitable distribution of randomly generated values to convene as much as possible with conditions on data defined later by data mining tasks. Although the evolutionary process will eventually converge to the right distribution, it is a waste of time to let evolution to re-define the whole histogram.

Due to its nature, the evolutionary algorithm needs a clear sign of direction from the random data to better and to the best ones. So it is not well equipped to deal with just a binary fitness function only (whether a data contains a pattern or does not). This was overcome by a fuzzy computation based on confidence values. Another important finding is to include negative constraints, i.e., to specify that no other patterns should be included, so only the truly desired patterns are present in data. These negative constraints could be defined either by the syntax of data mining tasks definition or by a negative weight the number of patterns potentially found contributes to fitness of each individual.

A graph of the history of evolution (see Fig. 16.2) allows for fine-tuning of tasks parameters (both conditions on data and evolution parameters) in an interactive way. Although it is possible to leave the evolution to find the desired solution in a long time, the experience suggests (in tune with the evolution algorithms theory) that the most important evolutionary changes are achieved during the first few steps. This behavior serves as a hint too, whether the case set-up and mainly the helper tasks are properly defined. The evolution parameters should be redefined, if no progress is seen within the first ten to twenty evolutionary steps.

16.5.6   Advantages and limitations

The most important advantage of the presented approach to generate artificial data is a definition of requirements on generated data in terms of the desired patterns that should be hidden in them. This corresponds to data mining tasks used to discover patterns. Although it is mainly left to the evolution process to find an acceptable solution, the means to compute the fitness function and thus acceptability of given datasets are using the same data mining tools as are used later for data analysis. It is guaranteed after a successful evolution that the generated dataset really meets the requirements, has the right properties and stores the desired patterns.

The proposed approach could be utilized even for automated preparation of test variants for exams where each student has a unique dataset with (slightly) different values but with desired properties and patterns still intact.

On the other hand, they are some visible limitations of this approach. First, no explicit statistical model is available to precisely describe all the properties of generated data. Moreover, as implied from the stochastic nature of evolutionary algorithms, there is no guarantee of the most optimal solution or even some solution at all. Solution times are difficult to estimate beforehand, and it could take significantly more time to find a solution after new requirements are included. A completely different solution could emerge too as the best fit after only a slight change in requirements.

Possible improvements could be in an upgrade of the evolutionary algorithm. Some measures aimed to escape from a local optimum or to improve decreased population diversity are studied. This could speed up evolution in the later stages mainly. But an important condition is to preserve the repeatability of the evolutionary process even after some of those measures are added.

Previous experiences with the ReverseMiner module proved that it is a feasible approach to generate demonstrational data and data for research and development purposes. It is also an interesting example of evolutionary algorithms utilization and could make classes more interesting for students. Instead of an analysis of data, more motivated students could try to generate them. A didactical goal is in the fact, that all the activities needed for data analysis are practiced in a very thorough manner in this way as well: business domain understanding, data understanding, data preprocessing, patterns understanding and recognition, and even how a fictional owner of data could deploy the found results in the real life. A significant advantage is that it is a student who chooses a business domain, usually depending on his interests. If a business domain is forced on someone and he/she has no sufficient prior knowledge, only poor results with only limited benefits for improvements of student skills could be expected.

Chapter 17

Applying Domain Knowledge

The goal of this chapter is to present three experiments with applications of items of domain knowledge. The first experiment described in Section 17.1 concerns the 4ft-Miner procedure and items of domain knowledge in a form of expert deduction rules. Expert deduction rules are logically incorrect deduction rules, which are, however, supported by indisputable facts justifying their applications. They are introduced in [124]. The second experiment concerns applications of items of domain knowledge in a form of formalized relations of two attributes. The application again concerns association rules and the 4ft-Miner procedure. It is described in Section 17.2. Possibilities of applications of expert deduction rules in data mining with histograms and the CF-Miner procedure are studied in the third experiment introduced in Section 17.3.

Additional experiments and considerations concerning applications of domain knowledge in data mining with the LISp-Miner system are described in [79, 87, 111, 116, 117, 118, 119, 120, 133, 134, 135, 136, 139, 149, 150, 152]. There are also thoughts on analytical reports summarising results of data mining with the GUHA procedures of the LISp-Miner, see [80, 129, 130, 132]. An additional approach to the interpretation of results of data mining with association rules is presented in [6, 7]. It is based on meta-learning and is applicable also to the results of the 4ft-Miner procedure.

17.1 Expert Deduction Rules and Association Rules

Our experiment with expert deduction rules and association rules concern the dataset *Adult* described in Section 2.2. Recall that there are several examples of applications of the 4ft-Miner procedure to this dataset described in Sections 7.2 and 7.3. We start with informal considerations in Section 17.1.1. An example of the application of expert deduction rules is introduced in Section 17.1.2.

17.1.1 Informal considerations

The goal of this section is to informally introduce expert deduction rules as a tool for filtering out uninteresting rules. We use the rule

$$\text{age}(\textit{Young}) \rightarrow_{0.6,0.05} \text{income}(\textit{Small}) \tag{17.1}$$

where $4ft(\text{age}(\textit{Young}), \text{income}(\textit{Small}), \textit{Adult}) = \langle 6297, 3330, 18423, 20792 \rangle$, see Fig. 17.1. This rule is true in the *Adult* dataset since it holds $\frac{6297}{6297+3330} = 0.65 \geq 0.6 \wedge \frac{6297}{6297+3330+18423+20792} = 0.129 \geq 0.05$.

Adult	income(*Small*)	¬ income(*Small*)
age(*Young*)	6297	3330
¬ age(*Young*)	18423	20792

Figure 17.1: $4ft(\text{age}(\textit{Young}), \text{income}(\textit{Small}), \textit{Adult})$.

Let us assume that the *Adult* matrix set has an additional attribute – column born with categories *Sunday, Monday, ..., Saturday* meaning a day of the week when a person was born. Let us have a rule

$$\text{age}(\textit{Young}) \wedge \text{born}(\textit{Sunday}) \rightarrow_{0.6,0.05} \text{income}(\textit{Small}) . \tag{17.2}$$

It is natural to understand this rule as a consequence of the rule (17.1). However, we show that the rule (17.2) is not a logical consequence of the rule (17.1). It can happen that a value of the attribute born is different from the value *Sunday* for each row of the data matrix *Adult*. Then it holds $4ft(\text{age}(\textit{Young}) \wedge \text{born}(\textit{Sunday}), \text{income}(\textit{Small}), \textit{Adult}) = \langle 0, 0, 24720, 24122 \rangle$, see Fig. 17.2.

Adult	income(*Small*)	¬ income(*Small*)
age(*Young*) ∧ born(*Sunday*)	0	24720
¬age(*Young*) ∧ born(*Sunday*)	0	24122

Figure 17.2: $4ft(\text{age}(\textit{Young}) \wedge \text{born}(\textit{Sunday}), \text{income}(\textit{Small}), \textit{Adult})$.

Thus, the rule (17.2) is false in the *Adult* data matrix. We can conclude that a deduction rule

$$\frac{\text{age}(\textit{Young}) \rightarrow_{0.6,0.05} \text{income}(\textit{Small})}{\text{age}(\textit{Young}) \wedge \text{born}(\textit{Sunday}) \rightarrow_{0.6,0.05} \text{income}(\textit{Small})} \qquad (17.3)$$

is not correct. Therefore, deduction rule (17.3) cannot serve as an argument that the rule (17.2) is a consequence of the rule (17.1).

Let us note that if $\varphi \approx \psi$ and $\varphi' \approx \psi'$ are association rules, then a deduction rule $\frac{\varphi \approx \psi}{\varphi' \approx \psi'}$ is correct if it holds for each data matrix $\mathcal{M}$: *If $\varphi \approx \psi$ is true in $\mathcal{M}$, then also $\varphi' \approx \psi'$ is true in $\mathcal{M}$* (see also Chapter 18).

Let us have a data matrix $\mathcal{M}$ such that

$$4\mathit{ft}(\text{age}(\textit{Young})), \text{income}(\textit{Small}), \text{M}) = \langle a_1, b_1, c_1, d_1 \rangle \text{ and}$$

$$4\mathit{ft}(\text{age}(\textit{Young}) \wedge \text{born}(\textit{Sunday}), \text{income}(\textit{Small}), \mathcal{M}) = \langle a_2, b_2, c_2, d_2 \rangle ,$$

see Fig. 17.3.

$\mathcal{M}$	income(*Small*)	¬ income(*Small*)
age(*Young*)	a_1	b_1
¬ age(*Young*)	c_1	d_1

$4\mathit{ft}(\text{age}(\textit{Young}), \text{income}(\textit{Small}), \mathcal{M})$ of the rule (17.1)

$\mathcal{M}$	income(*Small*)	¬ income(*Small*)
age(*Young*) ∧ born(*Sunday*)	a_2	b_2
¬ (age(*Young*) ∧ born(*Sunday*))	c_2	d_2

$4\mathit{ft}(\text{age}(\textit{Young}) \wedge \text{born}(\textit{Sunday}), \text{income}(\textit{Small}), \mathcal{M})$ of the rule (17.2)

Figure 17.3: 4ft-tables of rules (17.1) and (17.2) in a data matrix $\mathcal{M}$.

Let us further assume that both the rule (17.1) and the rule (17.2) are true in $\mathcal{M}$. We search for a reasonable argument that (17.2) is a consequence of the rule (17.1). We cannot use the deduction rule (17.3) since it is not correct. The core of incorrectness is the fact that the number of rows satisfying age(*Young*) ∧ born(*Sunday*) can be too small and thus $\frac{a_2}{a_2+b_2+c_2+d_2} < 0.05$ We have a true rule, thus this source of incorrectness does not work.

Our problem is that $\frac{a_2}{a_2+b_2} < 0.6$ can happen. We need a good reason to believe that if $\frac{a_1}{a_1+b_1} \geq 0.6$, then also $\frac{a_2}{a_2+b_2} \geq 0.6$. We consider as an indisputable fact that it holds for each data matrix $\mathcal{M}$:

(IF1): A relative frequency of persons satisfying born(*Sunday*) in the whole data matrix $\mathcal{M}$ is approximately $\frac{1}{7}$.

(IF2): relative frequencies of persons satisfying born(*Sunday*) in subsets of persons satisfying age(*Young*) ∧ income(*Small*) and age(*Young*) ∧¬ income(*Small*) are also approximately $\frac{1}{7}$.

These indisputable facts mean that it holds both $a_2 \doteq \frac{1}{7}a_1$ and $b_2 \doteq \frac{1}{7}b_1$ and thus

$$\frac{a_2}{a_2 + b_2} \doteq \frac{\frac{1}{7}a_1}{\frac{1}{7}a_1 + \frac{1}{7}b_1} \doteq \frac{a_1}{a_1 + b_1}.$$

This we understand as a good reason to consider $\frac{a_2}{a_2+b_2} \geq 0.6$ as a consequence of $\frac{a_1}{a_1+b_1} \geq 0.6$. It can, of course, happen that $\frac{a_2}{a_2+b_2} < 0.6$. However, if we consider bigger and bigger data matrices $\mathcal{M}$, then we can assume that $\frac{a_2}{a_2+b_2}$ is closer and closer to $\frac{a_1}{a_1+b_1}$.

This can be formalized by the assertion

$$\text{If} \quad \lim_{|\mathcal{M}| \to \infty} \frac{a_1}{a_1 + b_1} = L, \quad \text{then} \quad \lim_{|\mathcal{M}| \to \infty} \frac{a_2}{a_2 + b_2} = L, \qquad (17.4)$$

where $|\mathcal{M}| = a_1 + b_1 + c_1 + d_1 = a_2 + b_2 + c_2 + d_2$ is the number of rows of $\mathcal{M}$.

These considerations lead to an expert deduction rule

$$Exp \frac{\text{age}(\textit{Young}) \to_{0.6, 0.05} \text{income}(\textit{Small})}{\text{age}(\textit{Young}) \land \text{born}(\textit{Sunday}) \to_{0.6, 0.05} \text{income}(\textit{Small})}. \qquad (17.5)$$

We consider (17.5) as a correct expert deduction rule, if we can prove the validity of (17.4) from the suitable formalized indisputable facts (IF1) and (IF2) introduced above. Such correct expert deduction rule is used as an argument that the rule (17.2) is a consequence of the rule (17.1).

Expert deduction rules are introduced in [124]. The available theory related to expert deduction rules is also introduced in Chapter 18 together with challenges for further research. We show an example of the application of suitable expert deduction rules in the next section.

17.1.2 Applying expert deduction rules to association rules

We use expert deduction rules of the form

$$Exp \frac{\varphi \to_{p,s} \psi}{\varphi \land \text{Capital\_loss}(\textit{None}) \to_{p,s} \psi} \qquad (17.6)$$

where φ and ψ are Boolean attributes derived from the columns of the *Adult* data matrix. These deduction rules are introduced in [124] together with arguments for their correctness. The arguments are based on frequency analyses of the *Adult* data matrix, not on sociological research. The goal is to show that the application of correct expert deduction rules can remarkably decrease the number of output rules. The goal is not to provide a sociological analysis.

We use a run of the 4ft-Miner procedure to find all association rules

$$\mathscr{B}(\text{Personal}) \land \mathscr{B}(\text{Family}) \land \mathscr{B}(\text{Society}) \land \mathscr{B}(\text{Capital}) \to_{0.8, 0.01} \mathscr{B}(\text{Employment})$$

true in the *Adult* dataset. Here $\mathscr{B}$(Personal), $\mathscr{B}$(Family), $\mathscr{B}$(Society), $\mathscr{B}$(Capital), and $\mathscr{B}$(Employment) are Boolean characteristics of the groups of attributes Personal, Family, Society, Capital, and Employment. We use an $\mathscr{RP}$-expression Personal_1 defining a set of Boolean characteristics $\mathscr{B}$(Personal) of the group of attributes Personal presented in a tabular form in Fig. 17.4.

Name: Personal_1		Type: *Conjunction*		Length: $0-3$	
Attribute	Coefficients	Length	Gace	B/R	Class
sex	*Subsets*	$1-1$	*Pos*	B	—
age_exp	*Subsets*	$1-1$	*Pos*	B	—
education	*Sequences*	$1-4$	*Pos*	B	—

Figure 17.4: $\mathscr{RP}$-expression Personal_1.

Sets of Boolean characteristics $\mathscr{B}$(Family), $\mathscr{B}$(Society), and $\mathscr{B}$(Employment) are defined by the $\mathscr{RP}$-expressions Family, Society, and Employment presented already in Fig. 7.13. The set $\mathscr{B}$(Capital) of Boolean characteristics of the group Capital of attributes is defined by the $\mathscr{RP}$-expression Capital available in Fig. 17.5. The set of relevant antecedents is given by an $\mathscr{RC}$-expression $\mathscr{AE}$ introduced also in Fig. 17.5. A set of relevant succedents is given by the $\mathscr{RP}$-expression Employment already presented in the tabular form in Fig. 7.13.

Name: Capital		Type: *Conjunction*		Length: $0-3$	
Attribute	Coefficients	Length	Gace	B/R	Class
capital_gain	*Subsets*	$1-1$	*Pos*	B	—
capital_loss	*Subsets*	$1-1$	*Pos*	B	—

Name: $\mathscr{AE}$	Length: $1-10$
Set of partial cedents	Length
Personal_1	$0-3$
Family	$0-2$
Society	$0-3$
Capital	$0-2$

Figure 17.5: $\mathscr{RP}$-expression Capital and $\mathscr{RC}$-expression $\mathscr{AE}$.

We use again the secured approach to missing information introduced in Section 7.1.5. A run of the 4ft-Miner procedure with these parameters resulted in 285 rules in 119 seconds. More than $1.9 * 10^6$ rules were tested. The five rules with the highest confidence are displayed in a short form in Fig. 17.6.

PIM	Hypothesis
0.851	sex(*Male*) & Age(*Middle-aged*) & education(*Prof-school*, *Doctorate*) >-< occupation(*Prof-specialty*)
0.850	Age(*Middle-aged*) & education(*Prof-school*, *Doctorate*) & capital_gain(*None*) >-< occupation(*Prof-specialty*)
0.844	Age(*Middle-aged*) & education(*Prof-school*, *Doctorate*) & native_country(*United-States*) >-< occupation(*Prof-specialty*)
0.843	Age(*Middle-aged*) & education(*Prof-school*, *Doctorate*) & race(*White*) >-< occupation(*Prof-specialty*)
0.841	Age(*Middle-aged*) & education(*Prof-school*, *Doctorate*) & native_country(*United-States*) & race(*White*) >-< occupation(*Prof-specialty*)

Figure 17.6: 4ft-Miner output - five rules with the highest confidence.

The rule $Ant \rightarrow_{0.851,0.010}$ occupation(*Prof-specialty*) where
$Ant = $ sex(*Male*) $\wedge$ age(*Middle-aged*) $\wedge$ education(*Prof-school*, *Doctorate*) has
the highest confidence, i.e., 0.851.

We call a rule $\varphi \rightarrow_{p,s} \psi$ *expert prime according to the deduction rule (17.6)*
if $\varphi \rightarrow_{p,s} \psi$ is relevant and true and if it does follow from an other true relevant
rule according to expert deduction rules (17.6). In this case, this means that the
antecedent φ does not contain the Boolean attribute Capital_loss(*None*).

Two different succedents occur in output rules – hours-per-week(*Full-time*)
and occupation(*Prof-specialty*). The numbers of expert prime rules and the numbers of rules which can be filtered out by the expert deduction rules (17.6) are
presented in Table 17.1.

Table 17.1: The numbers of rules for particular succedents.

Succedent	the number of rules		
	Expert prime	Filtered out	Total
hours-per-week(*Full-time*)	139	135	274
occupation(*Prof-specialty*)	9	2	11
Total	148	137	285

Let us conclude that in our example, the application of correct expert deduction rules decreases the number of output rules by approximately 48 percent. This is a good reason to continue research in expert deduction rules. See
Chapter 18 for more information about achieved results and planned research
directions.

17.2 Items of Domain Knowledge and Association Rules

The GUHA association rule expresses a relation of two relatively general
Boolean attributes derived from columns of an analysed data matrix. However,
domain experts usually use more complex relations to express items of domain
knowledge. Such items can be formulated in a formal or semi-formal way. Two
examples of items of domain knowledge related to the *STULONG* dataset follow:

- BMI ↑↑ Diastolic meaning that if BMI of a patient increases, then his/here diastolic blood pressure increases too.

- Education ↑↓ Beer meaning that if education of a patient increases, then his/here beer consumption decreases.

Our goal is to present an example of the application of such items of domain knowledge in data mining with association rules. Let us emphasize that the presented example uses only basic medical knowledge. The goal is to present possibilities for how to deal with formalized items of domain knowledge, not to get new medical knowledge. We use the above mentioned item of domain knowledge BMI ↑↑ Diastolic. Principles of its application are introduced in Section 17.2.1. Details of particular steps of application are described in Sections 17.2.3– 17.2.6.

17.2.1 BMI ↑↑ Diastolic—principle of application

A principle of application of the item of domain knowledge BMI ↑↑ Diastolic is presented in Fig. 17.7. It holds:

1. The system makes possible to store and maintain formalized items of domain knowledge. The item BMI ↑↑ Diastolic is stored in the LISp-Miner system and presented in the top of Fig. 17.7 as displayed by the LISp-Miner.

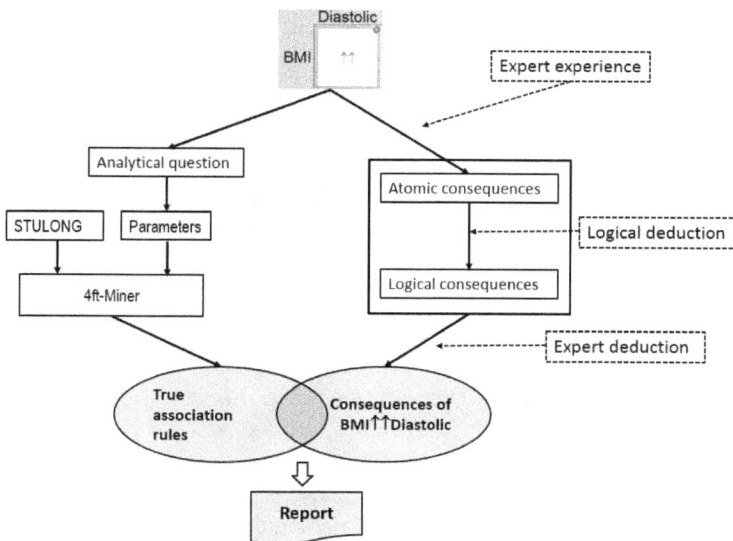

Figure 17.7: Principle of application of item of domain knowledge BMI ↑↑ Diastolic.

2. Various analytical questions can be formulated using stored items of domain knowledge. We will deal with the following analytical question:
In the data matrix STULONG, *are there any interesting relations between combinations of personal characteristics and results of measurements of patients on one side and systolic and diastolic blood pressures on the other side? However, we are not interested in consequences of the known fact that if BMI increases, then diastolic blood pressure increases too.*

3. Relations between combinations of personal characteristics and results of measurements of patients on one side and systolic and diastolic blood pressures on the other side will be understood as association rules.

4. All potentially interesting association rules will be produced as a result of a suitable application of the 4ft-Miner procedure. Details are presented in Section 17.2.2.

5. Expert experience will be used to map the item BMI ↑↑ Diastolic to a set of its atomic consequences – very simple association rules BMI(β) $\approx$ Diastolic(δ), see Section 17.2.3.

6. Logical deduction will be used to get all logical consequences of all atomic consequences of the item BMI ↑↑ Diastolic, see Section 17.2.4.

7. Expert deduction rules will be applied to a union of the set of atomic consequences and the set of all logical consequences of atomic consequences to get a set of all consequences of the item BMI ↑↑ Diastolic, see Section 17.2.5.

8. Output of the 4ft-Miner procedure will be compared with the set of all association rules – consequences of BMI ↑↑ Diastolic, see Section 17.2.6.

Let us note that the above described process is a part of FOFRADAR – a FOrmal FRAme for Data mining with Association Rules [122].

17.2.2 Applying 4ft-Miner

Recall that we solved the analytical question "*In the data matrix* STULONG, *are there any interesting relations between combinations of personal characteristics and results of measurements of patients on one side and systolic and diastolic blood pressures on the other side? However, we are not interested in consequences of the known fact that if BMI increases, then diastolic blood pressure increases too.*" introduced in Section 17.2.1.

Relations between combinations of personal characteristics and results of measurements of patients on one side and systolic and diastolic blood pressures on the other side are understood as association rules. We use attributes Marital-Status and Education as personal characteristics and attributes BMI, Subscapularis, and Triceps as results of measurement. Systolic and diastolic blood pressures are described by the attributes Systolic and Diastolic. We use 4ft-quantifier $\Rightarrow_{0.8,100}$ and we search for association rules

$$\mathscr{B}(\text{Personal}) \wedge \mathscr{B}(\text{Measurement}) \Rightarrow_{0.8,100} \mathscr{B}(\text{BloodPresure}) . \qquad (17.7)$$

The set of relevant antecedents for equation (17.7) is given by a $\mathscr{RC}$-expression "BMI ↑↑ Diastolic − Antecedent" presented in a long tabular form in Fig. 17.8. Note that this $\mathscr{RC}$-expression involves two $\mathscr{RP}$-expressions Personal and Measurement defining a set of Boolean characteristics $\mathscr{B}(\text{Personal})$ of the personal characteristics of patients and a set of Boolean characteristics $\mathscr{B}(\text{Measurement})$ of results of measurement respectively.

Name: BMI ↑↑ Diastolic − Antecedent		Length: $1-5$			
Name: Measurement		Type: *Conjunction*		Length: $0-3$	
Attribute	Coefficients	Length	Gace	B/R	Class
BMI	*Sequences*	$1-3$	*Pos*	*B*	−
Subscapularis	*Sequences*	$1-3$	*Pos*	*B*	−
Triceps	*Sequences*	$1-3$	*Pos*	*B*	−
Name: Personal		Type: *Conjunction*		Length: $0-2$	
MaritalStatus	*Subsets*	$1-1$	*Pos*	*B*	−
Education	*Subsets*	$1-1$	*Pos*	*B*	−

Figure 17.8: $\mathscr{RC}$-expression "BMI ↑↑ Diastolic − Antecedent".

The set of relevant succedents for equation (17.7) is given by a $\mathscr{RC}$-expression "BMI ↑↑ Diastolic − Succedent" presented in a long tabular form in Fig. 17.9. Note that this $\mathscr{RC}$-expression involves one $\mathscr{RP}$-expressions Blood-Pressure defining a set of Boolean characteristics $\mathscr{B}(\text{BloodPressure})$.

Name: BMI ↑↑ Diastolic − Succedent		Length: $1-5$			
Name: BloodPressure		Type: *Conjunction*		Length: $0-2$	
Attribute	Coefficients	Length	Gace	B/R	Class
Diastolic	*Sequences*	$1-3$	*Pos*	*B*	−
Systolic	*Sequences*	$1-4$	*Pos*	*B*	−

Figure 17.9: $\mathscr{RC}$-expression "BMI ↑↑ Diastolic − Succedent".

We used a run of the 4ft-Miner procedure with a set of relevant antecedents defined by the $\mathscr{RC}$-expression "BMI ↑↑ Diastolic − Antecedent", set of relevant antecedents defined by the $\mathscr{RC}$-expression "BMI ↑↑ Diastolic − Succedent", an empty set of relevant conditions, 4ft-quantifier $\Rightarrow_{0.8,100}$ and parameters *Missing = Secured* and *Prime = Yes*. A run of the procedure resulted in 164 rules, 1 004 172 rules were tested in 15 seconds. 12 rules with the highest confidence (denoted as PIM) are shown in Fig. 17.10.

Task run			
Start: 26.8.2020 19:15:21	Total time: 0h 0m 15s		
Number of verifications: 1004172			
Number of hypotheses: 164	Mode: Standard		Add group Del group Edit group

Actual group of hypotheses: All hypotheses			
Hypotheses in group: 164	Shown hypotheses: 164	Highlighted: 0	

Nr.	Id	PIM	Hypothesis
1	8	0.886	Subscapularis(< *14*) & BMI(*21..23*) & MaritalStatus(*married*) >÷< Diastolic(*<65;94>*)
2	76	0.886	BMI(<=*22*) >÷< Diastolic(*<65;94>*)
3	95	0.884	BMI(*22..24*) & Education(*university*) >÷< Diastolic(*<65;94>*)
4	82	0.880	BMI(*21..23*) & MaritalStatus(*married*) >÷< Diastolic(*<65;94>*)
5	114	0.877	BMI(*23..25*) & Education(*university*) >÷< Diastolic(*<65;94>*)
6	80	0.876	BMI(*21..23*) >÷< Diastolic(*<65;94>*)
7	6	0.875	Subscapularis(< *14*) & BMI(*21..23*) >÷< Diastolic(*<65;94>*)
8	78	0.873	BMI(*21,22*) >÷< Diastolic(*<65;94>*)
9	97	0.872	BMI(*23*) >÷< Diastolic(*<65;94>*)
10	109	0.872	BMI(*23..25*) & MaritalStatus(*married*) & Education(*university*) >÷< Diastolic(*<65;94>*)
11	107	0.871	BMI(*23..25*) & MaritalStatus(*married*) & Education(*secondary*) >÷< Diastolic(*<65;94>*)
12	123	0.869	BMI(*24,25*) & Education(*university*) >÷< Diastolic(*<65;94>*)

Figure 17.10: Rules before filtering out consequences of BMI ↑↑ Diastolic.

There is only one resulting rule with both attributes Diastolic and Systolic in succedent:

$$\text{BMI}(21,22,23) \wedge married \Rightarrow_{0.81,168} \text{Diastolic}\langle 65;94\rangle \wedge \text{Systolic}\langle 105;144\rangle$$

where we write *"married"* only instead of "MaritalStatus(*married*)". There are 132 resulting rules with attribute Diastolic only in succedent. A structure of antecedents for these rules is outlined in Fig. 17.11. We see that there are 20 rules attribute BMI only in antecedent, etc. An analogous information is in Fig. 17.12 for 31 resulting rules with attribute Systolic only in succedent.

17.2.3   Atomic consequences of BMI ↑↑ Diastolic

The first step in building a set of consequences of the item of domain knowledge BMI ↑↑ Diastolic is to define a set of atomic consequences of BMI ↑↑ Diastolic, see point 5 in Section 17.2.1. It is defined as a set of very simple association rules BMI(β) ≈ Diastolic(δ). We use a 4ft-quantifier $\Rightarrow_{0.8,100}$ applied in the run of the 4ft-Miner procedure described in Section 17.2.2.

Length	BMI	Subscapularis	Triceps	MaritalStatus	Education	Rules
1	X					20
1		X				2
1					X	2
1			X			1
2	X	X				21
2	X				X	17
2	X		X			14
2	X			X		13
2		X		X		4
2		X			X	2
2			X		X	2
2				X	X	2
3	X	X		X		14
3	X			X	X	10
3	X		X	X		8
			Total			132

Figure 17.11: Antecedent structure for Diastolic in succedent.

Length	BMI	Subscapularis	Triceps	MaritalStatus	Education	Rules
1	X					10
2	X				X	8
2	X			X		7
2	X	X				1
3	X			X	X	3
3	X	X		X		1
3	X		X	X		1
			Total			31

Figure 17.12: Antecedent structure for Systolic in succedent.

Thus, we are going to define a set $AC(\text{BMI} \uparrow\uparrow \text{Diastolic}, \Rightarrow_{0.8,100})$ of atomic consequences of BMI $\uparrow\uparrow$ Diastolic as a set of association rules $\text{BMI}(\beta) \Rightarrow_{p,Base} \text{Diastolic}(\delta)$ where $p \geq 0.8$, *Base* ≥ 100 and β and δ are suitable coefficients for attributes BMI and Diastolic. We show a way in which coefficients β and γ can be defined. Informally speaking, if $\text{BMI}(\beta)$ and $\text{Diastolic}(\delta)$ can be considered as saying "BMI is low" and "Diastolic is low" respectively, then the rule $\text{BMI}(\alpha) \Rightarrow_{0.8,100} \text{Diastolic}(\beta)$ can be considered as a simple consequence of BMI $\uparrow\uparrow$ Diastolic.

The attribute BMI has 14 categories $\leq 20, 21, 22, \ldots, 32, \geq 33$ and the attribute Diastolic has 7 categories $\leq 64, \langle 65; 74 \rangle, \ldots, \langle 105; 114 \rangle, \geq 115$. We can decide, in cooperation with a domain expert, that each basic Boolean attribute $\text{BMI}(\beta)$ satisfying condition $\beta \subseteq Low_{\text{BMI}}$ will be considered as saying "BMI is low" if $Low_{\text{BMI}} = \{\leq 20, 21, 22, \ldots, 26\}$. We can similarly decide that basic

Boolean attribute Diastolic(δ) will be considered as saying "Diastolic is low" if $\delta \subseteq Low_{\text{Diastolic}}$ where $Low_{\text{Diastolic}} = \{\leq 64, \langle 65; 74\rangle, \langle 75; 84\rangle\}$. Thus, we can say that the set of all rules BMI("is low") $\Rightarrow_{0.8,100}$ Diastolic("is low") is defined by a rectangle

$$Low_{\text{BMI}} \times Low_{\text{Diastolic}} = \{\leq 20, 21, 22, \ldots, 26\} \times \{\leq 64, \langle 65; 74\rangle, \langle 75; 84\rangle\}.$$

The LISp-Miner system makes possible to define the set $AC(\text{BMI} \uparrow\uparrow \text{Diastolic}, \Rightarrow_{0.8,100})$ by a union $\mathscr{A}_1 \times \mathscr{S}_1 \cup \ldots \cup \mathscr{A}_R \times \mathscr{S}_R$ of R similar, possibly overlapping, rectangles. The set $AC(\text{BMI} \uparrow\uparrow \text{Diastolic}, \Rightarrow_{0.8,100})$ is then considered as a set of all rules BMI(β) $\Rightarrow_{p,B}$ Diastolic(δ) satisfying $p \geq 0.8$ and $Base \geq 100$ for which there is $i \in \{1, \ldots, R\}$ such that $\beta \subseteq \mathscr{A}_i$ and $\delta \subseteq \mathscr{S}_i$. An example of such a definition is in Fig. 17.13, three rectangles are used.

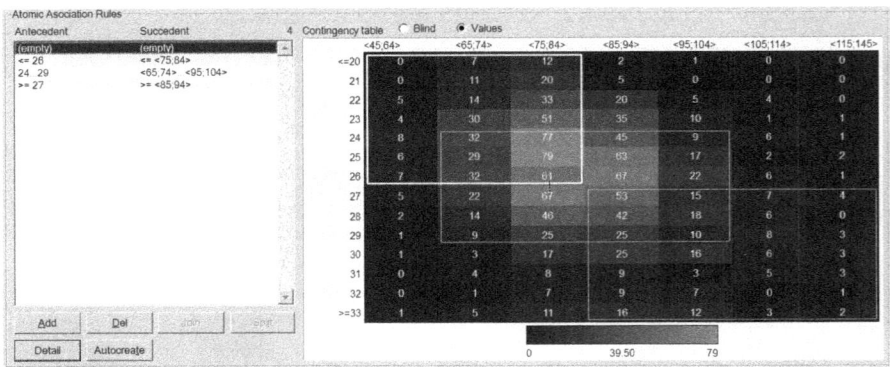

Figure 17.13: Definition of the set $AC(\text{BMI} \uparrow\uparrow \text{Diastolic}, \Rightarrow_{0.8,100})$.

We can say that the set $AC(\text{BMI} \uparrow\uparrow \text{Diastolic}, \Rightarrow_{0.8,100})$ is given by a union

$Low_{\text{BMI}} \times Low_{\text{Diastolic}} \cup Medium_{\text{BMI}} \times Medium_{\text{Diastolic}} \cup High_{\text{BMI}} \times High_{\text{Diastolic}}$

defined in Fig. 17.13. The set $AC(\text{BMI} \uparrow\uparrow \text{Diastolic}, \Rightarrow_{0.8,100})$ consists of all rules BMI(β) $\Rightarrow_{p,Base}$ Diastolic(δ) where $p \geq 0.8$, $Base \geq 100$ and one of the following assertions is true:

- $\beta \subseteq Low_{\text{BMI}}$ and $\delta \subseteq Low_{\text{Diastolic}}$
- $\beta \subseteq Medium_{\text{BMI}}$ and $\delta \subseteq Medium_{\text{Diastolic}}$
- $\beta \subseteq High_{\text{BMI}}$ and $\delta \subseteq High_{\text{Diastolic}}$.

17.2.4 Logical consequences of atomic consequence

Logical deduction is used to get all logical consequences of all atomic consequences of the item BMI $\uparrow\uparrow$ Diastolic, see point 6 in Section 17.2.1. In other

words, we are interested in all correct deduction rules $\dfrac{\mathsf{BMI}(\beta)\Rightarrow_{p,Base}\mathsf{Diastolic}(\delta)}{\varphi\Rightarrow_{p,Base}\psi}$
where $\varphi\Rightarrow_{p,Base}\psi$ is an association rule resulting from the application of the 4ft-Miner procedure described in Section 17.2.2. Each association rule resulting from this application of the 4ft-Miner is in one of the following forms

- $\omega\Rightarrow_{p,Base}\mathsf{Diastolic}(\delta)$

- $\omega\Rightarrow_{p,Base}\mathsf{Systolic}(\sigma)$

- $\omega\Rightarrow_{p,Base}\mathsf{Diastolic}(\delta)\wedge\mathsf{Systolic}(\sigma),$

see the end of Section 17.2.2. This fact and theoretical result presented later in Section 18.1.4 mean that only deduction rules $\dfrac{\mathsf{BMI}(\beta)\Rightarrow_{p,Base}\mathsf{Diastolic}(\delta)}{\mathsf{BMI}(\beta)\Rightarrow_{p,Base}\mathsf{Diastolic}(\delta')}$ where $\delta\subsetneqq\delta'$ can be applied.

A deduction rule $\dfrac{\mathsf{BMI}(21,22,23)\Rightarrow_{0.8,100}\mathsf{Diastolic}(\langle65;74\rangle,\langle75;84\rangle)}{\mathsf{BMI}(21,22,23)\Rightarrow_{0.8,100}\mathsf{Diastolic}(\langle65;74\rangle,\langle75;84\rangle,\langle85;94\rangle)}$ is a typical example of such correct deduction rule. We outline why this deduction rule is correct. Below we write "Diastolic$\langle65;84\rangle$" instead of "Diastolic$(\langle65;74\rangle,\langle75;84\rangle)$" and "Diastolic$\langle65;94\rangle$" instead of "Diastolic $(\langle65;74\rangle,\langle75;84\rangle),\langle85;94\rangle)$". 4ft-tables $4ft(\mathsf{BMI}(21,22,23),\mathsf{Diastolic}\langle65;84\rangle,$ $\mathcal{M})$ and $4ft(\mathsf{BMI}(21,22,23),\mathsf{Diastolic}\langle65;94\rangle,\mathcal{M})$ are shown in Fig. 17.14.

$\mathcal{M}$	Diastolic$\langle65;84\rangle$	¬Diastolic$\langle65;84\rangle$
BMI$(21,22,23)$	a	b
¬BMI$(21,22,23)$	c	d

$\mathcal{M}$	Diastolic$\langle65;94\rangle$	¬Diastolic$\langle65;94\rangle$
BMI$(21,22,23)$	a'	b'
¬BMI$(21,22,23)$	c'	d'

Figure 17.14: 4ft-tables for deduction rule $\dfrac{\mathsf{BMI}(21,22,23)\Rightarrow_{0.8,100}\mathsf{Diastolic}\langle65;84\rangle}{\mathsf{BMI}(21,22,23)\Rightarrow_{0.8,100}\mathsf{Diastolic}\langle65;94\rangle}.$

It sure holds

- $a'\geq a$ since each row of $\mathcal{M}$ satisfying BMI$(21,22,23)\wedge$ Diastolic$\langle65;84\rangle$ satisfies also BMI$(21,22,23)\wedge$ Diastolic$\langle65;94\rangle$

- $a+b=a'+b'$ since

$$a+b=\text{ the number of rows satisfiyng BMI(21,22,23) }=a'+b'$$

- $b'\leq b$ since both $a'\geq a$ and $a+b=a'+b'$.

It is evident that the validity of inequalities $a'\geq a$ and $b'\geq b$ implies that if $\frac{a}{a+b}\geq0.8\wedge a\geq100$ then also $\frac{a'}{a'+b'}\geq0.8\wedge a'\geq100$. This means that if the rule

BMI$(21,22,23)$ $\Rightarrow_{0.8,100}$ Diastolic$\langle 65;84 \rangle$ is true in a data matrix $\mathscr{M}$, then also the rule BMI$(21,22,23)$ $\Rightarrow_{0.8,100}$ Diastolic$\langle 65;94 \rangle$ is true in a data matrix $\mathscr{M}$.

We can conclude that the rule BMI$(21,22,23)$ $\Rightarrow_{0.8,100}$ Diastolic$\langle 65;94 \rangle$ is a logical consequence of the rule BMI$(21,22,23)$ $\Rightarrow_{0.8,100}$ Diastolic$\langle 65;84 \rangle$ which is an atomic consequence of the item of domain knowledge BMI $\uparrow\uparrow$ Diastolic. Let us note that the rule BMI$(21,22,23)$ $\Rightarrow_{0.8,100}$ Diastolic$\langle 65;94 \rangle$ is not an atomic consequence of BMI $\uparrow\uparrow$ Diastolic. This follows from Fig. 17.13.

17.2.5   Consequences of BMI $\uparrow\uparrow$ Diastolic

Expert deduction rules are applied to a union of the atomic consequence and the set of all logical consequences of atomic consequences to get a set of all consequences of the item of domain knowledge BMI $\uparrow\uparrow$ Diastolic, see point 7 in Section 17.2.1. Recall expert deduction rule (17.5) introduced in Section 17.1.1:

$$Exp \frac{\text{age}(Young) \rightarrow_{0.6,0.05} \text{income}(Small)}{\text{age}(Young) \wedge \text{born}(Sunday) \rightarrow_{0.6,0.05} \text{income}(Small)} .$$

This expert deduction rule is considered as correct because we assume that if the number of observed persons gets bigger and bigger, then a relative frequency of persons with small income among the young persons born on Sunday gets closer and closer to a relative frequency of persons with small income among all the young persons. A formal introduction of expert deduction rules is available in Section 18.2.2 together with their important properties.

Both atomic consequences and logical consequences of the item of domain knowledge BMI $\uparrow\uparrow$ Diastolic are in the form BMI(β) $\Rightarrow_{p,Base}$ Diastolic(δ), where $p \geq 0.8 \wedge Base \geq 100$, see Sections 17.2.3 and 17.2.4. This leads us to application of expert deduction rules

$$Exp \frac{\text{BMI}(\beta) \Rightarrow_{p,Base} \text{Diastolic}(\delta)}{\text{BMI}(\beta) \wedge \omega \Rightarrow_{p,Base} \text{Diastolic}(\delta)} \tag{17.8}$$

where BMI(α) $\Rightarrow_{p,Base}$ Diastolic(δ) is an atomic consequence of the item BMI $\uparrow\uparrow$ Diastolic or a logical consequence of such atomic consequence and one of the following (O1), (O2), (O3) is valid:

(O1): $\omega = $ MaritalStatus(κ)

(O2): $\omega = $ Education(λ)

(O3): $\omega = $ MaritalStatus(κ) $\wedge$ Education(λ).

This corresponds to our assumption that if the number of patients gets bigger and bigger, then a relative frequency of patients satisfying Diastolic(δ) among the patients satisfying BMI(β) $\wedge$ ω gets closer and closer to a relative frequency of patients satisfying Diastolic(δ) among the all patients satisfying BMI(β).

Very informally speaking, we can say that a relative frequency of patients satisfying Diastolic(δ) among patients satisfying BMI(β) does not depend either on marital status or on education.

We do not suppose a validity of analogous assumption concerning attributes Subscapularis and Triceps.

17.2.6 Interpreting results of 4ft-Miner

The considerations presented in Sections 17.2.3, 17.2.4 and 17.2.5 can be summarized such that a set of consequences of the item BMI $\uparrow\uparrow$ Diastolic of domain knowledge consists of the following association rules:

- Atomic consequences BMI(β) $\Rightarrow_{p,Base}$ Diastolic(δ) where $p \geq 0.8$, $Base \geq 100$ and one of (i), (ii), (iii) is true where

  (i): $\beta \subseteq Low_{\text{BMI}}$ and $\delta \subseteq Low_{\text{Diastolic}}$

  (ii): $\beta \subseteq Medium_{\text{BMI}}$ and $\delta \subseteq Medium_{\text{Diastolic}}$

  (iii): $\beta \subseteq High_{\text{BMI}}$ and $\delta \subseteq High_{\text{Diastolic}}$

  and Low_{BMI}, $Low_{\text{Diastolic}}$ $Medium_{\text{BMI}}$, $Medium_{\text{Diastolic}}$, $High_{\text{BMI}}$, and $High_{\text{Diastolic}}$ are defined in Fig. 17.13.

- Logical consequences BMI(β) $\Rightarrow_{p,Base}$ Diastolic(δ') of atomic consequences BMI(β) $\Rightarrow_{p,Base}$ Diastolic(δ) where $\delta \subsetneq \delta'$, see Section 17.2.4.

- Expert consequences BMI(β) $\wedge \omega \Rightarrow_{p,Base}$ Diastolic(δ) of atomic consequences and expert consequences BMI(β) $\wedge \omega \Rightarrow_{p,Base}$ Diastolic(δ') of logical consequences of atomic consequences where ω satisfies one of the (O1), (O2), (O3), see Section 17.2.5.

This makes it possible to filter out all the consequences of the item BMI $\uparrow\uparrow$ Diastolic of domain knowledge from the set of all the resulted rules with the attribute Diastolic in succedent. The structure of this set is described in Fig. 17.11. The filtering concerns only rules with antecedent BMI(β) and rules containing attribute BMI together with at least one of the attributes MaritalStatus and Education in the antecedent. There are 60 such rules, 45 can be filtered out as consequences of BMI $\uparrow\uparrow$ Diastolic. For details see Fig. 17.15.

We can summarize that the used tools made it possible to filter out 45 (i.e., 75 percent) of relevant 60 rules. However, let us emphasize that goal of the presented example is not to get a new medical knowledge. The goal is to present possibilities of developed tools for dealing with domain knowledge. Getting new medical knowledge requires participation of medical experts, and this is not so in the case of our example.

Rules

$$\text{BMI}(27) \Rightarrow_{0.81,142} \text{Diastolic}\langle 65; 94 \rangle \tag{17.9}$$

				Rules		
Length	BMI	MaritalStatus	Education	Total	Filtered out	Remaining
1	X			20	15	5
2	X		X	17	13	4
2	X	X		13	11	2
3	X	X	X	10	6	4
Total				60	45	15

Figure 17.15: Antecedent structure for Diastolic in succedent.

and

$$\mathsf{BMI}(26,27,28) \wedge \mathsf{MaritalStatus}(married) \Rightarrow_{0.8,345} \mathsf{Diastolic}\langle 65;94\rangle \quad (17.10)$$

are examples of remaining rules (cf. Fig. 17.13 and expert deduction rules 17.8).

Rules remaining after filtering consequences of BMI ↑↑ Diastolic can be understood as exceptions or as an impetus to revision of definition of the atomic consequences or revision of expert deduction rules. The same is true for additional applications of the presented tools for dealing with domain knowledge.

17.3 Expert Deduction Rules and Histograms

We again start with informal considerations, see Section 17.3.1. An example of application of expert deduction rules in data mining with histograms is introduced in Section 17.3.2.

17.3.1 *Expert deduction and histograms—considerations*

The goal of this section is to informally introduce expert deduction rules as a tool for filtering out uninteresting histograms from the output of the CF-Miner procedure. Note that ideas presented in this section are similar to that introduced in Section 17.1.1, see also [124] and [139]. Let us again assume that the *Adult* data matrix has an additional attribute – column born with categories *Sunday, Monday, ..., Saturday* meaning a day of the week when a person was born. Below, we again write "*Never-married*" instead of "marital_status(*Never-married*)" and "*Sunday*" instead of "born(*Sunday*)".

Let us consider CF-tables $CF(\mathsf{age\_exp}, Adult/Never\text{-}married)$ and $CF(\mathsf{age\_exp}, Adult/(Never\text{-}married \wedge Sunday))$, see Fig. 17.16.

Let us further consider a basic CF-quantifier $\approx_{Down3;3\%}$ defined as

$$\approx_{Down3;3\%} = \langle \mathsf{CFSt}:\ Down, 1, 4, 3, Yes, Rel, =, 3\rangle. \quad (17.11)$$

The parameter *Rel* means that this CF-quantifier defines a condition concerning histograms—heights of columns that correspond to the percentage of particular

Adult	Young	Middle-aged	Senior	Old	Σ
Never-married	n_1	n_2	n_3	n_4	n
True	m_1	m_2	m_3	m_4	m

CF-table $CF(\text{age\_exp}, Adult/Never\text{-}married)$

Adult	Young	Middle-aged	Senior	Old	Σ
Never-married ∧ Sunday	n_1'	n_2'	n_3'	n_4'	n'
True	m_1	m_2	m_3	m_4	m

CF-table $CF(\text{age\_exp}, Adult/(Never\text{-}married \wedge Sunday))$

Figure 17.16: Two CF-tables.

categories. The parameters "*Down*, 1, 4, 3" mean, that steps down in the range $\langle 1,4 \rangle$ with minimal height 3 per cent are required.

If we apply $\approx_{Down3;3\%}$ to the CF-tables $CF(\text{age\_exp}, Adult/Never\text{-}married)$ and $CF(\text{age\_exp}, Adult/(Never\text{-}married \wedge Sunday))$, we get conditions (17.12) and (17.13):

$$100\frac{n_1 - n_2}{n} \geq 3 \;\wedge\; 100\frac{n_2 - n_3}{n} \geq 3 \;\wedge\; 100\frac{n_3 - n_4}{n} \geq 3 \,, \qquad (17.12)$$

$$100\frac{n_1' - n_2'}{n'} \geq 3 \;\wedge\; 100\frac{n_2' - n_3'}{n'} \geq 3 \;\wedge\; 100\frac{n_3' - n_4'}{n'} - 100\frac{n_4'}{n'} \geq 3 \,. \quad (17.13)$$

We use also basic CF-quantifier $\approx_{Sum10000}$ defined in equation (17.14) and corresponding to conditions (17.15) and (17.16)

$$\approx_{Sum10000} \;=\; \langle \text{CFSF: } Sum, 1, 4, Abs, \geq, 10000 \rangle, \qquad (17.14)$$

$$n_1 + n_2 + n_3 + n_4 \geq 10000 \qquad (17.15)$$

$$n_1' + n_2' + n_3' + n_4' \geq 10000 \qquad (17.16)$$

for CF-tables introduced in Fig. 17.16. Finally, we define CF-quantifier $\approx_{Down3;3\%,10000}$ as

$$\approx_{Down3;3\%,10000} \;=\; \approx_{Down3;3\%} \wedge \approx_{Sum10000} \,. \qquad (17.17)$$

We will see in the next section that a CF-pattern

$$CF_{Nm} \;=\; \approx_{Down3;3\%,10000} \text{age\_exp}/Never\text{-}married$$

is true in the *Adult* data matrix. Assume that a CF-pattern

$$CF_{NmS} \;=\; \approx_{Down3;3\%,10000} \text{age\_exp}/(Never\text{-}married \wedge Sunday)$$

is also true in the enhanced *Adult* data matrix. A question is, if the CF-pattern CF_{NmS} can be considered as a consequence of the CF-pattern CF_{Nm}.

First, we show that the CF-pattern CF_{NmS} is not a logical consequence of CF_{Nm}. To show this, we need a data matrix $\mathcal{M}$ such that CF_{Nm} is true in $\mathcal{M}$ and CF_{NmS} is false in $\mathcal{M}$. The additional column born can have various values. We can assume that a value of the column born is Friday for each row. This implies that it holds $n_1' = n_2' = n_3' = n_4' = n' = 0$ for the CF-table $CF(\text{age\_exp}, Adult/(Never\text{-}married \wedge Sunday))$ for this extended *Adult*. However, this means that CF_{NmS} is false in this extended *Adult*. CF_{Nm} is true in the *Adult* data matrix and thus also in all extensions of *Adult*. We can conclude that CF_{NmS} is not a logical consequence of CF_{Nm}.

We can summarize that we have an extended data matrix *Adult* and that both CF_{Nm} and CF_{NmS} are true in *Adult*. We search for a reasonable argument that CF_{NmS} is a consequence of CF_{Nm}. We cannot use correct logical deduction. Our problem is not that $n_1' + n_2' + n_3' + n_4' < 10000$ (see (17.16)) can happen. We have a true pattern CF_{NmS} and this means that $n_1' + n_2' + n_3' + n_4' \geq 10000$. Our problem is that we need a good reason to believe that if (17.12), then also (17.13).

We consider as indisputable facts that it holds for each data matrix $\mathcal{M}$ with attributes born, marital and age_exp:

(CF1): A relative frequency of persons satisfying born(*Sunday*) in the whole data matrix is approximately $\frac{1}{7}$.

(CF2): A relative frequency of persons satisfying born(*Sunday*) in subsets of persons satisfying *Never-married* $\wedge$ age_exp(*Young*) is also approximately $\frac{1}{7}$. The same is true for *Never-married* $\wedge$ age_exp(*Middle-aged*), *Never-married* $\wedge$ age_exp(*Senior*), and *Never-married* $\wedge$ age_exp(*Old*).

Let us note that these indisputable facts are analogous to the indisputable facts (IF1) and (IF2) introduced in Section 17.1.1.

The indisputable fact (CF2) means that it holds:

$$n_1' \doteq \frac{1}{7} n_1, \quad n_2' \doteq \frac{1}{7} n_2, \quad n_3' \doteq \frac{1}{7} n_3, \quad n_4' \doteq \frac{1}{7} n_4 \quad n' \doteq \frac{1}{7} n, \tag{17.18}$$

which implies

$$\frac{n_1'}{n'} \doteq \frac{n_1}{n}, \quad \frac{n_2'}{n'} \doteq \frac{n_2}{n}, \quad \frac{n_3'}{n'} \doteq \frac{n_3}{n}, \quad \frac{n_4'}{n'} \doteq \frac{n_4}{n}. \tag{17.19}$$

If we consider bigger and bigger data extended matrices $\mathcal{M}$, then we can assume that $\frac{n_i'}{n'}$ is closer and closer to $\frac{n_i}{n}$ for $i = 1, \ldots, 4$. This can be formalized by the assertion

$$\lim_{|\mathcal{M}| \to \infty} \langle \frac{n_1}{n}, \frac{n_2}{n}, \frac{n_3}{n}, \frac{n_4}{n} \rangle = \lim_{|\mathcal{M}| \to \infty} \langle \frac{n_1'}{n'}, \frac{n_2'}{n'}, \frac{n_3'}{n'}, \frac{n_4'}{n'} \rangle, \tag{17.20}$$

where $|\mathcal{M}|$ is the number of rows of $\mathcal{M}$. The assertion (17.20) we consider as an indisputable fact following from indisputable fact (CF2). We will see in Chapter 18 that from indisputable fact (17.20) and an assumption

$$\lim_{|\mathcal{M}|\to\infty} 100\frac{n_1-n_2}{n} \geq 3 \wedge \lim_{|\mathcal{M}|\to\infty} 100\frac{n_2-n_3}{n} \geq 3 \wedge \lim_{|\mathcal{M}|\to\infty} 100\frac{n_3-n_4}{n} \geq 3$$

(17.21)

can be proved that it holds also

$$\lim_{|\mathcal{M}|\to\infty} 100\frac{n'_1-n'_2}{n'} \geq 3 \wedge \lim_{|\mathcal{M}|\to\infty} 100\frac{n'_2-n'_3}{n'} \geq 3 \wedge \lim_{|\mathcal{M}|\to\infty} 100\frac{n'_3-n'_4}{n'} \geq 3 .$$

(17.22)

This we understand as a good reason to consider the validity of (17.13) as a consequence of validity of (17.12).

These considerations lead to an expert deduction rule

$$Exp\frac{\approx_{Down3;3\%,10000} age\_exp/Never\text{-}married}{\approx_{Down3;3\%,10000} age\_exp/(Never\text{-}married \wedge Sunday)} .$$

(17.23)

We consider (17.23) as a correct expert deduction rule, if (17.20) can be understood as an indisputable fact. Then we can prove the validity of (17.22) from the assumption (17.21). This is a good reason to consider the validity of (17.13) as a consequence of the validity of (17.12). We use such correct expert deduction rule as an argument that the CF-pattern $\approx_{Down3;3\%,10000} age\_exp/(Never\text{-}married \wedge Sunday)$ is a consequence of the CF-pattern $\approx_{Down3;3\%,10000} age\_exp/Never\text{-}married$.

Expert deduction rules for CF-patterns are formally introduced in [139] together with their properties and examples of application. Available theory related to expert deduction rules for CF-patterns is also introduced in Chapter 18 together with challenges for further research. We show an example of the application of suitable expert deduction rules for CF-patterns in the next section.

17.3.2 Applying expert deduction rules to histograms

We use a task similar to that described in Section 8.2.2 to present possibilities of applications of the expert deduction rules introduced in the previous section. We use the CF-quantifier $\approx_{Down3;3\%,10000}$ further denoted as $\approx_E$ and defined in equation (17.17) instead of the $\approx_{Down3} \wedge \approx_{Sum5000}$ introduced in equation (8.10). A set of relevant conditions χ is defined in the same way as in Section 8.2.2. This means that relevant conditions are given by the $\mathcal{RC}$-expression $\mathcal{HCOND}$ introduced in Fig. 8.9.

Recall that we are interested in histograms $age\_exp/[\chi, Proc]$ at the data matrix *Adult* satisfying the condition given by the CF-quantifier $\approx_E$ and concerning the CF-table $CF(age\_exp, Adult/\chi)$ shown in Fig. 17.17. A histogram $age\_exp/[\chi, Proc]$ is a quadruple $\langle 100\frac{n_1}{n}, 100\frac{n_2}{n}, 100\frac{n_3}{n}, 100\frac{n_4}{n} \rangle$.

Adult	Young	Middle-aged	Senior	Old	Σ
χ	n_1	n_2	n_3	n_4	n
True	m_1	m_2	m_3	m_4	m

Figure 17.17: CF-table $CF(\text{age\_exp}, Adult/\chi$.

The condition given by the CF-quantifier $\approx_E$ is a conjunction of the condition $n_1 + n_2 + n_3 + n_4 \geq 10000$ given by the CF-quantifier $\approx_{Sum10000}$ (see equations (17.14) and (17.15)) and the condition

$$100\frac{n_1 - n_2}{n} \geq 3 \quad \wedge \quad 100\frac{n_2 - n_3}{n} \geq 3 \quad \wedge \quad 100\frac{n_3 - n_4}{n} \geq 3 \qquad (17.24)$$

given by the CF-quantifier $\approx_{Down3;3\%} = \langle \text{CFSt: } Down, 1, 4, 3, Yes, Rel, =, 3 \rangle$, see equation (17.11).

A run of the CF-Miner procedure with the above introduced parameters resulted in 31 CF-patterns $\approx_E$ age_exp/χ in less than 0.5 second. 8 162 CF-patterns were tested. The CF-pattern $\approx_E$ age_exp/*Never-married* is among the output patterns. There are 16 117 rows satisfying $\chi = Never\text{-}married$. A histogram age_exp/[*Never-married,Proc*] at a data matrix *Adult* is shown in Fig. 17.18.

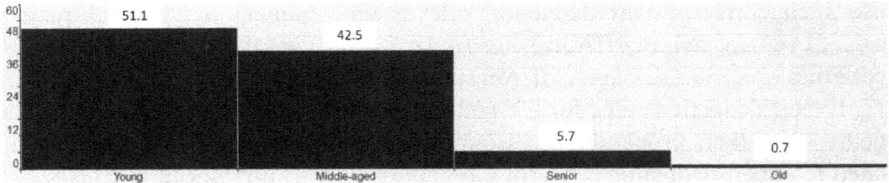

Figure 17.18: Histogram age_exp/[*Never-married, Proc*] at *Adult*.

All 30 remaining CF-patterns have a form *Never-married* $\wedge \kappa$ where κ is one of the following basic Boolean attributes: capital_loss(*None*), capital_gain(*None*), native_country(*United-States*), race(*White*), income(*Small*), workclass(*Private*), education(*11th,12th,HS-grad,Some-college*), education(*12th, HS-grad,Some-college,Assoc-voc*), education(*HS-grad,Some-college,Assoc-voc, Assoc-acdm*) or a conjunction of 2–4 of these basic Boolean attributes.

Histogram age_exp/[*Never-married, Proc*] and all the histograms age_exp/[*Never-married* $\wedge \kappa, Proc$] for κ – single basic Boolean attributes mentioned above are introduced in Table 17.2. We write "*United-States*" instead of "native_country(*United-States*)" only, etc.

We are interested in correct expert deduction rules

$$Exp\frac{\approx_E \text{ age\_exp}/Never\text{-}married}{\approx_E \text{ age\_exp}/(Never\text{-}married \wedge \kappa)} \,. \qquad (17.25)$$

Table 17.2: Histograms age_exp/[*Never-married* ∧ κ, *Proc*].

Adult	% of Σ				Σ
	Young	*Middle-aged*	*Senior*	*Old*	
Never-married	51.1	42.5	5.7	0.7	16 117
κ	*Young*	*Middle-aged*	*Senior*	*Old*	Σ
capital_loss(*None*)	51.5	42.2	5.6	0.7	15 629
capital_gain(*None*)	51.9	41.9	5.6	0.6	15 440
United-States	51.8	41.8	5.7	0.7	14 481
race(*White*)	53.2	40.5	5.5	0.7	13 218
workclass(*Private*)	52.3	42.2	4.9	0.6	12 243
11th,12th,HS-grad, Some-college	60.0	35.0	4.4	0.6	10 385
income(*Small*)	53.2	41.0	5.1	0.7	10 192
HS-grad,Some-college Assoc-voc,Assoc-acdm	55.4	39.5	4.6	0.5	10 185
12th,HS-grad, Some-college,Assoc-voc	57.0	37.9	4.6	0.5	10 011

Our goal is to use such correct expert deduction rules to filter out CF-patterns $\approx_E$ age_exp/(*Never-married* ∧ κ which can be considered as consequences of the simplest CF-pattern $\approx_E$ age_exp/*Never-married*.

Let us recall that the expert deduction rule (17.25) is considered correct if the assertion

$$\lim_{|\mathcal{M}|\to\infty} \langle \frac{n_1}{n}, \frac{n_2}{n}, \frac{n_3}{n}, \frac{n_4}{n} \rangle = \lim_{|\mathcal{M}|\to\infty} \langle \frac{n_1'}{n'}, \frac{n_2'}{n'}, \frac{n_3'}{n'}, \frac{n_4'}{n'} \rangle \qquad (17.26)$$

can be considered as an indisputable fact, see also assertion (17.20). Here n_1, n_2, n_3, n_4, n and n_1', n_2', n_3', n_4', and n' are frequencies defined in Table 17.3.

Table 17.3: Frequencies n_1, n_2, n_3, n_4, n and n_1', n_2', n_3', n_4', n'.

Adult	Young	Middle-aged	Senior	Old	Σ
marital(Never-married)	n_1	n_2	n_3	n_4	n
marital(Never-married) ∧κ	n_1'	n_2'	n_3'	n_4'	n'

According to values presented in Table 17.2, we will consider the validity of assertion (17.26) as an indisputable fact for κ equal to one of the basic Boolean attributes capital_loss(*None*), capital_gain(*None*), native_country(*United-States*) or to a conjunction of any two of them or to a conjunction to all of them. This way, we get six correct deduction rules, since the conjunction of all three basic Boolean attributes does not occur among output CF-patterns. Thus, this way we can decrease the number of output CF-patterns from 31 to 25.

Let us emphasize that this is only a simple example, we have chosen indisputable facts without necessary expertise. Note that we can search for further correct expert deduction rules to further decrease the number of output patterns. For example, we can consider expert deduction rules

$$Exp \frac{\approx_E \text{ age\_exp}/(\textit{Never-married} \wedge \text{race}(\textit{White}))}{\approx_E \text{ age\_exp}/(\textit{Never-married} \wedge \text{race}(\textit{White}) \wedge \kappa)} \ .$$

However, we will not do this due to lack of necessary expertise.

Chapter 18

Observational Calculi

Observational calculi and their role in mechanizing hypothesis formation are shortly introduced in Section 1.1.2. Observational calculi are logical calculi formulas which correspond to various assertions on analysed data. The patterns the described GUHA procedures deal with are examples of such assertions. Various results on observational calculi can be used when applying GUHA procedures.

Results introduced in Section 4.5 are used in the evaluation of all Boolean attributes in data with missing information. Results introduced in Sections 7.1.4–7.1.6 concern missing information and association rules. An example of their application is presented in Section 7.2.3. An example of the application of correct deduction rules is described in Section 7.3.4. Logical deduction is also used when applying domain knowledge in the interpretation of results of the procedures 4ft-Miner and CF-Miner. Incorrect deduction rules play a role in the definition of expert deduction rules concerning association rules and histograms, see Sections 17.2 and 17.3.

The first goal of this chapter is to give an overview of results related to missing information and deduction rules and relevant to the GUHA procedures described in this book. This is done in Section 18.1. The second goal is to summarize available results relevant to dealing with domain knowledge and expert deduction rules for association rules and histograms. This is done in Section 18.2 Relevant research challenges are also listed in both sections.

18.1   Definition and Overview of Results

The most of presented results concern GUHA association rules the 4ft-Miner procedure deals with. There are also some results concerning histograms the

CF-Miner procedure deals with. Let us emphasize that our goal is to present an overview of results, not details. All results are introduced in an exact mathematical way in the original publications. Most of the results are available in monographs [41] and [121]. Recent results are published in [123, 139]. Additional results and considerations relevant to observational calculi are available in [103, 104, 105, 106, 107, 108, 109, 110, 112, 113, 114, 115].

First, we present results relevant to association rules concerning GUHA association rules. Note that all these results are relevant also to "classical association rules" introduced in [2]. We present results relevant to dealing with missing information and to deduction rules. To do this, we need to introduce logical calculus of association rules and classes of association rules. This is done in an exact mathematical way in [121].

We will use a less formal way; however, still be enough precise to present relevant results in a suitable manner. Logical calculus of association rules is introduced in Section 18.1.1. Classes of association rules are described in Section 18.1.2. Results on missing information and deduction rules are presented in Sections 18.1.3 and 18.1.4 respectively. Logical calculus of histograms is introduced in Section 18.1.5. Several research challenges related to observational calculi are listed in Section 18.1.6.

18.1.1   *Logical calculus of association rules*

Observational calculi are defined in [41] in an exact mathematical way. Observational predicate calculus defined in Section II.2 in [41] is an example of observational calculi. Informally speaking, observational predicate calculus is a modification of classical predicate calculus – only finite models are allowed and generalized quantifiers are used instead of classical quantifiers $\forall$ and $\exists$. 4ft-quantifiers introduced in Section 7.1.2 are examples of generalized quantifiers.

The logical calculus of association rules can be seen as a special case of calculi with qualitative values introduced in Section III.4 in [41]. Association rules are understood as closed formulas of a suitable calculus with qualitative values. Calculus of association rules is defined in [121]. We use here a simplified definition. The simplification consists in using the same symbols $A_1, \ldots, A_K$ both as attributes—symbols of a language of association rules and as columns of data matrices in which the association rules are interpreted. This is used due to the definition of a data matrix introduced in Section 1.4.2 which is applied throughout the book. We believe that it is clear from a context, in what role the symbols $A_1, \ldots, A_K$ are used.

Note that the symbols $f_1, \ldots, f_K$ are used for columns of data matrices in [121]. The function f_i is then understood as an interpretation of an attribute A_i for $i = 1, \ldots, K$.

First, we define a type $\mathcal{T}$ of a data matrix and a set of data of data matrices of type $\mathcal{T}$. Then we introduce a language of a calculus of association rules, and finally we define a calculus of association rules.

Definition 18.1   Data matrices of type $\mathcal{T}$ are defined this way:

1. A *type of data matrices* is a K-tuple $\mathcal{T} = \langle t_1, \ldots, t_K \rangle$ where $K \geq 2$ is an integer and $t_i \geq 2$ are integers for $i = 1, \ldots, K$.

2. Let $\mathcal{T} = \langle t_1, \ldots, t_K \rangle$ be a type of data matrices. Then, a *data matrix of a type* $\mathcal{T}$ is a $K + 1$-tuple $\mathcal{M} = \langle M, A_1, \ldots, A_K \rangle$, where $M = \{o_1, \ldots, o_n\}$ is a non-empty finite set and A_i is a unary function from M to $\{1, \ldots, t_i\}$ for $i = 1, \ldots, K$.

3. Elements of M are called rows of $\mathcal{M}$ and a set M is a *set of rows* of data matrix $\mathcal{M}$. An expression $|\mathcal{M}|$ denotes the number of rows of $\mathcal{M}$. If $M = \{o_1, \ldots, o_n\}$, then $|\mathcal{M}| = n$.

4. A set of all data matrices of the type $\mathcal{T} = \langle t_1, \ldots, t_K \rangle$ is denoted as $\mathrm{M}_{\mathcal{T}}$.

Definition 18.2   A language $\mathcal{LBA}_{\mathcal{T}}$ of Boolean attributes of type $\mathcal{T} = \langle t_1, \ldots, t_K \rangle$ is given by attributes $A_1, \ldots, A_K$. We define:

1. A type of A_i is t_i.

2. If $\alpha \subsetneq \{1, \ldots t_i\}$, then $A(\alpha)$ is a basic Boolean attribute of $\mathcal{LBA}_{\mathcal{T}}$.

3. Each basic Boolean attribute of $\mathcal{LBA}_{\mathcal{T}}$ is a Boolean attribute of $\mathcal{LBA}_{\mathcal{T}}$.

4. The expression *True* is a Boolean attribute of $\mathcal{LBA}_{\mathcal{T}}$.

5. If φ and ψ are Boolean attributes of $\mathcal{LBA}_{\mathcal{T}}$, then $\neg\varphi$, $\varphi \wedge \psi$ and $\varphi \vee \psi$ are Boolean attributes $\mathcal{L}_{\mathcal{T}}$. Usual conventions on parentheses are valid.

6. If $\mathcal{M} = \langle M, A_1, \ldots, A_K \rangle$ is a data matrix of type $\mathcal{T}$, $o \in M$ and φ is a Boolean attribute of $\mathcal{LBA}_{\mathcal{T}}$, then a value $\varphi(o, \mathcal{M})$ of φ for a row o of $\mathcal{M}$ is given in a way described in Section 1.4.3.

7. If φ is a Boolean attribute of $\mathcal{LBA}_{\mathcal{T}}$, then $Atr(\varphi)$ denotes a set of all attributes A such that there is a basic Boolean attribute $A(\alpha)$ occurring in φ.

Definition 18.3   A logical calculus $\mathcal{LCAR}_{\mathcal{T}}$ of association rules of type $\mathcal{T} = \langle t_1, \ldots, t_K \rangle$ is given by:

■   A set $\mathrm{M}_{\mathcal{T}}$ of data matrices of the type $\mathcal{T} = \langle t_1, \ldots, t_K \rangle$.

■   A language $\mathcal{LBA}_{\mathcal{T}}$ of Boolean attributes of type $\mathcal{T} = \langle t_1, \ldots, t_K \rangle$ with attributes $A_1, \ldots, A_K$.

■ 4ft-quantifiers $\approx_1, \ldots, \approx_Q$; for each 4ft-quantifier $\approx$ its associated function $F_{\approx}$ is given. Each $F_{\approx}$ is a function mapping a set of all quadruples of $\langle a, b, c, d \rangle$ of non-negative integers into a set $\{0, 1\}$.

We define:

1. If φ and ψ are Boolean attributes of $\mathscr{LBA}_{\mathscr{T}}$ and if $\approx$ is a 4ft-quantifier, then $\varphi \approx \psi$ is an association rule of the logical calculus $\mathscr{LCAR}_{\mathscr{T}}$ of association rules.

2. If $\mathscr{M} \in M$ is a data matrix of the type $\mathscr{T} = \langle t_1, \ldots, t_K \rangle$ and $\varphi \approx \psi$ is an association rule, then a value $Val(\varphi \approx \psi, \mathscr{M})$ is defined as $F_{\approx}(a, b, c, d)$. Here, $\langle a, b, c, d \rangle = 4ft(\varphi, \psi, \mathscr{M})$ is a 4ft-table of φ and ψ in $\mathscr{M}$ introduced in Section 7.1.1.

3. If $Val(\varphi \approx \psi, \mathscr{M}) = 1$, then we say that $\varphi \approx \psi$ is true in $\mathscr{M}$. Otherwise we say that $\varphi \approx \psi$ is false in $\mathscr{M}$.

Let us note that we have introduced 4ft-quantifiers in Section 7.1.1. A 4ft-quantifier $\approx$ is used both as a symbol – a part of an association rule $\varphi \approx \psi$ and as a $\{0, 1\}$ - valued function defined for all 4ft-tables $\langle a, b, c, d \rangle$. In definition 18.3, we distinguish between a symbol and a related $\{0, 1\}$ - valued function A function related to a 4ft-quantifier $\approx$ is denoted as $F_{\approx}$ and called associated function of $\approx$, see point IV in the definition. This approach comes from [41] and it is used also in [121].

18.1.2 Classes of association rules

Properties of association rules $\varphi \approx \psi$ are determined by properties of a 4ft-quantifier $\approx$. This is the reason why classes of association rules are defined by classes of 4ft-quantifiers. Classes of 4ft-quantifiers are defined using TPC – truth preservation conditions [41, 121]. The following definition of TPC comes from [121].

Definition 18.4 A *truth preservation condition* (shortly TPC) is a $\{0, 1\}$-valued function $TPC(a, b, c, d, a', b', c', d')$ defined for all 8-tuples $\langle a, b, c, d, a', b', c', d' \rangle$ of non-negative integer numbers.

The *class of 4ft-quantifiers* $4ft[TPC]$ *defined by truth preservation condition* TPC is a set of all 4ft-quantifiers $\approx$ satisfying for each pair of 4ft-tables $\langle a, b, c, d \rangle$ and $\langle a', b', c', d' \rangle$ the condition

$$\text{if } F_{\approx}(a, b, c, d) = 1 \wedge TPC(a, b, c, d, a', b', c', d') = 1 \text{ then } F_{\approx}(a', b', c', d') = 1 \ .$$

More than ten classes of 4ft-quantifiers are defined and studied in [121]. We present definitions of six of them.

Definition 18.5   Classes of *implicational quantifiers, weakly implicational quanti-fiers, Σ-double implicational quantifiers, weakly Σ-double implicational quantifiers Σ-equivalence quantifiers, weakly Σ-equivalence quantifiers,* and *quantifiers with F-property* are defined by TPC introduced in Table 18.1.

Table 18.1: Classes of 4ft-quantifiers defined by TPC.

Class of	TPC	$TPC(a,b,c,d,a',b',c',d') = 1$ if and only if
implicational quantifiers	$TPC_{\Rightarrow}$	$a' \geq a \wedge b' \leq b$
weakly implicational quantifiers	$TPC_{\Rightarrow}^{\mathscr{M}}$	$a' \geq a \wedge b' \leq b \wedge$ $\wedge\, a+b+c+d = a'+b'+c'+d'$
Σ-double implicational quantifiers	$TPC_{\Sigma,\Leftrightarrow}$	$a' \geq a \wedge (b'+c') \leq (b+c)$
weakly Σ-double implicational quantifiers	$TPC_{\Sigma,\Leftrightarrow}^{\mathscr{M}}$	$a' \geq a \wedge (b'+c') \leq (b+c) \wedge$ $\wedge\, a+b+c+d = a'+b'+c'+d'$
Σ-equivalence quantifiers	$TPC_{\Sigma,\equiv}$	$(a'+d') \geq (a+d) \wedge$ $\wedge\, (b'+c') \leq (b+c)$
quantifiers with F-property	TPC_F	$a' = a \wedge d' = d \wedge X \vee Y$ $X: b \geq c-1 \wedge b' = b+1 \wedge c' = c-1$ $Y: b \geq c-1 \wedge b' = b-1 \wedge c' = c+1$

The association rule $\varphi \approx \psi$ belongs to a *class of implicational association rules* if a 4ft-quantifier $\approx$ belongs to a *class of implicational quantifiers*. We also say that the association rule $\varphi \approx \psi$ is an *implicational rule* and that the 4ft-quantifier $\approx$ is an *implicational quantifier*. This is the same for additional classes of association rules. We present several examples of 4ft-quantifiers belonging to the defined classes.

■ *Implicational quantifiers:*

  ■ 4ft-quantifier $\Rightarrow_{p,Base}$ introduced in Section 7.1.2 and defined by a condition $\frac{a}{a+b} \geq p \wedge a \geq Base$.

  ■ 4ft-quantifier $\rightarrow_{p,\alpha}^{!}$ of likely p-implication (i.e., upper critical im-plication) defined by a condition $\sum_{i=a}^{a+b} \binom{a+b}{i} p^i (1-p)^{a+b-i} \leq \alpha$. 4ft-quantifier $\rightarrow_{p,\alpha}^{!}$ is introduced in [41], see also [121]. It corre-sponds to the statistical test (on the level α) of the null hypothesis $H_0 : P(\psi|\varphi) \leq p$ against the alternative one $H_1 : P(\psi|\varphi) > p$. Here $P(\psi|\varphi)$ is the conditional probability of validity of ψ under the condition φ.

■ *Weakly implicational quantifiers:* 4ft-quantifier $\rightarrow_{p,s}$ introduced in Sec-tion 7.1.2 and defined by a condition $\frac{a}{a+b} \geq p \wedge \frac{a}{a+b+c+d} \geq s$.

- *Σ-double implicational quantifiers*: Jaccard 4ft-quantifier $\leftrightarrow_p$ defined by a condition $\frac{a}{a+b+c} \geq p \land a+b+c > 0$.

- *Weakly Σ-double implicational quantifiers*: 4ft-quantifier $\leftrightarrow_{p,s}$ defined by a condition $\frac{a}{a+b+c} \geq p \land \frac{a}{a+b+c+d} \geq s$.

- *Σ-equivalence quantifiers*: 4ft-quantifier p-equivalence $\equiv_p$ defined by a condition $\frac{a+d}{a+b+c+d} \geq q \land a+b+c+d > 0$.

- *Quantifiers with F-property*: 4ft-quantifier lift $\sim_q$ introduced in Section 7.1.2 and defined by a condition $\frac{a(a+b+c+d)}{(a+b)(a+c)} \geq q \land (a+b)(a+c) > 0$. Also 4ft-quantifiers corresponding to Fisher's test and χ^2 - test introduced in [41] are quantifiers with F-property.

Definitions of all these and additional 4ft-quantifiers, and proofs concerning their relations to particular classes, mutual relations among introduced classes of 4ft-quantifiers as well as definitions of additional classes of 4ft-quantifiers and description of their properties are available in [121].

18.1.3 Missing information in calculus of association rules

A specific approach to missing information called secured X-extension is developed in [41]. It is described in Section 4.5 and applied to Boolean attributes in all the GUHA procedures described in this book. This approach can be applied also to the evaluation of GUHA association rules, see Section 7.1.4. The application of the secured X-extension to GUHA association rules consists in the application of special 4ft-tables $\langle a_s, b_s, c_s, d_s \rangle$ called secured completions of nine-fold tables $9ft(\varphi, \psi, \mathcal{M}^X)$, for details see Section 7.1.5.

It is shown in Section 7.1.5, how the secured completions of nine-fold tables are computed for some of the 4ft-quantifiers. The goal of this section is to point out a relation of classes of association rules and a way of computing of secured completions of nine-fold tables.

It is shown in [121], that a secured completion $\langle a_s, b_s, c_s, d_s \rangle$ of nine-fold table a $9ft(\varphi, \psi, \mathcal{M}^X)$ depends on a class of 4ft-quantifiers in a way given in Table 18.2. We assume that $9ft(\varphi, \psi, \mathcal{M}^X)$ and $\langle a_s, b_s, c_s, d_s \rangle$ are in a form according to Fig. 18.1. Let us note that, except for quantifiers with F-property, there are more possibilities for how to compute $\langle a_s, b_s, c_s, d_s \rangle$. The secured completion $\langle a_s, b_s, c_s, d_s \rangle$ for implicational 4ft-quantifiers is given only by a_s and b_s. This is because a value of implicational 4ft-quantifiers depends only on frequencies a and b from 4ft-tables. For more details see [121].

$\mathcal{M}^X$	ψ	ψ_X	$\neg\psi$
φ	$f_{1,1}$	$f_{1,X}$	$f_{1,0}$
φ_X	$f_{X,1}$	$f_{X,X}$	$f_{X,0}$
$\neg\varphi$	$f_{0,1}$	$f_{0,X}$	$f_{0,0}$

$\mathcal{M}$	ψ	$\neg\psi$
φ	a_s	b_s
$\neg\varphi$	c_s	d_s

Figure 18.1: $9ft(\varphi,\psi,\mathcal{M}^X)$ and $4ft(\varphi,\psi,\mathcal{M})$ and $\langle a_s,b_s,c_s,d_s \rangle$.

Table 18.2: Secured completion $\langle a_s,b_s,c_s,d_s \rangle$ for classes of 4ft-quantifiers.

Class of	$\langle a_s,b_s,c_s,d_s \rangle$
implicational quantifiers	$\langle f_{1,1},\ f_{1,0}+f_{1,X}+f_{X,X}+f_{X,0},\ 0,\ 0 \rangle$
weakly implicational quantifiers	$\langle f_{1,1},\ f_{1,0}+f_{1,X}+f_{X,X}+f_{X,0},$ $f_{0,1}+f_{X,1}+f_{0,0},\ f_{0,0} \rangle$
Σ-double implicational quantifiers	$\langle f_{1,1},\ f_{1,0}+f_{1,X}+f_{X,X}+f_{X,0},$ $f_{0,1}+f_{X,1}+f_{0,X},\ 0 \rangle$
weakly Σ-double implicational quantifiers	$\langle f_{1,1},\ f_{1,0}+f_{1,X}+f_{X,X}+f_{X,0},$ $f_{0,1}+f_{X,1}+f_{0,X},\ f_{0,0} \rangle$
Σ-equivalence quantifiers quantifiers	$\langle f_{1,1},\ f_{1,0}+f_{1,X}+f_{X,X}+f_{X,0},$ $f_{0,1}+f_{X,1}+f_{0,X},\ f_{0,0} \rangle$
quantifiers with F-property	$\langle f_{1,1},\ f_{1,0}+f_{1,X}+f_{X,X,b}+f_{X,0},$ $f_{0,1}+f_{X,1}+f_{X,X,c}+f_{0,X},\ f_{0,0} \rangle;$ $f_{X,X,b},f_{X,X,c}$ chosen such that $|b_s-c_s|$ minimal

18.1.4 Deduction rules in calculus of association rules

We have shown in Section 7.3.4, that if the rule $\mathscr{A} \Rightarrow_{p,Base}$ income(*Large*) where

$\mathscr{A} =$ education(*Masters,Prof-school,Doctorate*) $\wedge$ marital(*Married-civ-spouse*)

is true in a data matrix $\mathcal{M}$, then also the rule

$$\mathscr{A} \Rightarrow_{p,Base} \text{gain\_positive}(\alpha) \vee \text{income}(Large)$$

is true in $\mathcal{M}$. This and analogous relations of two association rules are formalized by the following definition.

Definition 18.6 Let $\varphi \approx \psi$ and $\varphi' \approx \psi'$ are association rules of a logical calculus $\mathscr{LCAR}_{\mathscr{T}}$ of association rules of type $\mathscr{T} = \langle t_1,\ldots,t_K \rangle$. Then:

- A deduction rule $\frac{\varphi \approx \psi}{\varphi' \approx \psi'}$ is *sound* or *correct* if it holds: if $\mathcal{M}$ is a data matrix of type $\mathscr{T} = \langle t_1,\ldots,t_K \rangle$ and $Val(\varphi \approx \psi,\mathcal{M}) = 1$, then also $Val(\varphi' \approx \psi',\mathcal{M}) = 1$.

- A deduction rule $\frac{\varphi \approx \psi}{\varphi' \approx \psi'}$ is *not sound* or *not correct* if there is a data matrix $\mathcal{M}$ of type $\mathscr{T} = \langle t_1,\ldots,t_K \rangle$ such that $Val(\varphi \approx \psi,\mathcal{M}) = 1$ and $Val(\varphi' \approx \psi',\mathcal{M}) = 0$.

■ If a deduction rule $\frac{\varphi \approx \psi}{\varphi' \approx \psi'}$ is correct, then we also say that $\varphi' \approx \psi'$ *logically follows from* $\varphi \approx \psi$.

Correct deduction rules can be used to decrease the number of output rules, see Section 7.3.4. Incorrect deduction rules play a role in the definition of expert deduction rules, see Section 17.2. Thus it is natural to search for criteria making it possible to decide if a given deduction rule is correct or not. Various conditions related to the correctness of deduction rules $\frac{\varphi \approx \psi}{\varphi' \approx \psi'}$ are presented in [121]. There are both necessary and sufficient conditions of correctness of such deduction rules for some classes of association rules and also only sufficient conditions for additional classes.

A correctness of a deduction rule $\frac{\varphi \approx \psi}{\varphi' \approx \psi'}$ with implicational 4ft-quantifier $\approx$ depends on three propositional formulas $\Theta_1(\varphi, \psi, \varphi', \psi')$, $\Theta_2(\varphi, \psi, \varphi', \psi')$, and $\Theta_3(\varphi, \psi, \varphi', \psi')$ created from Boolean attributes $\varphi, \psi, \varphi', \psi'$; for details and a formal proof see [121]. The deduction rule $\frac{\varphi \approx \psi}{\varphi' \approx \psi'}$ is correct if and only if both $\Theta_1(\varphi, \psi, \varphi', \psi')$ and $\Theta_2(\varphi, \psi, \varphi', \psi')$ are tautologies or if $\Theta_3(\varphi, \psi, \varphi', \psi')$ is a tautology. The problem is that the propositional formulas $\Theta_1, \Theta_2, \Theta_3$ are very complex. Each category occurring in $\varphi, \psi, \varphi', \psi'$ corresponds to one propositional variable in Θ_1, Θ_2 and each category occurring in φ, ψ corresponds to one propositional variable in Θ_3. In addition, there is no similar criterion for weakly implicational 4ft-quantifiers. There is only a theorem saying that if both $\Theta_1(\varphi, \psi, \varphi', \psi')$ and $\Theta_2(\varphi, \psi, \varphi', \psi')$ are tautologies or if $\Theta_3(\varphi, \psi, \varphi', \psi')$ is a tautology, then a deduction rule $\frac{\varphi \approx \psi}{\varphi' \approx \psi'}$ with a weakly implicational 4ft-quantifier is correct.

Thus, we need a more effective criteria of correctness of implicational 4ft-quantifiers as well as at least some criteria equivalent to correctness for weakly implicational 4ft-quantifiers. There is a similar situation for additional 4ft-quantifiers.

Various results concerning the correctness of deduction rules $\frac{A(\alpha) \approx S(\sigma)}{\varphi \approx \psi}$ are presented in [123]. Here, $A(\alpha)$ and $S(\sigma)$ are basic Boolean attributes and $\varphi \approx \psi$ is a general association rule. Such deduction rules are used when dealing with items of domain knowledge in the way described in Section 17.2, see namely Section 17.2.4. Properties of these deduction rules depend on configuration of Boolean attributes $A(\alpha), S(\sigma)$, φ and ψ. Four types (incomplete, collapsed, direct, swapped) of deduction rules $\frac{A(\alpha) \approx S(\sigma)}{\varphi \approx \psi}$ and additional classes of 4ft-quantifiers (a-satisfiable, *db*-sensitive, *dc*-sensitive, *bc*-sensitive) are defined to study relevant configurations. The following theorem is an example of results achieved in [123].

Theorem 18.1
Let $A(\alpha) \approx S(\sigma)$ be an atomic association rule and let $\varphi \approx \psi$ be a relevant association rule. Then it holds: If $\approx$ is one of 4ft-quantifiers $\rightarrow_{p,s}$, $\Rightarrow_{p,Base}$, $\Rightarrow^!_{p,\alpha,Base}$ where

$0 < p \leq 1$, $0 < s \leq 1$, *Base* > 0, *and* $0 < \alpha \leq 0.5$, *then the only correct deduction rules* $\frac{A(\alpha)\approx S(\sigma)}{\varphi\approx\psi}$ *are* $\frac{A(\alpha)\approx S(\sigma)}{A(\alpha)\approx S(\sigma')}$ *and* $\frac{A(\alpha)\approx S(\sigma)}{A(\alpha)\approx S(\sigma')\vee\lambda_{Suc}}$ *where* $\sigma \subseteq \sigma'$.

Let us note that the 4ft-quantifiers $\rightarrow_{p,s}$, $\Rightarrow_{p,Base}$ are introduced in Section 7.1.2. The 4ft-quantifier $\Rightarrow^!_{p,\alpha,Base}$ is derived from the 4ft-quanitfier $\rightarrow^!_{p,\alpha}$ of likely p-implication (i.e., upper critical implication) introduced in Section 18.1.2. It is defined by the condition $\sum_{i=a}^{a+b}\binom{a+b}{i}p^i(1-p)^{a+b-i} \leq \alpha \wedge a \geq Base$. The notion of *relevant association rule* is introduced in [123]. All association rules $\varphi \approx \psi$ the 4ft-Miner procedure deals with are relevant association rules according to Theorem (18.1).

Let us emphasize that not all practically interesting questions concerning correctness of the deduction rules used when applying items of domain knowledge in data mining with association rules are solved. Important open related problems are listed in [123].

18.1.5   *Logical calculus of histograms*

A first attempt to define and study logical calculus of histograms is presented in [139]. The goal of this section is to introduce the definition of a calculus and the first available results concerning deduction rules. We use data matrices of type $\mathscr{T}$ and a language $\mathscr{LBA}_{\mathscr{T}}$ of Boolean attributes of type $\mathscr{T}$ introduced in Definitions 18.1 and 18.2. Recall that a histogram of an attribute A of type t with categories $1,\ldots,t$ is a K-tuple $\langle h_1,\ldots,h_t\rangle$, see Section 8.1.1.

The CF-Miner procedure deals with CF-patterns $\approx A/\chi$ where A is an attribute with t categories, χ is a Boolean attribute and $\approx$ is a CF-quantifier condition concerning all possible CF-tables $CF(A,\chi,\mathscr{M})$. Recall that a CF-table is a vector $\langle n_1,\ldots,n_t,m_1,\ldots,m_t\rangle$ of non-negative integers, see Section 4.2. A definition of a calculus of histograms follows.

Definition 18.7   A logical calculus $\mathscr{LCH}_{\mathscr{T}}$ of histograms of type $\mathscr{T} = \langle t_1,\ldots,t_K\rangle$ is given by:

- A set $\mathsf{M}_{\mathscr{T}}$ of data matrices of the type $\mathscr{T} = \langle t_1,\ldots,t_K\rangle$.

- A language $\mathscr{LBA}_{\mathscr{T}}$ of Boolean attributes of type $\mathscr{T} = \langle t_1,\ldots,t_K\rangle$ with attributes $A_1,\ldots,A_K$.

- CF-quantifiers $\approx_1,\ldots,\approx_Q$; for each CF-quantifier $\approx$ a set of relevant CF-tables $\langle n_1,\ldots,n_{t_i},m_1,\ldots,m_{t_i}\rangle$ is given together with its associated function $CF_{\approx}$. Each $CF_{\approx}$ is a function mapping a set of all CF-tables $\langle n_1,\ldots,n_{t_i},m_1,\ldots,m_{t_i}\rangle$ relevant to $\approx$ into a set $\{0,1\}$ for $i = 1,\ldots,Q$.

We define:

1. If A is an attribute, χ is a Boolean attribute, $\approx$ is a CF-quantifier of $\mathscr{L}_{\mathscr{T}}$ and a CF-table of A is relevant to $\approx$, then $\approx A/\chi$ is a CF-pattern of the logical calculus $\mathscr{LCH}_{\mathscr{T}}$ of histograms.

2. If $\mathscr{M} \in \mathsf{M}$ is a data matrix of the type $\mathscr{T} = \langle t_1, \ldots, t_K \rangle$ and $\approx A/\chi$ is a CF-pattern, then a value $Val(\approx A/\chi, \mathscr{M})$ is defined as $CF_{\approx}(n_1, \ldots, n_t, m_1, \ldots, m_t)$ where $\langle n_1, \ldots, n_t, m_1, \ldots, m_t \rangle$ is a CF-table $CF(A, \mathscr{M}/\chi)$.

3. If $Val(\approx A/\chi, \mathscr{M}) = 1$, then we say that $\approx A/\chi$ *is true in* $\mathscr{M}$. Otherwise we say that $\approx A/\chi$ *is false in* $\mathscr{M}$.

The CF-Miner procedure deals with simple frequencies CF-quantifiers $\langle\text{CFSF: } Sum, u, v, Abs, \circledast, Thr\rangle$ introduced in Section 8.1.4 and CF-quantifiers $\langle\text{CFSt: } Dir, u, v, MinH, Cons, TypeHist, \circledast, Thr\rangle$ concerning steps in histogram introduced in Section 8.1.5. Recall that $\langle u, v \rangle$ is a range of CF-quantifier, see Section 8.1.3. Let us note that we must be careful when using the range of CF-quantifiers $\langle u, v \rangle$. If a range $\langle u, v \rangle$ is used together with an attribute A of type t, then it must hold $1 \leq u < v \leq t$. (This can be formally solved by definition of particular CF-quantifiers for each t_i from $\mathscr{T} = \langle t_1, \ldots, t_K \rangle$.)

We need the following notions to present first available results on deduction rules in a calculus of histograms. Note that a *partial t-cedent* is a partial cedent in the sense of Section 4.4.4. However, *t-cedent* is generalisation of *cedent* which is only a conjunction (and not a disjunction) of partial cedents. Note that it holds for each attribute A of type t: $\neg A(\alpha)(o, \mathscr{M}) = A(\{1, \ldots, t\} \setminus \alpha)(o, \mathscr{M})$ for each row of each data matrix $\mathscr{M}$.

Definition 18.8 Let $\mathscr{LBA}_{\mathscr{T}}$ be a language of Boolean attributes of type $\mathscr{T} = \langle t_1, \ldots, t_K \rangle$ with attributes $A_1, \ldots, A_K$ We define:

1. A *partial t-cedent for the* $\mathscr{LBA}_{\mathscr{T}}$ *language* is each conjunction $A'_1(\alpha_1) \wedge \cdots \wedge A'_u(\alpha_u)$ and each disjunction $A'_1(\alpha_1) \vee \cdots \vee A'_u(\alpha_u)$ of basic Boolean attributes $A'_1(\alpha_1), \ldots, A'_u(\alpha_u)$ such that $u \geq 1$ and $A'_1, \ldots, A'_u$ are mutually distinct attributes belonging to the set $\{A_1, \ldots A_K\}$ of attributes.

2. A *t-cedent for the* $\mathscr{LBA}_{\mathscr{T}}$ *language* is each conjunction $\omega_1 \wedge \cdots \wedge \omega_v$ and each disjunction $\omega_1 \vee \cdots \vee \omega_v$ of partial t-cedents $\omega_1, \ldots, \omega_v$ such that $v \geq 1$ and all attributes occurring in $\omega_1, \ldots, \omega_v$ are mutually distinct.

Let us assume that we deal with an attribute A of type t and that we have a CF-table $CF(A, \chi, \mathscr{M}) = \langle n_1, \ldots, n_t, m_1, \ldots, m_t \rangle$. Then a histogram $Hist(CF(A, \chi, \mathscr{M}), Rel)$ is a vector $\langle 100 * \frac{n_1}{n}, \ldots, 100 * \frac{n_t}{n} \rangle$ where $n = n_1 + \cdots + n_t$, see Section 8.1.1.

A step CF-quantifier $\langle$CFSt: $Down, u, v, h, Yes, Rel, =, v - u + 1\rangle$ introduced in Section 8.1.5 means that the histogram $\langle 100 * \frac{n_1}{n}, \ldots, 100 * \frac{n_t}{n}\rangle$ decreases between the u-th and v-th categories with height of steps at least h. This can be expressed by the condition

$$100\frac{m_u - m_{u+1}}{m} \geq h \wedge 100\frac{m_{u+1} - m_{u+2}}{m} \geq h \wedge \cdots \wedge 100\frac{m_{v-1} - m_v}{m} \geq h \;\cdot$$

Let us denote a CF-quantifier $\langle$CFSt: $Down, u, v, h, Yes, Rel, =, v - u + 1\rangle$ by $\approx_{Dn,u,v,h}$. Then the following theorem is true; its proof is available in [139].

Theorem 18.2
Let $\mathscr{LCH}_{\mathscr{T}}$ of be a logical calculus of histograms of type $\mathscr{T} = \langle t_1, \ldots, t_K\rangle$ with CF-quantifier $\approx_{Dn,u,v,h}$ such that $v \geq u + 2$ and $h \leq \frac{100}{q}$ where $q = \frac{(v-u)(v-u+1)}{2}$. Let A be an attribute of type t such that $v \leq t$. Let χ and κ be t-cedents such that it holds $\{A\} \cap Atr(\chi) \cap Atr(\kappa) = \emptyset$. Then a deduction rule $\frac{\approx_{Dn,u,v,h}A/\chi}{\approx_{Dn,u,v,h}A/(\chi \wedge \kappa)}$ is not correct.

18.1.6 Research challenges to observational calculi

We have defined a logical calculus of association rules and presented results concerning deduction rules and dealing with missing information. The results are relevant to data mining with association rules. It is introduced in Section 18.1.4 that not all useful deduction rules relevant to association rules are known.

We have also defined a logical calculus of histograms and presented the first result concerning deduction rules in this calculus. Logical deduction is also used when applying domain knowledge in the interpretation of results of procedures 4ft-Miner and CF-Miner. Incorrect deduction rules play a role in the definition of expert deduction rules concerning both association rules and histograms, see Sections 17.2 and 17.3.

This situation can be understood as a challenge to research in the following directions:

- research for additional useful deduction rules $\frac{\varphi \approx \psi}{\varphi' \approx \psi'}$ concerning association rules

- research for additional useful deduction rules $\frac{\approx A/\chi}{\approx A/\chi'}$ in a calculus of histograms

- definition and study of logical calculi formulas which correspond to patterns the procedures KL-Miner, SD4ft-Miner, SDCF-Miner, SDKL-Miner, and Ac4ft-Miner deal with.

18.2   Expert Deduction Rules

We have presented informal considerations and applications of expert deduction rules in mining with association rules and histograms in Sections 17.1 and 17.3.

The first goal of this section is to present a formal definition of expert deduction rules for mining with association rules in a context of a logical calculus of association rule introduced in Section 18.1.1. Presentation of a summary of achieved theoretical results on expert deduction rules for association rules is also our goal. Introductory informal considerations on expert deduction rules and association rules are presented in Section 18.2.1. Definition of expert deduction rules for association rules is introduced in Section 18.2.2. Achieved theoretical results are summarized in Section 18.2.3.

Let us emphasize that the research of expert deduction rules is only at the beginning. This concerns both the association rules and histograms. The definition of expert deduction rules for histograms and first experience and theoretical results are available in [139]. These results can be presented analogously as for association rules. However, we will not do this since the expert deduction rules for histograms sufficiently presented in Section 17.3. The achieved results are promising. However, there are still many open problems and research challenges. They are listed in Section 18.2.4.

18.2.1   Informally on expert deduction and association rules

The idea of expert deduction rules for association rules comes from [124] and it is informally introduced in Section 17.1.1. We summarize the informal introduction in several points, and then we will discuss a question of expressing this idea by tools of observational calculus of association rules introduced in Section 18.1.1.

Let us have a data matrix $\mathcal{M}_0$ with the same attributes as the *Adult* data matrix with one additional attribute – column born with categories *Sunday, Monday, . . . , Saturday* meaning a day of the week when a person was born. Let us consider association rules

$$\text{age}(\textit{Young}) \rightarrow_{p,Base} \text{income}(\textit{Small}) \tag{18.1}$$

and

$$\text{age}(\textit{Young}) \wedge \text{born}(\textit{Sunday}) \rightarrow_{p,Base} \text{income}(\textit{Small}) \tag{18.2}$$

and let us assume that both rules are true in $\mathcal{M}$. It is natural to understand the rule (18.2) as a consequence of the (18.1) since we believe that to be born on Sunday does not influence a chance to have a small income. If we consider all data matrices $\mathcal{M}$ with the same columns as $\mathcal{M}_0$, then the fact that to be born on Sunday does not influence a chance to have a small income can be expressed by the assertion

$$\text{If} \quad \lim_{|\mathcal{M}| \to \infty} \frac{a_1}{a_1 + b_1} = L, \quad \text{then} \quad \lim_{|\mathcal{M}| \to \infty} \frac{a_2}{a_2 + b_2} = L, \tag{18.3}$$

which is the same as equation (17.4). Here $4ft(\text{age}(Young), \text{income}(Small), \mathscr{M}) = \langle a_1, b_1, c_1, d_1 \rangle$ and $4ft(\text{age}(Young) \wedge \text{born}(Sunday), \text{income}(Small), \mathscr{M}) = \langle a_2, b_2, c_2, d_2 \rangle$, see Fig. 18.2 (which is a copy of Fig. 17.3).

$\mathscr{M}$	income(*Small*)	¬ income(*Small*)
age(*Young*)	a_1	b_1
¬ age(*Young*)	c_1	d_1

$4ft(\text{age}(Young), \text{income}(Small), \mathscr{M})$ of rule (18.1)

$\mathscr{M}$	income(*Small*)	¬ income(*Small*)
age(*Young*) ∧ born(*Sunday*)	a_2	b_2
¬ (age(*Young*) ∧ born(*Sunday*))	c_2	d_2

$4ft(\text{age}(Young) \wedge \text{born}(Sunday), \text{income}(Small), \mathscr{M})$ of rule (18.2)

Figure 18.2: 4ft-tables of rules (18.1) and (18.2) in data matrix $\mathscr{M}$.

Recall 4ft-table in Fig. 17.2 in Section 17.1.1 which shows, that a deduction rule

$$\frac{\text{age}(Young) \rightarrow_{p,Base} \text{income}(Small)}{\text{age}(Young) \wedge \text{born}(Sunday) \rightarrow_{p,Base} \text{income}(Small)} \qquad (18.4)$$

is not correct. Therefore, this deduction rule cannot serve as an argument that the rule (18.2) is a consequence of the rule (18.1).

Our idea is to use an expert deduction rule

$$Exp\frac{\text{age}(Young) \rightarrow_{p,Base} \text{income}(Small)}{\text{age}(Young) \wedge \text{born}(Sunday) \rightarrow_{p,Base} \text{income}(Small)} . \qquad (18.5)$$

We consider (18.5) as a correct expert deduction rule, if we can prove validity of (18.3) from suitably formalized facts (IF1) and (IF2) concerning each data matrix $\mathscr{M}$:

(IF1): A relative frequency of persons satisfying born(*Sunday*) in the whole data matrix $\mathscr{M}$ is approximately $\frac{1}{7}$.

(IF2): Relative frequencies of persons satisfying born(*Sunday*) in subsets of persons satisfying age(*Young*) ∧ income(*Small*) and age(*Young*) ∧¬ income(*Small*) are approximately $\frac{1}{7}$.

We consider these facts as indisputable facts supported by a domain expert.

The indisputable facts (IF1) and (IF2) can be formalized in a natural way. In Table 18.3, there are all possible combinations of values of Boolean attributes age(*Young*), income(*Small*), and born(*Sunday*) in a data matrix $\mathscr{M}$. The first row corresponds to rows of $\mathscr{M}$ for which all these Boolean attributes are true. We assume that there are f_{111} such rows in the data matrix $\mathscr{M}$. The remaining rows of Table 18.3 are analogous.

Table 18.3: Frequencies of age(*Young*), income(*Small*), and born(*Sunday*).

age(*Young*)	income(*Small*)	born(*Sunday*)	# of rows
1	1	1	f_{111}
1	1	0	f_{110}
1	0	1	f_{101}
1	0	0	f_{100}
0	1	1	f_{011}
0	1	0	f_{010}
0	0	1	f_{001}
0	0	0	f_{000}

We assume that if we consider bigger and bigger data matrices $\mathscr{M}$, then all the mentioned relative frequencies of rows satisfying Born(*Sunday*) are getting closer and closer to $\frac{1}{7}$. The assertion (IF1) can be formalized by

$$\lim_{|\mathscr{M}| \to \infty} \frac{f_{111} + f_{101} + f_{011} + f_{001}}{|\mathscr{M}|} = \frac{1}{7} \qquad (18.6)$$

and the assertion (IF2) can be formalized by (18.7)

$$\lim_{|\mathscr{M}| \to \infty} \langle \frac{f_{111}}{f_{111} + f_{110}}, \frac{f_{101}}{f_{101} + f_{100}} \rangle = \langle \frac{1}{7}, \frac{1}{7} \rangle \qquad (18.7)$$

which is a concise form of $\lim_{|\mathscr{M}| \to \infty} \frac{f_{111}}{f_{111} + f_{110}} = \frac{1}{7}$ and $\lim_{|\mathscr{M}| \to \infty} \frac{f_{101}}{f_{101} + f_{100}} = \frac{1}{7}$. The assertion (18.3) can be proved as true if we assume that assertion (18.7) is true, see [124] and Section 18.2.3.

We can summarize that we start with assertion (IF2) which we consider as an item of expert domain knowledge. This expert domain knowledge is transformed into an expert correct deduction rule by the presented approach called *KRLR—Knowledge Representation by Limits of frequencies Ratios*. The transformation is done in three steps:

1. The item (IF2) of expert domain knowledge is formalised by the assertion (18.7): $\lim_{|\mathscr{M}| \to \infty} \langle \frac{f_{111}}{f_{111} + f_{110}}, \frac{f_{101}}{f_{101} + f_{100}} \rangle = \langle \frac{1}{7}, \frac{1}{7} \rangle$ which is an assertion on a logical calculus of association rules.

2. The assertion (18.3): "If $\lim_{|\mathscr{M}| \to \infty} \frac{a_1}{a_1 + b_1} = L$, then $\lim_{|\mathscr{M}| \to \infty} \frac{a_2}{a_2 + b_2} = L$" is proved as a consequence of the assertion (18.7). Note that both (18.3) and (18.7) are assertions talking on a logical calculus of association rules.

3. The assertion (18.3) is understood as a formalisation of the correct expert deduction rule (18.5): $Exp \frac{\text{age}(Young) \to_{p,Base} \text{income}(Small)}{\text{age}(Young) \wedge \text{born}(Sunday) \to_{p,Base} \text{income}(Small)}$ (and of many others correct deduction rules).

Note that two conditions must be satisfied by the expert deduction rule $Exp\dfrac{\text{age}(Young)\rightarrow_{p,Base}\text{income}(Small)}{\text{age}(Young)\wedge\text{born}(Sunday)\rightarrow_{p,Base}\text{income}(Small)}$ to be a correct expert deduction rule:

- A deduction rule $\dfrac{\text{age}(Young)\rightarrow_{p,Base}\text{income}(Small)}{\text{age}(Young)\wedge\text{born}(Sunday)\rightarrow_{p,Base}\text{income}(Small)}$ must be incorrect.

- The assertion (18.3) must be in a mathematical sense a consequent of the assertion (18.7).

The three steps of the *KRLR* approach are shown also in Fig. 18.3. A formal definition of expert deduction rules is presented in the following section.

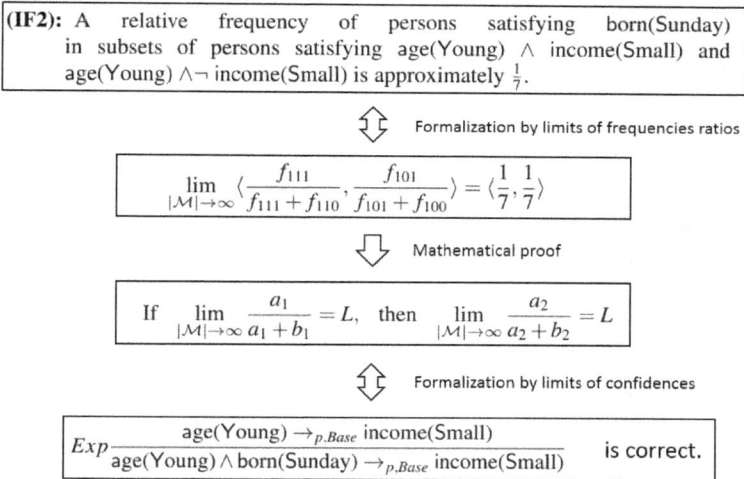

$$\boxed{\begin{array}{l}\textbf{(IF2):}\ \text{A\quad relative\quad frequency\quad of\quad persons\quad satisfying\quad born(Sunday)}\\ \text{in subsets of persons satisfying age(Young) $\wedge$ income(Small) and}\\ \text{age(Young) $\wedge\neg$ income(Small) is approximately $\frac{1}{7}$.}\end{array}}$$

⇕ Formalization by limits of frequencies ratios

$$\boxed{\lim_{|\mathcal{M}|\to\infty}\left\langle\frac{f_{111}}{f_{111}+f_{110}},\frac{f_{101}}{f_{101}+f_{100}}\right\rangle=\left\langle\frac{1}{7},\frac{1}{7}\right\rangle}$$

⇓ Mathematical proof

$$\boxed{\text{If}\ \lim_{|\mathcal{M}|\to\infty}\frac{a_1}{a_1+b_1}=L,\ \text{then}\ \lim_{|\mathcal{M}|\to\infty}\frac{a_2}{a_2+b_2}=L}$$

⇕ Formalization by limits of confidences

$$\boxed{Exp\dfrac{\text{age}(Young)\rightarrow_{p,Base}\text{income}(Small)}{\text{age}(Young)\wedge\text{born}(Sunday)\rightarrow_{p,Base}\text{income}(Small)}}\ \text{is correct.}$$

Figure 18.3: Three steps of the *KRLR* approach.

18.2.2 Expert deduction rules for association rules

We used expert deduction rules of the form

$$Exp\dfrac{\varphi\rightarrow_{p,s}\psi}{\varphi\wedge\text{Capital\_loss}(None)\rightarrow_{p,s}\psi}$$

in Section 17.1.2. In this section we deal with generalizations of such expert deduction rules. We are interested in expert deduction rules $Exp\dfrac{\varphi\approx\psi}{\varphi\wedge\chi\approx\psi}$, $Exp\dfrac{\varphi\approx\psi}{\varphi\approx\psi\wedge\chi}$, $Exp\dfrac{\varphi\approx\psi}{\varphi\vee\chi\approx\psi}$, $Exp\dfrac{\varphi\approx\psi}{\varphi\approx\psi\vee\chi}$. A formal definition of a general expert deduction rule in a logical calculus of association rules follows.

Definition 18.9 Let $\mathscr{LCAR}$ be a logical calculus of association rules. Let $\kappa \approx \lambda$ and $\varphi \approx \psi$ be association rules of $\mathscr{LCAR}$ and let $\frac{\kappa \approx \lambda}{\varphi \approx \psi}$ be an incorrect deduction rule in $\mathscr{LCAR}$. Then $Exp\frac{\kappa \approx \lambda}{\varphi \approx \psi}$ is an *expert deduction rule in* $\mathscr{LCAR}$.

Recall the indisputable fact (IF2):
Relative frequencies of persons satisfying born(Sunday) *in subsets of persons satisfying* age(Young) $\wedge$ income(Small) *and* age(Young) $\wedge \neg$ income(Small) *are approximately* $\frac{1}{7}$.
This is formalized by the assertion (18.7): $\lim_{|\mathscr{M}| \to \infty} \langle \frac{f_{111}}{f_{111}+f_{110}}, \frac{f_{101}}{f_{101}+f_{100}} \rangle = \langle \frac{1}{7}, \frac{1}{7} \rangle$.
The indisputable fact (IF2) is generalized for Boolean attributes φ, ψ and χ of a logical calculus $\mathscr{LCAR}$ of association rules by an indisputable fact:
Relative frequencies of rows satisfying χ *in subsets of rows satisfying* $\varphi \wedge \psi$ *and* $\varphi \wedge \neg\psi$ *are approximately* f.
This fact is denoted as $\mathscr{LCAR}(Impl : \chi, \varphi, \psi; f)$ and formalized by an assertion on a logical calculus $\mathscr{LCAR}$, see Definition 18.10.

Definition 18.10 Let $\mathscr{LCAR}$ be a logical calculus of association rules. Let φ, ψ, and χ be Boolean attributes of $\mathscr{LCAR}$. Let $\mathscr{M}$ be a data matrix and let combinations of values of φ, ψ, and χ and their frequencies be according to Table 18.4. The first row of the left part of the table corresponds to rows of $\mathscr{M}$ for which all these Boolean attributes are true. We assume that there are f_{111} such rows. Additional rows in Table 18.4 are analogous.

Table 18.4: Combinations of values of Boolean attributes φ, ψ and χ in a matrix $\mathscr{M}$.

φ	ψ	χ	# of rows	φ	ψ	χ	# of rows
1	1	1	f_{111}	0	1	1	f_{011}
1	1	0	f_{110}	0	1	0	f_{010}
1	0	1	f_{101}	0	0	1	f_{001}
1	0	0	f_{100}	0	0	0	f_{000}

Let f be a real number satisfying $0 < f < 1$. We say that *a relative frequency of* χ *is I-uniform for* φ, ψ, *and* f *in the logical calculus* $\mathscr{LCAR}$ if it holds

$$\lim_{|\mathscr{M}| \to \infty} \langle \frac{f_{111}}{f_{111}+f_{110}}, \frac{f_{101}}{f_{101}+f_{100}} \rangle = \langle f, f \rangle . \tag{18.8}$$

Then we write $\mathscr{LCAR}(Impl : \chi, \varphi, \psi; f)$.

Let us emphasize that the assertion 18.8 is not mathematically provable or even provable in the calculus $\mathscr{LCAR}$. It is an expert opinion about calculus $\mathscr{LCAR}$. However, it can be used to prove additional assertions about calculus $\mathscr{LCAR}$. Correctness and incorrectness of expert deduction rules are formalized by Definition 18.11 inspired by considerations presented in Section 18.2.1.

Definition 18.11   Let $Exp\frac{\varphi\approx\psi}{\varphi'\approx\psi'}$ be one of the expert deduction rules $Exp\frac{\varphi\approx\psi}{\varphi\wedge\chi\approx\psi}$, $Exp\frac{\varphi\approx\psi}{\varphi\approx\psi\wedge\chi}$, $Exp\frac{\varphi\approx\psi}{\varphi\vee\chi\approx\psi}$, and $Exp\frac{\varphi\approx\psi}{\varphi\approx\psi\vee\chi}$ in a logical calculus of association rules $\mathcal{LCAR}$. Let $\mathcal{M}$ be a data matrix belonging to $\mathcal{LCAR}$. In addition, let $4ft(\varphi,\psi,\mathcal{M}) = \langle a,b,c,d \rangle$ and $4ft(\varphi',\psi',\mathcal{M}) = \langle a_1,b_1,c_1,d_1 \rangle$ be 4ft-tables according to Fig. 18.4.

$\mathcal{M}$	ψ	$\neg\psi$		$\mathcal{M}$	ψ'	$\neg\psi'$
φ	a	b		φ'	a_1	b_1
$\neg\varphi$	c	d		$\neg\varphi'$	c_1	d_1

Figure 18.4: 4ft-tables $4ft(\varphi,\psi,\mathcal{M})$ and $4ft(\varphi',\psi',\mathcal{M})$.

Let there are 4ft-quantifiers $\approx_M$ and $\approx_s$ such that $F_\approx = F_{\approx_M} * F_{\approx_s}$ and $F_{\approx_M}(a,b,c,d) = 1$ if and only if $MI(a,b) \geq p$ where $MI(a,b)$ is a real-value function defined for all integers $a > 0$, $b \geq 0$ and p is a real number. Let $\mathcal{LC}(Impl : \chi,\varphi,\psi;f)$, see Definition 18.10. Then the expert deduction rule $Exp\frac{\varphi\approx\psi}{\varphi'\approx\psi'}$:

- *is correct according to* $\mathcal{LC}(Impl : \chi,\varphi,\psi;f)$ *and* $\approx_M$ *if there is a real L such that* $\lim_{|\mathcal{M}|\to\infty} MI_\approx(a,b) = L$ *implies* $\lim_{|\mathcal{M}|\to\infty} MI_\approx(a_1,b_1) = L$.

- *is incorrect according to* $\mathcal{LC}(Impl : \chi,\varphi,\psi;f)$ *and* $\approx_M$ *if there is a real L such that* $\lim_{|\mathcal{M}|\to\infty} MI_\approx(a,b) = L$ *and it does not hold* $\lim_{|\mathcal{M}|\to\infty} MI_\approx(a_1,b_1) = L$.

Remark 18.1   Let us give three comments to Definition 18.11:

- The function $MI(a,b)$ introduced in Definition 18.11 is of course inspired by the confidence measure of interest of association rules defined by the fraction $\frac{a}{a+b}$, see Table 7.1 in Section 7.1.2.

- The 4ft-quantifiers $\approx$, $\approx_M$ and $\approx_s$ satisfying $F_\approx = F_{\approx_M} * F_{\approx_s}$ are inspired by the 4ft-quantifiers $\to_p$, $\odot_s$, and $\oplus_{Base}$ introduced in Table 7.1 in Section 7.1.2 and by the 4ft-quantifiers $\Rightarrow_{p,Base}$ and $\to_{p,s}$ introduced at the end of Section 7.1.2.

- It holds $F_{\Rightarrow_{p,Base}} = F_{\to_p} * F_{\oplus_{Base}}$ and $F_{\to_{p,s}} = F_{\to_p} * F_{\odot_s}$. Note that in Table 7.1 we write $\approx(a,b,c,d)$ instead of $F_\approx(a,b,c,d)$. Here we distinguish $F_\approx$ from $\approx$.

■

18.2.3   Results on expert deduction rules for association rules

The goal of this section is to present examples of useful results on the correctness of expert deduction rules. We use expert deduction rules $Exp\frac{\varphi\to_{p,s}\psi}{\varphi\wedge\chi\to_{p,s}\psi}$ and

$Exp \frac{\varphi \to_{p,s} \psi}{\varphi \to_{p,s} \psi \wedge \chi}$ for a 4ft-quantifier $\to_{p,s}$ of confidence-support defined by the condition $\frac{a}{a+b} \geq p \wedge \frac{a}{a+b+c+d} \geq s$ for $0 < p \leq 1$ and $0 < s \leq 1$. We assume that φ, ψ, χ are sound Boolean attributes according to Definition 18.12.

Definition 18.12   Let φ be a Boolean attribute of a calculus $\mathscr{LCAR}$ of association rules. We say that φ is a *sound Boolean attribute* if there is a data matrix $\mathscr{M}$ of $\mathscr{LCAR}$ and rows o_1, o_2 of $\mathscr{M}$ such that it holds $\varphi(o_1, \mathscr{M}) = 1$ and $\varphi(o_2, \mathscr{M}) = 0$.

First, we have to ensure that $Exp \frac{\varphi \to_{p,s} \psi}{\varphi \wedge \chi \to_{p,s} \psi}$ and $Exp \frac{\varphi \to_{p,s} \psi}{\varphi \to_{p,s} \psi \wedge \chi}$ are expert deduction rules according to Definition 18.9. This means to show that $\frac{\varphi \to_{p,s} \psi}{\varphi \wedge \chi \to_{p,s} \psi}$ and $\frac{\varphi \to_{p,s} \psi}{\varphi \to_{p,s} \psi \wedge \chi}$ are not correct deduction rules in a calculus of association rules. This is formulated by Theorem 18.3, which is proved as Lemma 1 in [124]. Examples of useful results on correctness of expert deduction rules are presented in Theorem 18.4, which is proved as Theorem 1 in [124].

Theorem 18.3
Let $\mathscr{LCAR}$ be a logical calculus of association rules. Let φ, ψ and χ be sound Boolean attributes of $\mathscr{LC}$ such that such that $Atr(\varphi) \cap Atr(\psi) \cap Atr(\chi) = \emptyset$. Then $Exp \frac{\varphi \to_{p,s} \psi}{\varphi \wedge \chi \to_{p,s} \psi}$ and $Exp \frac{\varphi \to_{p,s} \psi}{\varphi \to_{p,s} \psi \wedge \chi}$ where $0 < p \leq 1$ and $0 < s \leq 1$ are expert deduction rules.

Theorem 18.4
Let φ, ψ and χ be sound Boolean attributes of a logical calculus $\mathscr{LCAR}$ of association rules. Let $0 < f < 1$ and let $\to_{p,s}$ be a 4ft-quantifier of confidence-support defined by the condition $\frac{a}{a+b} \geq p \wedge \frac{a}{a+b+c+d} \geq s$ for $0 < p \leq 1$ and $0 < s \leq 1$. Then it holds that:

1. *An expert deduction rule $Exp \frac{\varphi \to_{p,s} \psi}{\varphi \wedge \chi \to_{p,s} \psi}$ is correct according to $\mathscr{LC}(Impl : \chi, \varphi, \psi; f)$ and $\to_p$.*

2. *An expert deduction rule $Exp \frac{\varphi \to_{p,s} \psi}{\varphi \to_{p,s} \psi \wedge \chi}$ is incorrect according to $\mathscr{LC}(Impl : \chi, \varphi, \psi; f)$ and $\to_p$.*

18.2.4   Open problems and challenges

There is promising experience with applications of domain knowledge in data mining with association rules and with histograms, see Chapter 17 and [122, 124, 139]. This led to the formal definition of expert deduction rules and to the study of their theoretical properties as well as the study of possibilities of applications of correct deduction rules in dealing with items of domain knowledge, see Sections 18.2.1–18.2.3 and [122, 124, 139, 121, 123].

However, this has brought new open problems and challenges. We briefly outline three of them related both to Sections 18.2.1–18.2.3 and 17.3 and also to results presented in [139].

CH1 Correctness and incorrectness of expert deduction rules are formalized by definition 18.11 which is inspired by the 4ft-quantifiers $\rightarrow_{p,s}$ and $\Rightarrow_{p,Base}$, see also Remark 18.1. A question is if additional 4ft-quantifiers can be used for definition of useful correct expert deduction rules. 4ft-quantifiers $\sim_q$ of lift, $\equiv_p$ of p-equivalence and Jaccard 4ft-quantifier $\leftrightarrow_p$ introduced in Section 18.1.2 are examples of 4ft-quantifiers interesting from this point of view.

CH2 We are interested in expert deduction rules $Exp\frac{\varphi\approx\psi}{\varphi\wedge\chi\approx\psi}$, $Exp\frac{\varphi\approx\psi}{\varphi\approx\psi\wedge\chi}$, $Exp\frac{\varphi\approx\psi}{\varphi\vee\chi\approx\psi}$, $Exp\frac{\varphi\approx\psi}{\varphi\approx\psi\vee\chi}$, see Section 18.2.2. Only expert deduction rules $Exp\frac{\varphi\rightarrow_{p,s}\psi}{\varphi\wedge\chi\rightarrow_{p,s}\psi}$ and $Exp\frac{\varphi\rightarrow_{p,s}\psi}{\varphi\rightarrow_{p,s}\psi\wedge\chi}$ were studied, see Theorem 18.4. Study of properties of expert deduction rules $Exp\frac{\varphi\approx\psi}{\varphi\vee\chi\approx\psi}$ and $Exp\frac{\varphi\approx\psi}{\varphi\approx\psi\vee\chi}$ is thus still a challenge as well as a study of expert deduction rules $Exp\frac{\varphi\approx\psi}{\varphi\wedge\chi\approx\psi}$ and $Exp\frac{\varphi\approx\psi}{\varphi\approx\psi\wedge\chi}$ for 4ft-quantifiers different from $\rightarrow_{p,s}$.

CH3 The definition of expert deduction rules for histograms introduced in Section 17.3.1 is tailored to considerations presented in Section 17.3. This definition can be generalized in various ways and relevant practically important expert deduction rules have to be investigated.

References

[1] Charu C. Aggarwal. *Data Mining - The Textbook*. Springer, 2015.

[2] Rakesh Agrawal, Tomasz Imielinski, and Arun N. Swami. Mining association rules between sets of items in large databases. In *Proceedings of the 1993 ACM SIGMOD International Conference on Management of Data, Washington, DC, USA, May 26-28, 1993*, pages 207–216, 1993.

[3] Martin Atzmueller. Subgroup discovery. *WIREs Data Mining and Knowledge Discovery*, 5(1):35–49, 2015.

[4] Stefan Behnel, Robert Bradshaw, Craig Citro, Lisandro Dalcin, Dag Sverre Seljebotn, and Kurt Smith. Cython: The best of both worlds. *Computing in Science & Engineering*, 13(2):31–39, 2011.

[5] Petr Berka. Practical aspects of data mining using LISp-Miner. *Comput. Informatics*, 35(3):528–554, 2016.

[6] Petr Berka and Jan Rauch. Meta-learning for post-processing of association rules. In Torben Bach Pedersen, Mukesh K. Mohania, and A Min Tjoa, editors, *Data Warehousing and Knowledge Discovery, 12th International Conference, DAWAK 2010, Bilbao, Spain, August/September 2010. Proceedings*, volume 6263 of *Lecture Notes in Computer Science*, pages 251–262. Springer, 2010.

[7] Petr Berka and Jan Rauch. Mining and post-processing of association rules in the atherosclerosis risk domain. In Sami Khuri, Lenka Lhotská, and Nadia Pisanti, editors, *Information Technology in Bio- and Medical Informatics, ITBAM 2010, First International Conference, Bilbao, Spain, September 1-2, 2010. Proceedings*, volume 6266 of *Lecture Notes in Computer Science*, pages 110–117. Springer, 2010.

[8] Petr Berka, Jan Rauch, and Marie Tomečková. Lessons learned from the ECML/PKDD Discovery Challenge on the atherosclerosis risk factors data. *Comput. Informatics*, 26(3):329–344, 2007.

[9] Mario Boley, Bryan R. Goldsmith, Luca M. Ghiringhelli, and Jilles Vreeken. Identifying consistent statements about numerical data with dispersion-corrected subgroup discovery. *CoRR*, abs/1701.07696, 2017.

[10] Pete Chapman, Julian Clinton, Randy Kerber, Thomas Khabaza, Thomas Reinartz, Colin Shearer, and Rudiger Wirth. Crisp-dm 1.0 step-by-step data mining guide. Technical report, The CRISP-DM consortium, August 2000.

[11] David Chudán. *Association rule mining as a support for OLAP*. PhD thesis, Prague University of Economics and Business, 2015.

[12] Alonzo Church. *Introduction to mathematical logic, Volume I*. Princeton University Press, 1956.

[13] Agnieszka Dardzinska. *Action Rules Mining*, volume 468 of *Studies in Computational Intelligence*. Springer, 2013.

[14] Dheeru Dua and Casey Graff. UCI machine learning repository, 2017.

[15] A. W. F. Edwards. The measure of association in a 2 × 2 table. *Journal of the Royal Statistical Society. Series A (General)*, 126(1):109–114, 1963.

[16] Patrik Eklund, Johan Karlsson, Jan Rauch, and Milan Šimůnek. On the logic of medical decision support. In Harrie C. M. de Swart, Ewa Orlowska, Gunther Schmidt, and Marc Roubens, editors, *Theory and Applications of Relational Structures as Knowledge Instruments II, International Workshops of COST Action 274, TARSKI, 2002-2005, Selected Revised Papers*, volume 4342 of *Lecture Notes in Computer Science*, pages 50–59. Springer, 2006.

[17] T. S. Ferguson. *Probability and Mathematical Statistics*. Academic Press, New York, 1967.

[18] Johannes Fürnkranz and Tomáš Kliegr. A brief overview of rule learning. In Nick Bassiliades, Georg Gottlob, Fariba Sadri, Adrian Paschke, and Dumitru Roman, editors, *Rule Technologies: Foundations, Tools, and Applications - 9th International Symposium, RuleML 2015, Berlin, Germany, August 2-5, 2015, Proceedings*, volume 9202 of *Lecture Notes in Computer Science*, pages 54–69. Springer, 2015.

[19] Michael Hahsler, Sudheer Chelluboina, Kurt Hornik, and Christian Buchta. The arules R-package ecosystem: Analyzing interesting patterns

from large transaction data sets. *Journal of Machine Learning Research*, 12:2021–2025, 2011.

[20] Michael Hahsler, Bettina Gruen, Kurt Hornik, and Christian Buchta. Mining association rules and frequent itemsets. R package version 1.3-1.

[21] Petr Hájek. The problem of a general conception of the GUHA method. *Kybernetika*, 4(6):505–515, 1968.

[22] Petr Hájek. Automatic listing of important observational statements. I. *Kybernetika*, 9(3):187–205, 1973.

[23] Petr Hájek. Automatic listing of important observational statements. II. *Kybernetika*, 9(4):251–271, 1973.

[24] Petr Hájek. Some logical problems of automated research. In *Mathematical Foundations of Computer Science: Proceedings of Symposium and Summer School, Štrbské Pleso, High Tatras, Czechoslovakia, September 3-8, 1973*, pages 85–93. Mathematical Institute of the Slovak Academy of Sciences, 1973.

[25] Petr Hájek. Automatic listing of important observational statements. III. *Kybernetika*, 10(2):95–124, 1974.

[26] Petr Hájek. Observationsfunktorenkalküle und die logik der automatisierten forschung. *J. Inf. Process. Cybern.*, 12(4/5):181–186, 1976.

[27] Petr Hájek. Decision problems of some statistically motivated monadic modal calculi. *International Journal of Man-Machine Studies*, 15(3):351–358, 1981.

[28] Petr Hájek. Applying artificial intelligence to data analysis. In *5th European Conference on Artificial Intelligence, ECAI 1982, Paris, 1982, Proceedings*, pages 149–150, 1982.

[29] Petr Hájek. The new version of the GUHA procedure ASSOC. In *COMPSTAT 1984, Proceedings in Computational Statistics*, pages 360–365, 1984.

[30] Petr Hájek. Relations in GUHA style data mining. In Harrie C. M. de Swart, editor, *Relational Methods in Computer Science, 6th International Conference, RelMICS 2001, and 1st Workshop of COST Action 274 TARSKI Oisterwijk, The Netherlands, October 16-21, 2001, Revised Papers*, volume 2561 of *Lecture Notes in Computer Science*, pages 81–87. Springer, 2001.

[31] Petr Hájek. On generalized quantifiers, finite sets and data mining. In Mieczyslaw A. Klopotek, Slawomir T. Wierzchon, and Krzysztof Trojanowski, editors, *Intelligent Information Processing and Web Mining, Proceedings of the International IIS: IIPWM'03 Conference held in Zakopane, Poland, June 2-5, 2003*, Advances in Soft Computing, pages 489–496. Springer, 2003.

[32] Petr Hájek. Relations and GUHA-style data mining II. In Rudolf Berghammer, Bernhard Möller, and Georg Struth, editors, *Relational and Kleene-Algebraic Methods in Computer Science: 7th International Seminar on Relational Methods in Computer Science and 2nd International Workshop on Applications of Kleene Algebra, Bad Malente, Germany, May 12-17, 2003, Revised Selected Papers*, volume 3051 of *Lecture Notes in Computer Science*, pages 163–170. Springer, 2003.

[33] Petr Hájek. Logics for data mining. In Oded Maimon and Lior Rokach, editors, *Data Mining and Knowledge Discovery Handbook, 2nd ed*, pages 541–551. Springer, 2010.

[34] Petr Hájek, Kamila Bendová, and Zdenek Renc. The GUHA method and the three-valued logic. *Kybernetika*, 7(6):421–435, 1971.

[35] Petr Hájek, Marie Hájková, Tomáš Havránek, and Milan Daniel. The expert system shell EQUANT-PC: brief information. *Kybernetika*, 25(7):4–9, 1989.

[36] Petr Hájek, Ivan Havel, and Metoděj Chytil. The GUHA method of automatic hypotheses determination. *Computing*, 1(4):293–308, 1966.

[37] Petr Hájek, Ivan Havel, and Metoděj K. Chytil. GUHA - the method of systematical hypotheses searching. *Kybernetika*, 2(1):31–47, 1966.

[38] Petr Hájek, Ivan Havel, and Metoděj K. Chytil. The GUHA method of systematical hypotheses searching. II. *Kybernetika*, 3(5):430–437, 1967.

[39] Petr Hájek and Tomáš Havránek. On generation of inductive hypotheses. *International Journal of Man-Machine Studies*, 9(4):415–438, 1977.

[40] Petr Hájek and Tomáš Havránek. The GUHA method – its aims and techniques (twenty-four questions and answers). *International Journal of Man-Machine Studies*, 10(1):3–22, 1978.

[41] Petr Hájek and Tomáš Havránek. *Mechanising Hypothesis Formation - Mathematical Foundations for a General Theory*. Springer, 1978.

[42] Petr Hájek and Martin Holeňa. Formal logics of discovery and hypothesis formation by machine. *Theor. Comput. Sci.*, 292(2):345–357, 2003.

[43] Petr Hájek, Martin Holeňa, and Jan Rauch. The GUHA method and foundations of (relational) data mining. In Harrie C. M. de Swart, Ewa Orlowska, Gunther Schmidt, and Marc Roubens, editors, *Theory and Applications of Relational Structures as Knowledge Instruments, COST Action 274, TARSKI, Revised Papers*, volume 2929 of *Lecture Notes in Computer Science*, pages 17–37. Springer, 2003.

[44] Petr Hájek, Martin Holeňa, and Jan Rauch. The GUHA method and its meaning for data mining. *J. Comput. Syst. Sci.*, 76(1):34–48, 2010.

[45] Petr Hájek and Jiří Ivánek. Artificial intelligence and data analysis. In H. Caussinus, P. Ettinger, and R. Tomassone, editors, *COMPSTAT 1982 5th Symposium held at Toulouse 1982*, pages 54–60, Heidelberg, 1982. Physica-Verlag HD.

[46] Petr Hájek, Jan Rauch, David Coufal, and Thomas Feglar. The GUHA method, data preprocessing and mining. In Rosa Meo, Pier Luca Lanzi, and Mika Klemettinen, editors, *Database Support for Data Mining Applications: Discovering Knowledge with Inductive Queries*, volume 2682 of *Lecture Notes in Computer Science*, pages 135–153. Springer, 2004.

[47] Petr Hájek, Anna Sochorová, and Jana Zvárová. GUHA for personal computers. *Computational Statistics & Data Analysis*, 19(2):149–153, 1995.

[48] Jaroslava Hálová, Oldřich Štrouf, Přemysl Žák, Anna Sochorová, Noritaka Uchida, Hiroshi Okimoto, Tomoaki Yuzuri, Kazuhisa Sakakibara, and Minoru Hirota. Computer aided hypothesis based drug discovery using CATALYST RTM and PC GUHA software systems (A case study of catechol analogs against malignant melanoma). In Setsuo Arikawa and Hiroshi Motoda, editors, *Discovery Science, First International Conference, DS '98, Fukuoka, Japan, December 14-16, 1998, Proceedings*, volume 1532 of *Lecture Notes in Computer Science*, pages 447–448. Springer, 1998.

[49] Mohamed Said Hamani, Ramdane Maamri, Yacine Kissoum, and Maamar Sedrati. Unexpected rules using a conceptual distance based on fuzzy ontology. *Journal of King Saud University - Computer and Information Sciences*, 26(1):99 – 109, 2014.

[50] Jiawei Han, Jian Pei, and Yiwen Yin. Mining frequent patterns without candidate generation. *SIGMOD Rec.*, 29(2):1–12, May 2000.

[51] Tomáš Havránek. The statistical interpretation and modification of GUHA method. *Kybernetika*, 7(1):13–21, 1971.

[52] Tomáš Havránek. The computation of characteristic vectors of logical-probabilistic expressions. *Kybernetika*, 10(2):80–94, 1974.

[53] Tomáš Havránek. The approximation problem in computational statistics. In Jiří Bečvář, editor, *Mathematical Foundations of Computer Science 1975, 4th Symposium, Mariánské Lázně, Czechoslovakia, September 1-5, 1975, Proceedings*, volume 32 of *Lecture Notes in Computer Science*, pages 258–265. Springer, 1975.

[54] Tomáš Havránek. Statistics and computability. *Kybernetika*, 12(5):303–315, 1976.

[55] Tomáš Havránek. Enumeration calculi and rank methods. *International Journal of Man-Machine Studies*, 10(1):59–65, 1978.

[56] Tomáš Havránek. Statistics of multidimensional contingency tables and the guha method. *International Journal of Man-Machine Studies*, 10(1):87–93, 1978.

[57] Tomáš Havránek. An alternative approach to missing information in the GUHA method. *Kybernetika*, 16(2):145–155, 1980.

[58] Tomáš Havránek. Formal systems for mechanized statistical inference. *International Journal of Man-Machine Studies*, 15(3):333–350, 1981.

[59] Tomáš Havránek. The GUHA method in the context of data analysis. *International Journal of Man-Machine Studies*, 15(3):265–282, 1981.

[60] Tomáš Havránek. The present state of the GUHA software. *International Journal of Man-Machine Studies*, 15(3):253–264, 1981.

[61] Tomáš Havránek, Miloš Chyba, and Dan Pokorný. Processing sociological data by the GUHA method - an example. *International Journal of Man-Machine Studies*, 9(4):439–447, 1977.

[62] Tomáš Havránek and Dan Pokorný. GUHA-style processing of mixed data. *International Journal of Man-Machine Studies*, 10(1):47–57, 1978.

[63] Tomáš Havránek and Jiří Vosáhlo. A GUHA procedure with correlational quantifiers. *International Journal of Man-Machine Studies*, 10(1):67–74, 1978.

[64] Francisco Herrera, Cristóbal J. Carmona, Pedro González, and María José del Jesús. An overview on subgroup discovery: foundations and applications. *Knowl. Inf. Syst.*, 29(3):495–525, 2011.

[65] Martin Holeňa. Fuzzy hypotheses for GUHA implications. *Fuzzy Sets Syst.*, 98(1):101–125, 1998.

[66] Martin Holeňa. Observational logic integrates data mining based on statistics and neural networks. In Djamel A. Zighed, Henryk Jan Komorowski, and Jan M. Zytkow, editors, *Principles of Data Mining and Knowledge Discovery, 4th European Conference, PKDD 2000, Lyon, France, September 13-16, 2000, Proceedings*, volume 1910 of *Lecture Notes in Computer Science*, pages 440–445. Springer, 2000.

[67] Martin Holeňa. Extraction of logical rules from data by means of piecewise-linear neural networks. In Steffen Lange, Ken Satoh, and Carl H. Smith, editors, *Discovery Science, 5th International Conference, DS 2002, Lübeck, Germany, November 24-26, 2002, Proceedings*, volume 2534 of *Lecture Notes in Computer Science*, pages 192–205. Springer, 2002.

[68] Martin Holeňa, Lukáš Bajer, and Martin Scavnicky. Using copulas in data mining based on the observational calculus. *IEEE Trans. Knowl. Data Eng.*, 27(10):2851–2864, 2015.

[69] Cindi Howson. *Successful business intelligence: unlock the value of BI & big data*. McGraw-Hill Education, 2014.

[70] Roberto Ierusalimschy, Luiz Henrique de Figueiredo, and Waldemar Celes Filho. Lua an extensible extension language. *Software: Practice and Experience*, 26(6):635–652, 1996.

[71] Jiří Ivánek. Some examples of transforming ordinal data to an input for GUHA-procedures. *International Journal of Man-Machine Studies*, 15(3):309–318, 1981.

[72] Jiří Ivánek. On the correspondence between classes of implicational and equivalence quantifiers. In Jan M. Zytkow and Jan Rauch, editors, *Principles of Data Mining and Knowledge Discovery, Third European Conference, PKDD '99, Prague, Czech Republic, September 15-18, 1999, Proceedings*, volume 1704 of *Lecture Notes in Computer Science*, pages 116–124. Springer, 1999.

[73] Jiří Ivánek. Construction of implicational quantifiers from fuzzy implications. *Fuzzy Sets Syst.*, 151(2):381–391, 2005.

[74] Jiří Ivánek. Combining implicational quantifiers for equivalence ones by fuzzy connectives. *Int. J. Intell. Syst.*, 21(3):325–334, 2006.

[75] Petr Jirků and Tomáš Havránek. On verbosity levels in cognitive problem solvers. In *Proceedings of the 9th International Conference on Computational Linguistics, COLING '82, Prague, Czechoslovakia, July 5-10, 1982*, pages 142–145. ACADEMIA, Publishing House of the Czechoslovak Academy of Sciences, 1982.

[76] Tomáš Karban, Jan Rauch, and Milan Šimůnek. SDS-rules and association rules. In Hisham Haddad, Andrea Omicini, Roger L. Wainwright, and Lorie M. Liebrock, editors, *Proceedings of the 2004 ACM Symposium on Applied Computing (SAC), Nicosia, Cyprus, March 14-17, 2004*, pages 520–524. ACM, 2004.

[77] Ralph Kimball and Margy Ross. *The Data Warehouse Toolkit: The Definitive Guide to Dimensional Modeling, 3rd Edition*, volume 3. Wiley, 2013.

[78] Stephen Cole Kleene. *Introduction to Metamathematics*. Van Nostrand, 1952.

[79] Tomáš Kliegr, David Chudán, Andrej Hazucha, and Jan Rauch. SEWEBAR-CMS: A system for postprocessing data mining models. In Monica Palmirani, M. Omair Shafiq, Enrico Francesconi, and Fabio Vitali, editors, *Proceedings of the RuleML-2010 Challenge, at the 4th International Web Rule Symposium, Washington, DC, USA, October, 21-23, 2010*, volume 649 of *CEUR Workshop Proceedings*. CEUR-WS.org, 2010.

[80] Tomáš Kliegr, Martin Ralbovský, Vojtěch Svátek, Milan Šimůnek, Vojtěch Jirkovský, Jan Nemrava, and Jan Zemánek. Semantic analytical reports: A framework for post-processing data mining results. In Jan Rauch, Zbigniew W. Ras, Petr Berka, and Tapio Elomaa, editors, *Foundations of Intelligent Systems, 18th International Symposium, ISMIS 2009, Prague, Czech Republic, September 14-17, 2009. Proceedings*, volume 5722 of *Lecture Notes in Computer Science*, pages 88–98. Springer, 2009.

[81] Tomáš Kliegr and Jan Rauch. An XML format for association rule models based on the GUHA method. In Mike Dean, John Hall, Antonino Rotolo, and Said Tabet, editors, *Semantic Web Rules - International Symposium, RuleML 2010, Washington, DC, USA, October 21-23, 2010. Proceedings*, volume 6403 of *Lecture Notes in Computer Science*, pages 273–288. Springer, 2010.

[82] Tomáš Kliegr, Vojtěch Svátek, Martin Ralbovský, and Milan Šimůnek. SEWEBAR-CMS: semantic analytical report authoring for data mining results. *J. Intell. Inf. Syst.*, 37(3):371–395, 2011.

[83] Douglas B. Lenat. AM: An artificial intelligence approach to discovery in mathematics as heuristic search. Technical Report STAN-CS-76-570, Stanford Artificial Intelligence Laboratory, July 1976.

[84] Douglas B. Lenat. Computers and thought lecture: The ubiquity of discovery. In *IJCAI*, 1977.

[85] Václav Lín, Petr Dolejší, Jan Rauch, and Milan Šimůnek. The KL-Miner procedure for datamining. *Neural Network World*, 14(5):411–420, 2004.

[86] Bing Liu, Wynne Hsu, Lai-Fun Mun, and Hing-Yan Lee. Finding interesting patterns using user expectations. *IEEE Transactions on Knowledge and Data Engineering*, 11(6):817–832, Nov 1999.

[87] Viktor Nekvapil. Data mining with trusted knowledge. In Maria Ganzha, Leszek A. Maciaszek, and Marcin Paprzycki, editors, *Communication Papers of the 2017 Federated Conference on Computer Science and Information Systems, FedCSIS 2017, Prague, Czech Republic, September 3-6, 2017*, volume 13 of *Annals of Computer Science and Information Systems*, pages 9–16, 2017.

[88] Balaji Padmanabhan. A belief-driven method for discovering unexpected patterns. In *Proceedings of the Fourth International Conference on Knowledge Discovery and Data Mining*, KDD'98, pages 94–100. AAAI Press, 1998.

[89] Petr Máša. *Cleverminer: Beyond apriori*. Petr Máša.

[90] G. D. Plotkin. A further note on inductive generalization. In *Machine Intelligence 6*, pages 101–124. American Elsevier, 1971.

[91] Dan Pokorný. The GUHA method and desk calculators. *International Journal of Man-Machine Studies*, 10(1):75–86, 1978.

[92] Jaroslav Pokorný and Jan Rauch. The GUHA-DBS data base system. *International Journal of Man-Machine Studies*, 15(3):289–298, 1981.

[93] Laurel Powell, Anna Gelich, and Zbigniew W. Ras. The construction of action rules to raise artwork prices. In Denis Helic, Gerhard Leitner, Martin Stettinger, Alexander Felfernig, and Zbigniew W. Ras, editors, *Foundations of Intelligent Systems - 25th International Symposium, ISMIS 2020, Graz, Austria, September 23-25, 2020, Proceedings*, volume 12117 of *Lecture Notes in Computer Science*, pages 11–20. Springer, 2020.

[94] Pavel Pudlák. Polynomially complete problems in the logic of automated discovery. In Jiří Bečvář, editor, *Mathematical Foundations of Computer Science 1975, 4th Symposium, Mariánské Lázně, Czechoslovakia, September 1-5, 1975, Proceedings*, volume 32 of *Lecture Notes in Computer Science*, pages 358–361. Springer, 1975.

[95] Pavel Pudlák and Frederick N. Springsteel. Complexity in mechanized hypothesis formation. *Theor. Comput. Sci.*, 8:203–225, 1979.

[96] Python Core Team. *Python: A dynamic, open source programming language*. Python Software Foundation, 2021.

[97] R Core Team. *R: A Language and Environment for Statistical Computing*. R Foundation for Statistical Computing, Vienna, Austria, 2021.

[98] Michael O. Rabin. Theoretical impediments to artificial intelligence. In Jack L. Rosenfeld, editor, *Information Processing, Proceedings of the 6th IFIP Congress 1974, Stockholm, Sweden, August 5-10, 1974*, pages 615–619. North-Holland, 1974.

[99] Martin Ralbovský and Tomáš Kuchař. Using disjunctions in association mining. In Petra Perner, editor, *Advances in Data Mining. Theoretical Aspects and Applications, 7th Industrial Conference, ICDM 2007, Leipzig, Germany, July 14-18, 2007, Proceedings*, volume 4597 of *Lecture Notes in Computer Science*, pages 339–351. Springer, 2007.

[100] Jan Rauch. Ein Beitrag zu der GUHA Methode in der dreiwertigen Logik. *Kybernetika*, 11(2):101–113, 1975.

[101] Jan Rauch. Some remarks on computer realizations of GUHA procedures. *International Journal of Man-Machine Studies*, 10(1):23 – 28, 1978.

[102] Jan Rauch. Main problems and further possibilities of the computer realization of GUHA procedures. *International Journal of Man-Machine Studies*, 15(3):283–287, 1981.

[103] Jan Rauch. Logical calculi for knowledge discovery in databases. In Henryk Jan Komorowski and Jan M. Zytkow, editors, *Principles of Data Mining and Knowledge Discovery, First European Symposium, PKDD '97, Trondheim, Norway, June 24-27, 1997, Proceedings*, volume 1263 of *Lecture Notes in Computer Science*, pages 47–57. Springer, 1997.

[104] Jan Rauch. Classes of four-fold table quantifiers. In Jan M. Zytkow and Mohamed Quafafou, editors, *Principles of Data Mining and Knowledge Discovery, Second European Symposium, PKDD '98, Nantes, France, September 23-26, 1998, Proceedings*, volume 1510 of *Lecture Notes in Computer Science*, pages 203–211. Springer, 1998.

[105] Jan Rauch. Four-fold table calculi for discovery science. In Setsuo Arikawa and Hiroshi Motoda, editors, *Discovery Science, First International Conference, DS '98, Fukuoka, Japan, December 14-16, 1998, Proceedings*, volume 1532 of *Lecture Notes in Computer Science*, pages 405–406. Springer, 1998.

[106] Jan Rauch. Deduction in logic of association rules. In P. S. Thiagarajan and Roland H. C. Yap, editors, *Advances in Computing Science - ASIAN'99, 5th Asian Computing Science Conference, Phuket, Thailand, December 10-12, 1999, Proceedings*, volume 1742 of *Lecture Notes in Computer Science*, pages 386–387. Springer, 1999.

[107] Jan Rauch. Logic of association rules. *Appl. Intell.*, 22(1):9–28, 2005.

[108] Jan Rauch. Definability of association rules in predicate calculus. In Tsau Young Lin, Setsuo Ohsuga, Churn-Jung Liau, and Xiaohua Hu, editors, *Foundations and Novel Approaches in Data Mining*, volume 9 of *Studies in Computational Intelligence*, pages 23–40. Springer, 2006.

[109] Jan Rauch. Many sorted observational calculi for multi-relational data mining. In *Workshops Proceedings of the 6th IEEE International Conference on Data Mining (ICDM 2006), 18-22 December 2006, Hong Kong, China*, pages 417–422. IEEE Computer Society, 2006.

[110] Jan Rauch. Observational calculi, classes of association rules and F-property. In *2007 IEEE International Conference on Granular Computing, GrC 2007, San Jose, California, USA, 2-4 November 2007*, pages 287–293. IEEE Computer Society, 2007.

[111] Jan Rauch. Project SEWEBAR – considerations on Semantic Web and data mining. In Bhanu Prasad, editor, *Proceedings of the 3rd Indian International Conference on Artificial Intelligence, Pune, India, December 17-19, 2007*, pages 1763–1782. IICAI, 2007.

[112] Jan Rauch. Classes of association rules: An overview. In Tsau Young Lin, Ying Xie, Anita Wasilewska, and Churn-Jung Liau, editors, *Data Mining: Foundations and Practice*, volume 118 of *Studies in Computational Intelligence*, pages 315–337. Springer, 2008.

[113] Jan Rauch. Remarks to logical aspects of measures of interestingness of association rules. In *Workshops Proceedings of the 8th IEEE International Conference on Data Mining (ICDM 2008), December 15-19, 2008, Pisa, Italy*, pages 599–608. IEEE Computer Society, 2008.

[114] Jan Rauch. Logic of discovery, data mining and Semantic Web - position paper. In Ana L. N. Fred and Joaquim Filipe, editors, *KDIR 2010 - Proceedings of the International Conference on Knowledge Discovery and Information Retrieval, Valencia, Spain, October 25-28, 2010*, pages 342–351. SciTePress, 2010.

[115] Jan Rauch. Logical aspects of the measures of interestingness of association rules. In Jacek Koronacki, Zbigniew W. Ras, Slawomir T. Wierzchon, and Janusz Kacprzyk, editors, *Advances in Machine Learning II, Dedicated to the Memory of Professor Ryszard S. Michalski*, volume 263 of *Studies in Computational Intelligence*, pages 175–203. Springer, 2010.

[116] Jan Rauch. Modifying logic of discovery for dealing with domain knowledge in data mining. In Marzena Kryszkiewicz and Sergei A. Obiedkov, editors, *Proceedings of the 7th International Conference on Concept Lattices and Their Applications, Sevilla, Spain, October 19-21, 2010*, volume

672 of *CEUR Workshop Proceedings*, pages 175–186. CEUR-WS.org, 2010.

[117] Jan Rauch. Consideration on a formal frame for data mining. In Tzung-Pei Hong, Yasuo Kudo, Mineichi Kudo, Tsau Young Lin, Been-Chian Chien, Shyue-Liang Wang, Masahiro Inuiguchi, and Guilong Liu, editors, *2011 IEEE International Conference on Granular Computing, GrC-2011, Kaohsiung, Taiwan, November 8-10, 2011*, pages 562–569. IEEE Computer Society, 2011.

[118] Jan Rauch. Domain knowledge and data mining with association rules - A logical point of view. In Li Chen, Alexander Felfernig, Jiming Liu, and Zbigniew W. Ras, editors, *Foundations of Intelligent Systems - 20th International Symposium, ISMIS 2012, Macau, China, December 4-7, 2012. Proceedings*, volume 7661 of *Lecture Notes in Computer Science*, pages 11–20. Springer, 2012.

[119] Jan Rauch. EverMiner: consideration on knowledge driven permanent data mining process. *Int. J. Data Min. Model. Manag.*, 4(3):224–243, 2012.

[120] Jan Rauch. Formalizing data mining with association rules. In Tsau Young Lin, Xiaohua Hu, Zhaohui Wu, Arbee L. P. Chen, Andrei Z. Broder, Howard Ho, and Shuliang Wang, editors, *2012 IEEE International Conference on Granular Computing, GrC 2012, Hangzhou, China, August 11-13, 2012*, pages 406–411. IEEE Computer Society, 2012.

[121] Jan Rauch. *Observational Calculi and Association Rules*, volume 469 of *Studies in Computational Intelligence*. Springer, 2013.

[122] Jan Rauch. Formal framework for data mining with association rules and domain knowledge - overview of an approach. *Fundam. Inform.*, 137(2):171–217, 2015.

[123] Jan Rauch. Logical aspects of dealing with domain knowledge in data mining with association rules. *Fundam. Informaticae*, 148(1-2):1–33, 2016.

[124] Jan Rauch. Expert deduction rules in data mining with association rules: a case study. *Knowledge and Information Systems*, 59(1):167–195, 2019.

[125] Jan Rauch and Milan Šimůnek. Mining for 4ft association rules. In Setsuo Arikawa and Shinichi Morishita, editors, *Discovery Science, Third International Conference, DS 2000, Kyoto, Japan, December 4-6, 2000, Proceedings*, volume 1967 of *Lecture Notes in Computer Science*, pages 268–272. Springer, 2000.

[126] Jan Rauch and Milan Šimůnek. Mining for association rules by 4ft-Miner. In *Proceedings of the 14th International Conference on Applications of Prolog, INAP 2001, University of Tokyo, Tokyo, Japan, October 20-22, 2001*, pages 285–295. The Prolog Association of Japan, 2001.

[127] Jan Rauch and Milan Šimůnek. An alternative approach to mining association rules. In *Foundations of Data Mining and Knowledge Discovery*, pages 211–231. 2005.

[128] Jan Rauch and Milan Šimůnek. GUHA method and Granular Computing. In Xiaohua Hu, Qing Liu, Andrzej Skowron, Tsau Young Lin, Ronald R. Yager, and Bo Zhang, editors, *2005 IEEE International Conference on Granular Computing, Beijing, China, July 25-27, 2005*, pages 630–635. IEEE, 2005.

[129] Jan Rauch and Milan Šimůnek. Semantic Web presentation of analytical reports from data mining - preliminary considerations. In *2007 IEEE / WIC / ACM International Conference on Web Intelligence, WI 2007, 2-5 November 2007, Silicon Valley, CA, USA, Main Conference Proceedings*, pages 3–7. IEEE Computer Society, 2007.

[130] Jan Rauch and Milan Šimůnek. LAREDAM - considerations on system of local analytical reports from data mining. In Aijun An, Stan Matwin, Zbigniew W. Ras, and Dominik Slezak, editors, *Foundations of Intelligent Systems, 17th International Symposium, ISMIS 2008, Toronto, Canada, May 20-23, 2008, Proceedings*, volume 4994 of *Lecture Notes in Computer Science*, pages 143–149. Springer, 2008.

[131] Jan Rauch and Milan Šimůnek. Action rules and the GUHA method: Preliminary considerations and results. In Jan Rauch, Zbigniew W. Ras, Petr Berka, and Tapio Elomaa, editors, *Foundations of Intelligent Systems, 18th International Symposium, ISMIS 2009, Prague, Czech Republic, September 14-17, 2009. Proceedings*, volume 5722 of *Lecture Notes in Computer Science*, pages 76–87. Springer, 2009.

[132] Jan Rauch and Milan Šimůnek. Dealing with background knowledge in the SEWEBAR project. In Bettina Berendt, Dunja Mladenic, Marco de Gemmis, Giovanni Semeraro, Myra Spiliopoulou, Gerd Stumme, Vojtěch Svátek, and Filip Železný, editors, *Knowledge Discovery Enhanced with Semantic and Social Information*, volume 220 of *Studies in Computational Intelligence*, pages 89–106. 2009.

[133] Jan Rauch and Milan Šimůnek. Applying domain knowledge in association rules mining process - first experience. In Marzena Kryszkiewicz, Henryk Rybinski, Andrzej Skowron, and Zbigniew W. Ras, editors, *Foundations of Intelligent Systems - 19th International Symposium, ISMIS*

2011, Warsaw, Poland, June 28-30, 2011. Proceedings, volume 6804 of *Lecture Notes in Computer Science*, pages 113–122. Springer, 2011.

[134] Jan Rauch and Milan Šimůnek. Dealing with domain knowledge in association rules mining - several experiments. In Ji Zhang and Giovanni Livraga, editors, *2011 International Conference on Data and Knowledge Engineering, ICDKE 2011, Milano, Italy, September 6, 2011*, pages 13–17. IEEE, 2011.

[135] Jan Rauch and Milan Šimůnek. Using domain knowledge in association rules mining - case study. In Kecheng Liu, Ana L. N. Fred, and Joaquim Filipe, editors, *KDIR/KMIS 2013 - Proceedings of the International Conference on Knowledge Discovery and Information Retrieval and the International Conference on Knowledge Management and Information Sharing, Vilamoura, Algarve, Portugal, 19 - 22 September, 2013*, pages 104–111. SciTePress, 2013.

[136] Jan Rauch and Milan Šimůnek. Learning association rules from data through domain knowledge and automation. In Antonis Bikakis, Paul Fodor, and Dumitru Roman, editors, *Rules on the Web. From Theory to Applications - 8th International Symposium, RuleML 2014, Co-located with the 21st European Conference on Artificial Intelligence, ECAI 2014, Prague, Czech Republic, August 18-20, 2014. Proceedings*, volume 8620 of *Lecture Notes in Computer Science*, pages 266–280. Springer, 2014.

[137] Jan Rauch and Milan Šimůnek. Data mining with histograms - A case study. In Floriana Esposito, Olivier Pivert, Mohand-Saïd Hacid, Zbigniew W. Ras, and Stefano Ferilli, editors, *Foundations of Intelligent Systems - 22nd International Symposium, ISMIS 2015, Lyon, France, October 21-23, 2015, Proceedings*, volume 9384 of *Lecture Notes in Computer Science*, pages 3–8. Springer, 2015.

[138] Jan Rauch and Milan Šimůnek. Apriori and GUHA - comparing two approaches to data mining with association rules. *Intell. Data Anal.*, 21(4):981–1013, 2017.

[139] Jan Rauch and Milan Šimůnek. Data mining with histograms and domain knowledge - case studies and considerations. *Fundam. Inform.*, 166(4):349–378, 2019.

[140] Jan Rauch, Milan Šimůnek, and Václav Lín. Mining for patterns based on contingency tables by *KL-Miner* - first experience. In Tsau Young Lin, Setsuo Ohsuga, Churn-Jung Liau, and Xiaohua Hu, editors, *Foundations and Novel Approaches in Data Mining*, volume 9 of *Studies in Computational Intelligence*, pages 155–167. Springer, 2006.

[141] Z. Renc, K. Kubát, and J. Kouřim. An application of the GUHA method in medicine. *International Journal of Man-Machine Studies*, 10(1):29–35, 1978.

[142] Zdeněk Renc. On interpretation of GUHA results. *International Journal of Man-Machine Studies*, 10(1):37–46, 1978.

[143] James Rumbaugh, Mike Blaha, Bill Premerlani, Fred Eddy, and Bill Lorense. *Object-oriented Analysis and Design*. OMT Methodology, Prentice-Hall, 1996.

[144] Rajiv Sabherwal and Irma Becerra-Fernandez. *Business Intelligence, Practices, Technologies and Management*. John Wiley and Son, 2011.

[145] Paul D. Scott and Elwood Wilkins. Evaluating data mining procedures: techniques for generating artificial data sets. *Inf. Softw. Technol.*, 41(9):579–587, 1999.

[146] Avi Silberschatz and Alexander Tuzhilin. What makes patterns interesting in knowledge discovery systems. *IEEE Transactions on Knowledge and Data Engineering*, 8(6):970–974, Dec 1996.

[147] Frederic Springsteel. Complexity of hypothesis formation problems. *International Journal of Man-Machine Studies*, 15(3):319–332, 1981.

[148] Robert Stackowiak, Joseph Rayman, and Rick Greenwald. *Oracle data warehousing and business intelligence solutions*. Wiley Pub., 2007.

[149] Petr Strossa and Jan Rauch. Converting association rules into natural language - an attempt. In Mieczyslaw A. Klopotek, Slawomir T. Wierzchon, and Krzysztof Trojanowski, editors, *Intelligent Information Processing and Web Mining, Proceedings of the International IIS: IIPWM'03 Conference held in Zakopane, Poland, June 2-5, 2003*, Advances in Soft Computing, pages 383–392. Springer, 2003.

[150] Petr Strossa, Zdeněk Černý, and Jan Rauch. Reporting data mining results in a natural language. In Tsau Young Lin, Setsuo Ohsuga, Churn-Jung Liau, Xiaohua Hu, and Shusaku Tsumoto, editors, *Foundations of Data Mining and Knowledge Discovery*, volume 6 of *Studies in Computational Intelligence*, pages 347–361. Springer, 2005.

[151] Einoshin Suzuki. Undirected exception rule discovery as local pattern detection. In *Local Pattern Detection, International Seminar, Dagstuhl Castle, Germany, April 12-16, 2004, Revised Selected Papers*, pages 207–216, 2004.

[152] Vojtěch Svátek, Jan Rauch, and Martin Ralbovský. Ontology-enhanced association mining. In Markus Ackermann, Bettina Berendt, Marko Grobelnik, Andreas Hotho, Dunja Mladenic, Giovanni Semeraro, Myra Spiliopoulou, Gerd Stumme, Vojtěch Svátek, and Maarten van Someren, editors, *Semantics, Web and Mining, Joint International Workshops, EWMF 2005 and KDO 2005, Porto, Portugal, October 3 and 7, 2005, Revised Selected Papers*, volume 4289 of *Lecture Notes in Computer Science*, pages 163–179. Springer, 2005.

[153] David Taniar, J. Wenny Rahayu, Vincent C. S. Lee, and Olena Daly. Exception rules in association rule mining. *Applied Mathematics and Computation*, 205(2):735–750, 2008.

[154] IBM Cloud team. Python vs. r: What's the difference? blog, March 23, 2021 [Online].

[155] Esko Turunen. Using GUHA data mining method in analyzing road traffic accidents occurred in the years 2004-2008 in Finland. *Data Sci. Eng.*, 2(3):224–231, 2017.

[156] Esko Turunen and Klara Dolos. Revealing driver's natural behavior – a GUHA data mining approach. *Mathematics*, 9(15), 2021.

[157] V. Vlček. An application of the GUHA method to chemical engineering. *International Journal of Man-Machine Studies*, 15(3):299–307, 1981.

[158] Milan Šimůnek. Academic KDD project LISp-Miner. In Ajith Abraham, Katrin Franke, and Mario Köppen, editors, *Intelligent Systems Design and Applications*, pages 263–272, Berlin, Heidelberg, 2003. Springer Berlin Heidelberg.

[159] Milan Šimůnek. LISp-Miner Control Language – description of scripting language implementation. *Journal of systems integration [online]*, 5(2):28–44, 2014.

[160] Milan Šimůnek and Jan Rauch. EverMiner prototype using LISp-Miner Control Language. In Troels Andreasen, Henning Christiansen, Juan Carlos Cubero Talavera, and Zbigniew W. Ras, editors, *Foundations of Intelligent Systems - 21st International Symposium, ISMIS 2014, Roskilde, Denmark, June 25-27, 2014. Proceedings*, volume 8502 of *Lecture Notes in Computer Science*, pages 113–122. Springer, 2014.

[161] Milan Šimůnek and Teppo Tammisto. Distributed data-mining in the LISp-Miner system using Techila grid. In Filip Zavoral, Jakub Yaghob, Pit Pichappan, and Eyas El-Qawasmeh, editors, *Networked Digital Technologies - Second International Conference, NDT 2010, Prague, Czech*

Republic, July 7-9, 2010. Proceedings, Part I, volume 87 of *Communications in Computer and Information Science*, pages 15–20. Springer, 2010.

[162] Thomas Weise. *Global Optimization Algorithm: Theory and Application.* 2011.

[163] Qiang Yang and Xindong Wu. 10 challenging problems in data mining research. *Int. J. Inf. Technol. Decis. Mak.*, 5(4):597–604, 2006.

[164] M. J. Zaki, S. Parthasarathy, M. Ogihara, and W. Li. New algorithms for fast discovery of association rules. In *Proceedings of the Third International Conference on Knowledge Discovery and Data Mining*, KDD'97, page 283–286. AAAI Press, 1997.

Index

For Product Safety Concerns and Information please contact our EU
representative GPSR@taylorandfrancis.com
Taylor & Francis Verlag GmbH, Kaufingerstraße 24, 80331 München, Germany